KENNETH A. HASSALL

THE CHEMISTRY OF PESTICIDES

Other Related Books

ISBN	Title	Author
978-81-7233-566-3	A Manual of Practical Entomology	*Trigunayat, M.M.*
978-81-7233-712-4	Agricultural Insect Pests & their Control	*Awasthi, V.B.*
81-7233-430-3	Animal Biodiversity : Patterns and Processes	*Ananthakrishnan*
978-81-7233-562-5	Bamboo Flowering & Rodent Outbreaks	*Sridhara, S.*
978-81-7233-797-1	Biopesticides in Environment and Food Security: *Issues and Strategies*	*Koul, Opendra*
81-7233-402-8	Changing Faunal Ecology in the Thar Desert	*Tyagi, B.K.*
978-81-7233-740-7	Dynamics of Insect Behaviour	*Ananthakrishnan*
978-81-7233-599-1	Ecodynamics of Insect Communities	*Ananthakrishnan*
978-81-7233-509-0	Ecological Entomology: Insect Life in Odd Environments	*Ananthakrishnan,*
81-7233-401-X	Ecology of Small Mammals of Desert and Montane Ecosystems	*Prakash, I.*
978-81-7233-727-8	Entomology : Ecology and Biodiversity	*Tyagi, B.K.*
978-81-7233-694-3	Fresh Water Aquaculture (3rd Ed.)	*Rath, R.K.*
978-81-7233-641-7	Insect Biodiversity : Functional Dynamics and Ecological Perspectives	*Ananthakrishnan,*
81-7233-390-0	Insect Neuroendocrines	*Awasthi, V.B.*
978-81-7233-597-7	Introduction to General & Applied Entomology (3rd Ed)	*Awasthi, V.B.*
81-7233-351-X	Medical Entomology	*Tyagi, B.K.*
81-7233-482-6	Odonata : Biology of Dragonflies	*Tyagi, B.K.*
978-81-7233-532-8	Pests of Forest Importance & their Management	*Tyagi, B.K.*
978-81-7233-540-3	Photographic Atlas of Entomology and Guide to Insect Identification	*Castner, J.L.*
———	Principles of Insect Behaviour	*Awasthi, V.B.*
978-81-7233-583-0	Scientists : A unique 'species' and the backbone of a nation's economic and intellectual prosperity	*Tyagi, B.K.*
81-7233-365-X	The Invincible Deadly Mosquitoes	*Tyagi, B.K.*
978-81-7233-760-5	Theory and Practice of Integrated Pest Management	*Arora, R.*
81-7233-436-2	Vertebrate Pests in Agriculture-The Indian Scenario	*Sridhara, S.*

for more details visit www.scientificpub.com

KENNETH A. HASSALL

THE CHEMISTRY OF PESTICIDES

Their Metabolism, Mode of Action and Uses in Crop Protection

SCIENTIFIC
PUBLISHERS (INDIA)

Reprint in India, 2013

Published by:

Scientific Publishers (India)
5 A, New Pali Road, P.O. Box 91
Jodhpur 342 001 (India)

Scientific Publishers (India)
4806/24, Ansari Road, Daryaganj
New Delhi - 110 002 (India)

E-mail: info@scientificpub.com
Website: www.scientificpub.com

Print : 2021

ISBN: 978-81-7233-819-0

Printed in India

To
C.H.E.

Contents

Preface

Few aspects of agriculture have advanced so rapidly or generated so much controversy as has the use of chemicals for crop protection. Such chemicals include insecticides, fungicides and herbicides, as well as several minor groups of compounds. All of these are here collectively termed *pesticides* – an unsatisfactory but useful umbrella term, from 'pest: anything noxious, mischievous or destructive'. In its early days the subject of pesticides involved little more than an enumeration of the toxic compounds used in agriculture, with lists of their chemical properties and practical uses. Such has been the progress in the subject that nowadays it is evident that an understanding of metabolism, persistence and mode of action is as essential for the safe and efficient use of pesticidal substances as it is for poisons and drugs of importance in medical toxicology.

A consequence of this progress has been that the subject has become more fascinating as it has become more complex. This book attempts to portray something of this fascination by demonstrating the multidisciplinary nature of the science of pesticides, with specific reference to substances used in crop protection. The agriculturist or his advisor normally has to handle and apply insecticides, fungicides and herbicides and will find it convenient for these three major types of agricultural pesticides to be considered under one cover. This (regrettably unusual) event is highly desirable scientifically and educationally since so many of the principles involved transcend the type of compound under discussion – the scientific aspects of formulation, application, disposition, metabolism and even mode of action, are extremely similar for insecticides, fungicides and herbicides, and one may suspect academic demarcation often separates what should naturally be united.

At least three per cent of the market value of agricultural crops in technologically advanced countries is spent on toxic chemicals and their application. This large sum is spent in an effort to minimise the effects of the army of pests which find large areas of monoculture a paradise for explosive reproduction. Some of us may regret the necessity for the use of pesticides on crops, but it is a necessity arising from the fact that man has opted out of the usual constraints placed upon species equilibration; man neither wishes to limit his reproduction to two children per family nor wishes to take a cut in his (worldwide) standard of eating. Remove the chemicals and, in the present world situation, food will become scarcer for the poor nations and dearer for the rich. On the other hand, Rachel Carson and many others have pointed out that the use of toxic chemicals on crops can help to create some of the problems which perpetuate their use (the development of resistance and the destruction of competitors are two examples). They can also pollute the environment if used in the wrong place or at the wrong time, or in the wrong amounts.

Let me make quite clear that the comments above are not to be construed as an apologia; even though I am sometimes proud to wear a conservationist's hat, I have dedicated much of my life to teaching this subject and several well-known names were former students at Reading. In short, in my considered judgment chemical crop protection is both inevitable and desirable for the forseeable future, albeit increasingly as a part of integrated pest control. So, if such substances are to be used it is imperative that all who use them, or advise about their use, are aware of the hazards which can arise and of the metabolic factors which determine their persistence, activation or degradation.

A major purpose of this book is to meet the need just described, whilst demonstrating that the underlying principles are often of intrinsic interest to the pure scientist. It is thus designed to meet the needs of those in academic, commercial or international organisations who, while expert in some other discipline, have neither the time nor the inclination to sieve through advanced texts for various types of basic information. In addition, most students studying agricultural disciplines now receive courses in Crop Protection Chemistry. Pesticides also provide excellent examples of how molecules can be designed to fulfil predetermined objectives (Biochemical and Molecular Pharmacology) and of the problems arising from the widespread use of toxic chemicals (Environmental and Pollution Chemistry).

From undergraduate teaching in both Britain and East Africa, I have found the subject both intelligible and interesting to those with a standard in chemistry approximately equivalent to Advanced Level in Britain (i.e. school chemistry to about eighteen years of age). Clearly the book is not intended to deal with this wide subject exhaustively, for merely to list the *titles* of the articles on pesticide chemistry which have appeared in the last ten years would occupy a space larger than this book. On the other hand, where chemistry (or biochemistry) is an essential vehicle to convey the rationale of structure, enzymic attack, or mode of action of a particular pesticide, chemical or biochemical discussion has not been side-stepped. However, recognising that long chemical names can break the continuity of presentation (and perhaps be rather off-putting to some non-chemists), chemical formulae have, wherever possible, been kept out of the main text but provided for reference as a part of the legend to Figures. For rather similar reasons latin names do not always accompany trivial names in the text although they have been used where ambiguity might arise. For others, cross-reference is possible through the Index.

Substances used for crop protection vary from year to year but the fundamentals governing their choice, use and metabolism evolve much more slowly. Of the hundreds of chemicals in current use as pesticides only a relatively small number have been selected to illustrate these general principles, for a lengthy catalogue would obscure rather than enlighten – and would be, one imagines, profoundly tedious. The same would apply to lists of insect pests, pathogens and weeds together with the chemicals which have been used to control them. This is, in fact, something that is best done at the local level anyway, and in many countries Agricultural Ministries provide literature indicating how best to deal with local problems. In addition, excellent commercial literature exists on the practicalities of pest control and should always be sought and carefully read before handling any crop protection chemical. In the present book the principles already outlined are illustrated by reference to substances selected as representatives of groups or of sub-groups of pesticides and have been chosen to illustrate the widest range of structure–toxicity relationships and metabolic patterns as is reasonably possible within the space available.

No doubt these subjective choices will merit criticism. Nevertheless it is my hope that this book will be a stimulus to all who are concerned with scientific agriculture, whether as growers, governmental advisors or as representatives of commercial firms. In addition, I should regard it as a welcome bonus if its contents provide an appreciation of the subject of molecular biology, the fascination of which grows with each new discovery, and which often lies at the heart of pesticide metabolism, mechanism of action and the development of resistant strains of organisms.

Reading, May, 1981 K. A. H.

Introduction, Acknowledgments and a Disclaimer

Pesticide science is a multi-disciplinary subject and readers of this book are likely to be approaching it from a variety of scientific backgrounds. Consequently, although the book is an introductory text which is suitable for use by undergraduates for whom the subject forms part of a wider course, it is also likely that specialists in such fields as chemistry, biochemistry, agriculture, biology, toxicology and soil science may turn to it to superimpose a knowledge of pesticides upon their pre-existing expertise. Thus, for any one reader, 'introductory' is not necessarily the same as 'elementary' through all facets of the subject. This heterogeneity of background imposes a number of problems to both author and reader since a book such as this must, of necessity, assume the existence of a certain familiarity with a range of supporting sciences. To minimise such difficulties the first section of the Bibliography at the end of the book (p. 333) lists recommended texts which will, where the need arises, help individual readers quickly to acquire basic but essential knowledge in one or more basic disciplines. The latter could include entomology, genetics, plant anatomy and mycology as well as structural and metabolic biochemistry.

Other books in the Bibliography provide a wider insight into the ecological controversy which often surrounds pesticide usage, and yet others are reference books which are indispensable to all concerned with the practical use of pesticides (the present book is seldom concerned with the timing or dosage of a particular pesticide application, nor with the best application volume or spray pattern to employ). The references which occur at the end of each chapter serve a different purpose, for they give guidance to those wishing to pursue selected subjects in greater depth and to those who wish to consult original articles or more detailed reviews.

It is a pleasure to acknowledge the help provided by several colleagues in my own and other departments of the University of Reading, as well as that given by staff of the British Agrochemicals Association and members of several Research Institutes. Numerous reference books and review articles have proved invaluable as sources of information and, more importantly, have enabled me to make decisions with regard to balance and priorities of subject matter. Many of these are listed in the Bibliography at the end of the book and others appear in the references at the ends of the chapters. I should like to single out for special mention the *List of Approved Products for Farmers and Growers* (Ref. 20, p. 334).

the *British Insecticide and Fungicide Handbook* (Ref. 21) and the *British Weed Control Handbook* (Ref. 22). In addition, the book on insecticides by Matsumura (Ref. 12) and that edited by Wilkinson (Ref. 13) are valuable sources of more detailed information and list many hundreds of references to original work. The same is true of the text on antifungal compounds edited by Siegel and Sisler (Ref. 15) and those on the chemistry of herbicides edited by Audus (Ref. 16) and by Kearney and Kaufman (Ref. 17).

A few of the tables and figures in this book are based on data or diagrams of other authors. These are acknowledged where they occur, but there are a few authors to whom I am particularly indebted. Figures in table 1.1 have been selected from several tables which appear in *Chemicals for Crop Protection and Pest Control* (1977) by Green, Hartley and West (reference list, chapter 1). Many of the toxicity data in table 1.2 appeared in *World Review of Pest Control* (1966) and were collated by Edson, Sanderson and Noakes (Bibliography, p. 335, Ref. 28). Some of the data in table 5.1 are quoted from *Carbamate Insecticides: Chemistry, Biochemistry and Toxicology* (1976) by Kuhr and Dorough (reference list, chapter 5) and some of those in table 6.2 are from a review in *Environmental Toxicology of Pesticides* (1972) by Kenaga (reference list, chapter 6). I have also used several diagrams and tables from two of my earlier works, namely, *World Crop Protection: vol. 2, Pesticides* (Iliffe, 1969) and *Biochemistry and Physiology of the Cell* (Edwards and Hassall, 1980, McGraw-Hill (UK) Ltd).

Finally, because of the nature of the subject a disclaimer is necessary. This book is for educational information, not for trade; although every care has been taken to get facts right and to avoid printing errors, the ultimate responsibility in pesticide usage rests with the user, whose job it is to read the label on the container, to consult commercial or official advisors and to be acquainted with local rules and regulations. Consequently, notwithstanding any statement made in this book, *neither the author nor the publisher accept any responsibility for any damage or accident of any sort whatsoever, arising from the use of any of the pesticidal compounds mentioned in this book.*

1 General Considerations

1.1 SOME HISTORICAL LANDMARKS

The use of chemicals to control insects possibly dates back to classical Greece and Rome, for Homer mentioned the fumigant value of burning sulphur and Pliny the Elder advocated the insecticidal use of arsenic and referred to the use of soda and olive oil for the seed treatment of legumes. Pliny experienced the non-selective action of sulphurous fumes for he died when Vesuvius erupted in A.D. 79 ('sulphur and the gross air', in the words of his nephew, Pliny the Younger). The Chinese were employing moderate amounts of arsenicals as insecticides by the sixteenth century and not long afterwards nicotine, in the form of tobacco extracts, was used to control the plum curculio. By the nineteenth century, both pyrethrum and soap had been used for insect control and Forsyth described a combined wash of tobacco, sulphur and lime to combat insects and fungi (see Horsfall, 1956; frontispiece).

As in the twentieth century, chemical pest control was not without its competitors even in ancient times. Some mystical and legal anecdotes are entertainingly described by Ordish (1976). A method described by Mouffet and apocryphally attributed to Pliny would undoubtedly still be popular if it were totally reliable – caterpillars can allegedly be enticed off infected trees by persuading an unadorned maiden of appropriate charm to dance around the orchards. Those who favour the wisdom of age to the sallies of youth may prefer a legal indictment of 1485, when the High Vicar of Valencia 'commanded caterpillars to appear before him, gave them a defence counsel and, in judgment, condemned them to leave the area' (Ordish, 1976, p. 44).

The middle of the nineteenth century marked the beginning of the first systematic scientific studies into the use of crop protection chemicals. Work on arsenic compounds led to the introduction in 1867 of Paris green, an impure copper arsenite. It was used in the USA to check the spread of the Colorado beetle and by 1900 its use was so widespread that it led to the introduction of what was probably the first pesticide legislation in the world. Bordeaux mixture was in use in 1885 to control the downy mildew of vine and by the turn of the century lime sulphur was employed both in Europe and California as a fungicide for use against diseases of fruit trees.

In 1896 a French grape grower, applying Bordeaux mixture to his vines, observed that the leaves of yellow charlock growing nearby turned black. This chance observation demonstrated the possibility of chemical weed control and shortly afterwards it was found that iron sulphate, when sprayed on to a mixture of cereal and dicotyledonous weeds, killed the weeds without damaging the crop. Within a decade several other inorganic substances had been shown to act selectively at appropriate concentrations. Another important landmark before the First World War was the introduction in 1913 in Germany of the first organomercury seed dressings (prior to this organomercury compounds had been used medically to combat syphilis).

In the years between the two world wars both the number and the complexity of crop protection chemicals increased. Tar oil was, and still is, used to control eggs of aphids on dormant trees. Dinitro-orthocresol was patented in France in 1932 for the control of weeds in cereals, and in 1934 thiram, the first of several important dithiocarbamate fungicides, was patented in the USA.

The cosmopolitan nature of contributions to chemical pest control was even more evident after 1939. During the Second World War, the insecticidal potential of DDT was discovered in Switzerland and insecticidal organophosphorus compounds were developed in Germany. At about the same time, work was in progress in Britain that was to lead to the commercial production of herbicides of the phenoxyalkanoic acid group. In 1945 the first soil-acting carbamate herbicides were discovered by British workers and the organochlorine insecticide, chlordane, was introduced in the USA and in Germany. Shortly afterwards, the insecticidal carbamates were developed in Switzerland.

In 1950–1955, herbicidal urea derivatives were developed in the USA, the fungicides captan and glyodin appeared, and malathion was introduced. Between 1955 and 1960, newcomers included herbicidal triazines (Switzerland) and quaternary ammonium herbicides (Britain). Relatively few new *groups* of crop protection compounds have appeared since the introduction of dichlobenil, trifluralin and bromoxynil in 1960–1965, but the important systemic fungicide, benomyl, was described in 1968 and the soil-acting herbicide glyphosate was discovered in the USA soon afterwards. Workers in Britain and Japan are currently working on light-stable pyrethroid insecticides and research is continuing in several countries in an attempt to develop new systemic fungicides.

1.2 GENERAL LITERATURE ON PESTICIDES

Pesticide science is a multi-disciplinary field. It encompasses the chemistry of synthesis and analysis, the physics of formulation and the mechanics of application. Toxicology and statistics contribute to assessment of safety hazards and biochemistry is relevant to an understanding of the mode of action of pesticides and of their degradation. Biology contributes a knowledge of life cycles of insects and of fungi and provides essential data on the growth habits of weeds. Economic considerations enter into their development and marketing and determine return on outlay on the part of farmers.

Workers in this subject may therefore enter with a range of backgrounds and, while possessing expertise in other fields, may not always be familiar with sources of general information on crop protection chemicals. Some of the more useful of these sources are listed in the Bibliography. Any choice of a couple of dozen items must of necessity be subjective but the area to which each chosen reference contributes has been indicated. There are, in addition, numerous specialised texts and monographs which are primarily the concern of research specialists. Although omitted from the Bibliography, some appear in the reference lists at the end of individual chapters of this book.

1.3 THE SIZE AND SCOPE OF THE PESTICIDE MARKET

The end of the war in 1945 was marked by a couple of decades of rapid industrial expansion and a concomitant increase in prosperity in many industrialised countries. This was reflected in the market value of good quality agricultural produce and it became financially worthwhile for a farmer to protect his crops from the predations of insects and the effects of fungi and weeds. This happened at about the time that effective and relatively cheap organic pesticides were becoming readily available. The situation was to change in that second-generation pesticides were often more costly than the early ones – most pyrethroid and carbamate insecticides, for instance, are more expensive than organophosphorus compounds, and most of these are dearer than some organochlorine compounds (compare, for example, USDA, 1978, Table 670). Furthermore, the petrochemical industry provides many of the precursors from which pesticides are made, and, in consequence, oil-price inflation since 1973 has inevitably led to spiralling costs. However, the success of first-generation pesticides had convinced farmers of the benefits accruing from the use of crop protection chemicals. It would, perhaps, be expedient to start enquiring soon about the source of raw materials for pesticides when oil becomes a precious commodity some time between 1995 and 2020.

The dramatic expansion of the world pesticides industry is illustrated by the histograms in figure 1.1, which are based on tonnages quoted by Green *et al.* (1977). The growth is the more remarkable when it is recalled that cheap but heavily-applied inorganic chemicals have been in considerable measure replaced during this period by organic chemicals of higher unit cost but often applied at a rate of only a fraction of a kilogram per hectare. In financial terms, it is estimated that, at grower prices, the world crop protection market has grown from £500 million in 1960 to £4200 million in 1978 (Braunholtz, 1979).

The British Agrochemicals Association (BAA) report that (excluding inflation) the total sales of pesticides of all kinds by the British pesticides industry rose from £192 million in 1976 to £367 million in 1979 (BAA, 1980). About £210 million of these sales were to the home market, comprising herbicides, insecticides and fungicides (figure 1.2) as well as smaller amounts of seed dressings, herbicide/fertilizer mixtures and other products. The predominant contribution

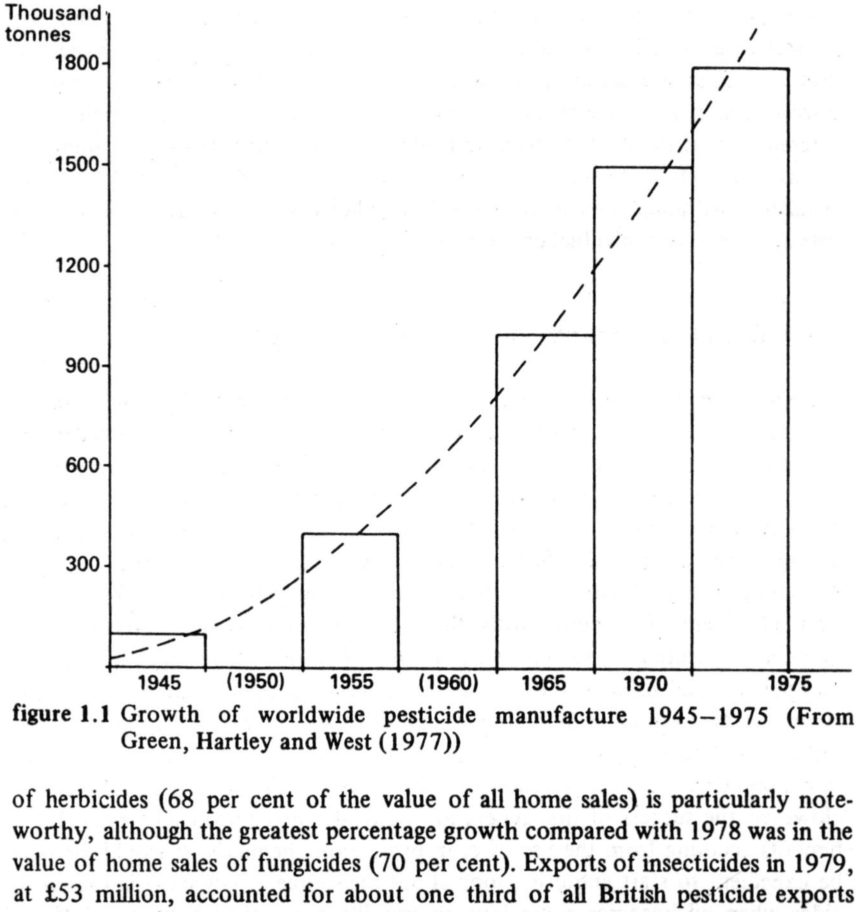

figure 1.1 Growth of worldwide pesticide manufacture 1945–1975 (From Green, Hartley and West (1977))

of herbicides (68 per cent of the value of all home sales) is particularly noteworthy, although the greatest percentage growth compared with 1978 was in the value of home sales of fungicides (70 per cent). Exports of insecticides in 1979, at £53 million, accounted for about one third of all British pesticide exports (£157 million); herbicides accounted for most of the remainder.

It would be instructive to know the world production figures for individual pesticides but surprisingly little reliable information appears to exist. Instead, it is necessary to turn to production data provided by individual countries. Since USA production represents about one third of total known world production, the breakdown of data from the USA to the level of groups of pesticides is probably a fair reflection of the world pattern. Even for the USA, however, the picture is becoming unsatisfactory, for commercial secrecy has led to an increasing tendency for figures to be witheld. For example, the Agricultural Statistics of the United States Department of Agriculture (USDA, 1977) show that the last declared figures for DDT and 2,4-D were in 1970 and for the aldrin-toxaphene group were in 1974. The USA production figures which appear in table 1.1 are quoted with permission from Green *et al.*, 1977.

The great post-war bonanza in the development of new pesticides has probably ended, the reason being in part economic. It has been estimated that for every 8000 substances synthesised, only one is likely to be successfully marketed.

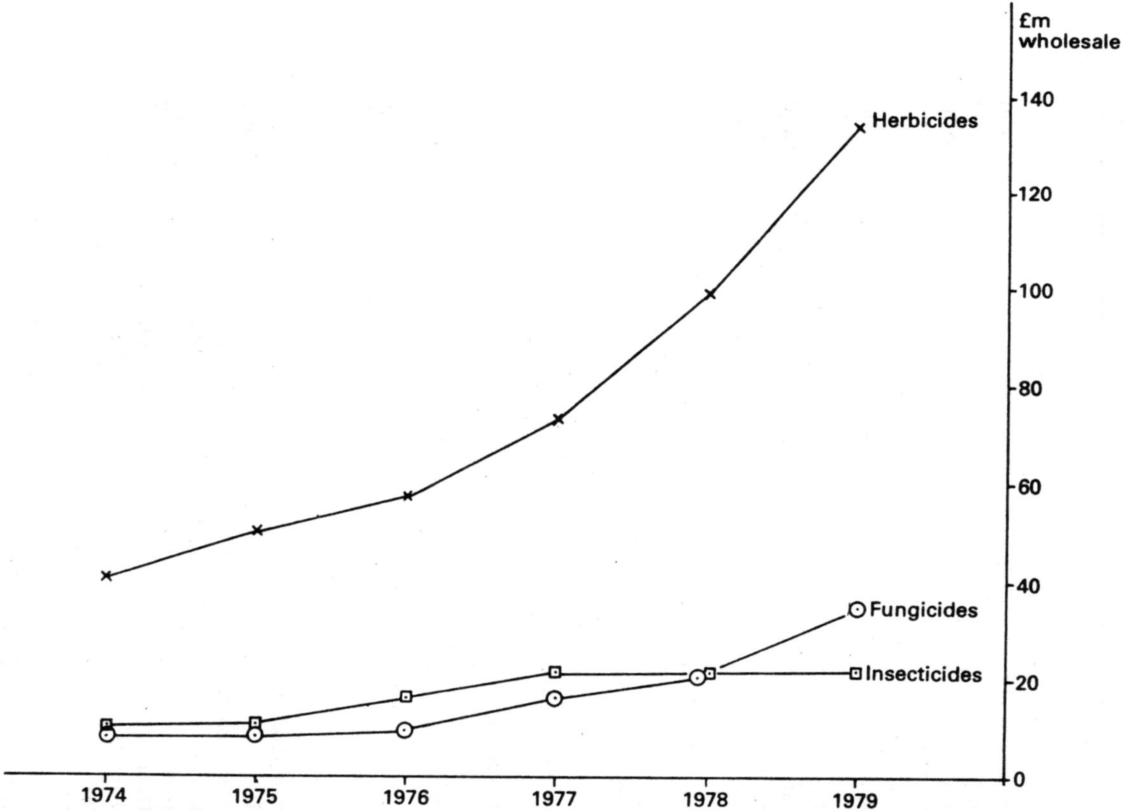

figure 1.2 Wholesale value of UK sales of crop protection chemicals by British Manufacturers 1974–1979 (It is a pleasure to acknowledge the co-operation of Mr. C. S. Major, Director of the British Agrochemicals Association and his staff, in providing the figures on which this graph is based)

Table 1.1 Tonnage of some major groups of pesticides manufactured in the USA in 1971 (From Green, Hartley and West (1977))

	Total (tonnes)	Individual contributions
A. Insecticides		
1. *Organophosphorus compounds*	67 900	
methyl parathion		20 400
malathion		15 900
parathion		6800
dichlorvos		4100
2. *Organochlorine compounds*	55 300	
toxaphene		22 700
DDT		20 400
3. *Carbamates*	27 800	
carbaryl		20 400
carbofuran		3600
4. *Others*	700	
Total insecticides	151 700	
B. Fungicides		
1. *Sulphur*	68 000	
2. *Copper compounds*	6500	
3. *Organic compounds*	35 100	
dithiocarbamates		18 000
phthalimides		9100
others		8000
Total fungicides	109 600	
C. Herbicides		
1. Phenoxyalkanoic acids	23 400	
2,4D		20 400
2. Inorganic compounds	13 600	
3. Triazines	43 900	
atrazine		40 900
4. Trifluralin	11 400	
5. Carbamates	10 000	
6. Ureas	6600	
diuron		2700
7. Propachlor	10 700	
8. Others (including dalapon, alachlor and dinitrophenols)	52 800	
Total herbicides	172 400	

In addition, about 34 000 tonnes of petroleum oil were used by US farmers for insect control and 66 100 tonnes for weed control.

Even when a winner is discovered, the cost of development is such that the innovating manufacturer is unlikely to break even financially for about a decade. Furthermore, the money tied up for these years is very great – nowadays it could approach £10 million. Consequently, only the largest organisations are able to support the necessary research, development, toxicological studies and field trials essential to meet the stringent requirements for safety (and to demonstrate efficiency) demanded by governments in all the major pesticide markets in the world.

1.4 ECONOMIC ASPECTS OF PESTICIDE USAGE

The current value of agricultural produce in the United Kingdom is approximately £6500 million, and expenditure on agrochemicals exceeds £210 million. Thus the money spent by farmers on these chemicals is about 3.3 per cent of the total annual value of their produce. In addition, farmers spend money purchasing and maintaining spray equipment and on the labour necessary for pesticide application.

On a global level it is probable that the current expenditure by farmers on pesticides is of the order of £5000 million. In view of the fact that farmers have a reputation for hard-headedness, it must be assumed that they are convinced this outlay provides an acceptable return on capital. Various estimates have been made which appear to demonstrate that this conviction is justified. Commonly-quoted figures for benefit to cost ratio are 4-fold overall in the USA, 3-fold for cereal herbicides in W. Germany and 8-fold for the control of aphids on sugar beet in W. Germany (Braunholtz, 1979). Unfortunately, it is not easy to check the validity of such figures in individual cases, for each farmer's activities provide an uncontrolled experiment, and assumptions have to be made when predicting what losses a particular crop would have sustained had pesticides not been used.

Despite these uncertainties, there are some useful guidelines to the economics of pesticide utilisation. The cost of *additional* pest control rises rapidly as the pest population falls; to control the final 10 per cent of a population costs about as much as it does to control the first 90 per cent. In practice, this means that, for many pests, an uninformed farmer could double his costs further to reduce a population that was already acceptably low. Since the rate of increase in the yield of a crop rapidly falls once the pest population has decreased to manageable proportions, there is little chance that in such a case the farmer would recoup his additional expenditure with the proceeds from extra marketed produce. The break-even point, where money accruing from increase in crop yield is balanced by the increased cost of chemical control measures, is known as the **economic threshold** (Headley, 1972).

In an abnormally bad season, a pesticide may save the crop from being a write-off, and thus save the farmer from bankruptcy. Consequently, some farmers tend to regard money spent on pesticides as being in the same category as money spent on insurance. A slightly damaged crop (e.g. apples infested with codling moth) may be almost unmarketable, so that 'write-off', from a farmer's viewpoint, does not necessarily mean 'zero yield'. This is especially likely to be true where agreements exist for contract growing.

In the case of herbicides, it is often possible to quantify the efficacy of chemicals by comparing the total cost of purchase and application with the labour costs of hoeing. Some studies on the comparative costs and the consequent return per unit cost of herbicide are mentioned by Green *et al.* (1977); in various instances the advantage was between two-fold and thirteen-fold in favour of the use of weedkillers.

Pesticides, together with mechanisation, fertilisers and irrigation, have helped to increase the productivity of farm workers. In Europe and in the USA, one farm worker grows enough to feed about 60 people, so the labour costs distribute themselves over the price paid for the food by numerous consumers. In consequence, the typical European or American spends only 15–30 per cent of his income on food, offering prospects of fulfilled life by deploying the purchasing power of the remainder. In many a less mechanised country an agricultural worker may produce enough food for only 4–10 people. When 50–70 per cent of income is spent on food, the apathy and frustration inherent in the situation can undermine the political fabric of the nation.

1.5 TOXICOLOGICAL CONSIDERATIONS

Any chemical substance may evoke one or both of two toxic effects. The first, the **acute effect**, is the one more readily comprehensible to the layman, and normally occurs shortly after contact with a *single dose* of poison. The magnitude of the effect depends on the innate toxicity of the substance and upon its method of application to a particular organism. Thus, a smaller dose of arsenious oxide than of sodium chloride will produce toxic symptoms in most animal species, and a drop of sulphuric acid is less dangerous on the skin than it is in the eye. Acute toxicity very often results from the disruption of an identifiable biochemical or physiological system and, in consequence, acute toxic responses are usually readily quantifiable.

A **chronic effect**, on the other hand, sometimes occurs when an organism is exposed to *repeated small and non-lethal doses* of a potentially harmful substance. Well-known chronic responses to various irritants include silicosis, lung cancer, brain damage and necrosis of the liver or kidneys.

A frequently used measure of acute toxicity is the 50 per cent lethal dose, or LD_{50}. It is the amount of poison which kills half the organisms in a randomly chosen batch of a named species when applied in a particular way under stated experimental conditions. It is usually expressed in terms of mg poison per kg body weight of the experimental animals. Such a method of representation eliminates the variation of body weight between species. This is important, for most toxicological tests are done vicariously – the experimenter is not usually directly interested in the LD_{50} of a pesticide to, say, adult male rats, but employs this test species to obtain some indication of what the toxicity might have been had the substance been administered to man or to his livestock.

It will have been noted that in the previous paragraph the example referred to 'adult male' rats. This illustrates another important point, for sometimes male and female, or young and old, of a single species may vary greatly in their susceptibility to a particular poison. In the case of the rat, for example, the susceptibilities of males may sometimes be regarded as forming one distribution and those of the females a somewhat different one. Female rats are, in fact, on average more susceptible than males to a number of foreign compounds (or xenobiotics, as they are sometimes called). A useful survey of age, sex and species differences in the responses of animals to pesticides and other poisons has been provided by Brown (1980).

The LC_{50} is a somewhat similar concept to the LD_{50}. It is the concentration, usually in the air or water surrounding experimental animals, which causes 50 per cent mortality. It is therefore a convenient index for studying vapours inhaled by mammals or water-soluble pesticides in contact with fish; in neither of these cases is the amount of poison taken up by individual test organisms evident without additional investigation. Related terms are ED_{50} and EC_{50}, which are, respectively, the 50 per cent effective dose and 50 per cent effective concentration and are applicable when some criterion of toxicity other than death is adopted. The I_{50} is similarly the concentration which inhibits by 50 per cent an *in vitro* enzyme system.

To determine the LD_{50} of a substance, batches of animals are exposed to a range of doses and the percentage of organisms dying in each batch is recorded. The results can usually be represented by the (idealised) sigmoid curve shown in figure 1.3a. The curve is steepest in the region of 50 per cent response so that a

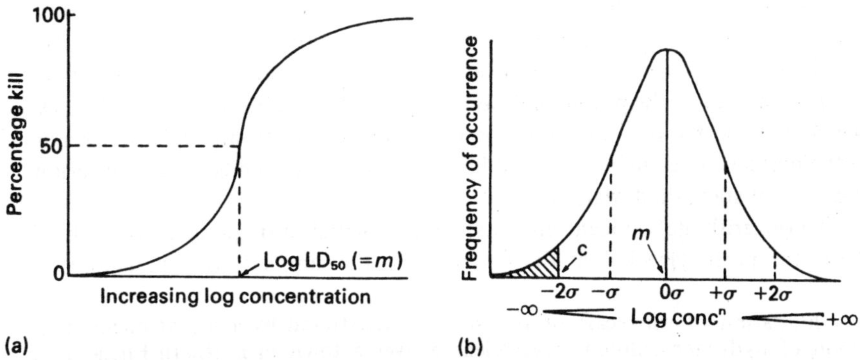

figure 1.3 Statistical distribution of tolerances of individuals in a population to increasing amounts of a poison. (a) Relation of percentage response to the log of toxic concentration. (b) Normal distribution of responses to log concentrations

small change in concentration in this region of the curve causes a large change in percentage kill. The dose which kills half the organisms is thus a more sensitive index of toxicity than any other dose, and this is why the LD_{50} is usually adopted as a standard for comparing the relative toxicity of substances.

Acute toxicity data for numerous pesticides have been provided by several authors, including Edson, Sanderson and Noakes (1966) and by Fairchild (1977). Some selected examples appear in table 1.2. It should be noted that compounds with *small* LD_{50}s are more toxic than those with large LD_{50}s, and that mg/kg (or μg/g) is equivalent to expressing a toxic dose as parts per million (p.p.m.) body weight.

An empirical scale of assessment of levels of toxicity was introduced by Du Bois and Geiling (1959). It is a useful rule-of-thumb description so long as it is not raised to the level of a dogma*. On the scale they proposed a substance with an LD_{50} of less than 1 mg/kg body weight is described as being *extremely* toxic and one with an LD_{50} lying between 1 and 50 mg/kg is *highly* toxic. The corresponding range for a *moderately* toxic material is 50 to 500 mg/kg and, for a *slightly* toxic substance, 500 to 5000 mg/kg.

In common with many other variable factors of a biological nature, the susceptibility of individuals of a population to a poison approximates to a normal distribution when the variable – in this case the applied concentration – is plotted on a logarithmic scale (figure 1.3b). As the concentration increases from $-\infty$ to $+\infty$, more and more organisms respond. For example, at -2σ, few respond, but at zero σ from the mean, half the population responds. In each case, the proportion of the whole population responding is the area to the left of the designated σ value as a fraction of the total area under the curve. Thus the sigmoid curve obtained in dosage-response tests and shown in figure 1.3a is merely the summation of the areas under the normal curve for each concentration as the latter increases progressively from zero to infinity.

The normal distribution of tolerances allows statistical treatment of toxicity data. It enables, for example, the best estimate of the LD_{50} for a particular population to be calculated from the available data and permits confidence limits to be ascribed to it. The **probit analysis** method of Bliss, later extended by Finney (1964) is one of several methods commonly employed. In principle, the x-axis of the normal curve can be expressed in alternative units – units of concentration and units of standard deviation.

The (logarithmic) concentrations are experimental in origin and, for each of these, the percentage kill of the organisms in the sample is known. The other

* A limitation of this scale of assessment is illustrated by a recent incident. A cloud of hydrogen sulphide was released over a town in northern England. A police statement broadcast by the BBC said (quite rightly) that it was not dangerous. But it is interesting to speculate what the recipients of this comfort would have thought if they had been told (equally rightly) that hydrogen sulphide has an LD_{50} half that of the dreaded prussic acid, HCN! Why, then is H_2S less dangerous than HCN in such circumstances?

Table 1.2 Acute oral LD_{50}s of some pesticides to male rats (mg/kg) (From Green, Hartley and West (1977))

Insecticides		Fungicides		Herbicides	
Mevinphos	3	Bordeaux	700	2,4-D	300
Dichlorvos	25	Sulphur	(not toxic)	MCPA, MCPB	700
Parathion	5	Ethylmercury chloride	30		
Malathion	1400	Phenylmercury chloride	60	Paraquat	112
Fenitrothion	250	Thiram	375	Aminotriazole	1100
Azinphos methyl	10				
Phorate	2	Zineb	1000	Dalapon	6600
Dimethoate	200	Captan	9000	Barban	600
Chlorfenvinphos	10	Dinocap	1000	Glyphosate	4320
Bromophos	4000	Binapacryl	150	Bromoxynil	190
Coumaphos	100	Quintozene	1650	Dinoseb	50
Fenchlorphos	1250	Chloranil	4000	Diuron	3200
Carbaryl	400	Dicloran	1500	Linuron	1500
Other carbamates, see Table 5.1		Dodine	1000	Simazine	5000
DDT	250	Benomyl	100†	Desmetryne	1630
BHC (HCH)	125	Ethirimol	4000	Bromacil	5200
Methoxychlor	5000	Tridemorph	650	Chlorpropham	3800
Aldrin	40	Carboxin	3200	Diallate	395
Pyrethrin	570	Pyrazophos	140	Trifluralin	5000
Nicotine	70			Propachlor	710
				Chlorthiamid	760

Substances are listed in order of appearance in this book. Data (mostly for adult male rats) from several sources, including commercial literature, Edson *et al.* (1966), Klingman and Ashton (1975), Fairchild (1977) and Brown (1980). Where discrepancies exist, lowest values are given.
† For wild birds.

experimental quantity, which is the percentage kill, can be expressed in units of standard deviation, σ. For example, 2.25 per cent of organisms lie in the tail to the left of -2σ from the mean. These are the organisms that would be killed if a poison were applied in a toxicity test at a (logarithmic) concentration, shown as c in figure 1.3b, to an idealised batch of test organisms (i.e. one in which the tolerances were exactly normally distributed). By the use of a table of σ, each observed percentage kill can similarly be expressed in σ units (table 1.3). A 'probit' (probability unit) is merely the σ value plus 5, and is a device to avoid the use of negative values of σ. Figure 1.4 shows a typical probit-log concentration graph plotted from experimental results (Hassall, 1955).

In contrast to acute toxic effects which normally follow ingestion or contact with a single dose of appropriate size*, the long-term presence of small amounts

* There is really no such thing as a poisonous *substance* but only of a poisonous *dose*. It is possible to kill oneself with salt if one has the mind to do so.

Table 1.3 Transformation of percentage response to probits (simplified; for the full table, see Finney, 1952)

Percentage	Probit	Percentage	Probit	Percentage	Probit
6	3.45	36	4.64	66	5.41
8	3.59	38	4.69	68	5.47
10	3.72	40	4.75	70	5.52
12	3.82	42	4.80	72	5.58
14	3.92	44	4.85	74	5.64
16	4.01	46	4.90	76	5.71
18	4.08	48	4.95	78	5.77
20	4.16	50	5.00	80	5.84
22	4.23	52	5.05	82	5.92
24	4.29	54	5.10	84	5.99
26	4.36	56	5.15	86	6.08
28	4.42	58	5.20	88	6.18
30	4.48	60	5.25	90	6.28
32	4.53	62	5.31	92	6.41
34	4.59	64	5.36	94	6.55

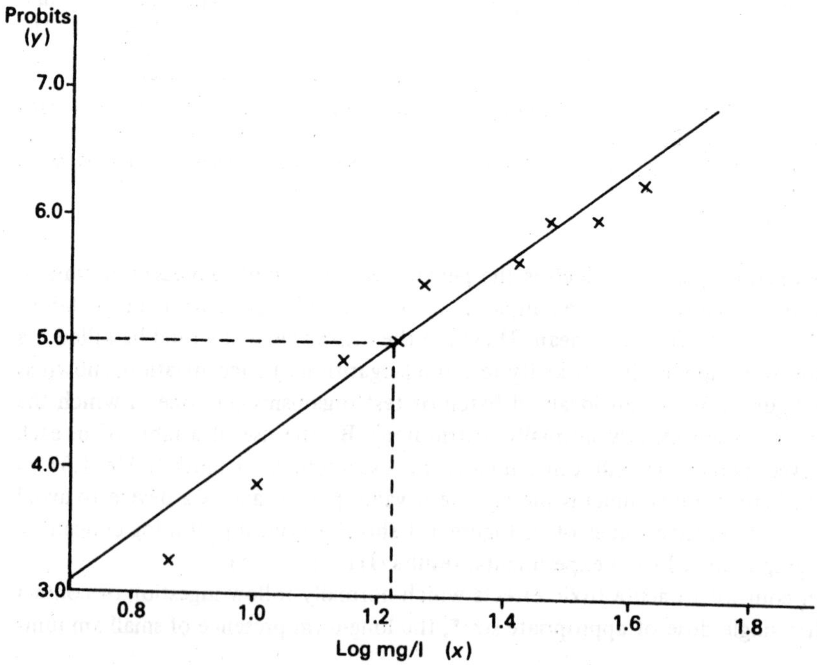

figure 1.4 Dosage-response data for *sec*-amyl iodide used as a fumigant to the grain weevil (Hassall, 1955). ($y = 3.624x + 0.568$. When $y = 5$, x the log of LD_{50} is 1.225 ± 0.025)

of a poison in a living organism may (but does not necessarily) lead eventually to the appearance of chronic effects. The debilitating effects of blue asbestos or of coal dust are well-known occupational hazards where contact with these minerals causes physical damage to lung tissue. The uptake of lead from the atmosphere is believed by some to be responsible for mental disorders and behavioural changes (Bryce-Smith and Waldron, 1974). For pesticidal chemicals, chronic effects in human beings are most likely to occur when the substance is stable enough to survive the processing or cooking of food, and then is able to persist in the body tissues of the consumers.

Much work has been done to discover whether repetitive intake of small quantities of highly stable and lipid-soluble substances results in indefinite linear accumulation with time. Without exception, at all doses of a magnitude at all relevant to the residue problem, a nearly constant internal level is eventually reached when an organism is exposed to a constant small daily intake. This implies that the various mechanisms which lead to detoxication and removal sooner or later balance the uptake, with the establishment of a **steady state** between uptake and elimination (figure 1.5). It is, in fact, not unusual for the presence of a poison to stimulate detoxifying systems and thus to accelerate the rate of its own decomposition or elimination. When this occurs, the internal level may actually fall after a certain period of time, even though the daily intake remains unchanged.

It is therefore evident that a constant daily intake does not cause continued accumulation until a dose capable of producing an acute response is reached. The main problem in predicting chronic toxic hazard relates to assessing the possible harmful effects of different steady state concentrations. It is here that the scientific dilemma begins. All that can usually be said with reasonable certainty

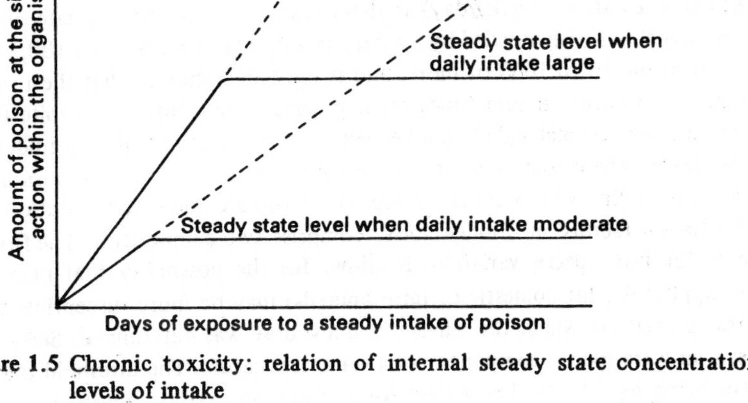

figure 1.5 Chronic toxicity: relation of internal steady state concentrations to levels of intake

is that a higher internal steady state concentration is likely to be more hazardous to a greater percentage of the population than is a lower internal concentration. Furthermore, when considering chronic effects that are only manifest after years of exposure, it is usually impossible to say whether or not there is a 'no-effect' level. That is to say, it is not possible to assert that there is a given internal concentration (and therefore a corresponding external daily dose) below which no damage whatsoever occurs in any one individual of the population. Science, in such matters, cannot demonstrate a negative. However, even if such a dose could be demonstrated to exist for that one individual, for statistical reasons it could not be assumed that the same dose would be safe in another individual.

By the very nature of a long-term effect, it is difficult to quantify the sort of damage that is, or might be happening in any particular circumstance. If we all died when exactly 70 years old, it could be said authoritatively that someone who died at 68 years had had his lifespan shortened. Even then, however, it might not be possible to say that this truncation resulted exclusively from, say, liver necrosis. Even if liver necrosis was the cause of death it would probably not be possible to disentangle the individual contributions of the multifarious environmental and self-inflicted pollutants to which the deceased person had been exposed over much of a life-time.

In order to assess the potential chronic toxic hazard of a pesticide, individuals of a suitable animal species are fed varying amounts of the substance and the highest concentration having 'no effect' on any of the organisms in that batch of animals is determined. This is known as the threshold daily dose. Numerous observations are made both during the long-duration feeding trial and by autopsy examination when the animals are eventually sacrificed. The kinds of data it is desirable to collect in chronic toxicity tests have been described by Benitz (1970). These include changes in body weight, in the intake of food and in the composition of blood, urine and faeces. The activities of enzymes in serum, liver and other tissues are also investigated. Behavioural changes are often studied and, after death, histological examinations are made of liver, kidney, lung and brain.

In order to arrive at a figure for the **'acceptable daily intake'** (ADI) for man, a **'threshold daily dose'** (mg/kg/day) is determined using an appropriate animal species, and this is then divided by a **safety factor**. The latter is often 100, but it may vary from 10 to 1000 or more, and it must be admitted that the reasons given for such variation sometimes seem pragmatic or arbitrary. Despite these limitations, the acceptable daily intake figures are of practical value as guidelines and are taken into consideration by the scientific advisers of many governments when tolerance limits for pesticide residues on crops are under consideration.

The 'safety factors' mentioned above comprise two components. The first of these is for inter-species variation; it allows for the possibility that man (or, where applicable, his domestic or farm animals) may be more susceptible than the test species on which the threshold daily dose was determined. Secondly, there is an intra-species component of a statistical nature. The threshold for one human being may be far lower than for another, and safety factors must take into account possible adverse effects on highly susceptible individuals rather than the response of people of average susceptibility. The ADIs of a number of pesticides have been published jointly by the World Health Organisation and by the Food and Agriculture Division of the United Nations (WHO, 1973). The list has been amended and updated yearly thereafter. Safety factors, ADIs and other aspects of the evaluation of health hazards have been reviewed by Sharratt (1977) and Vettorazzi (1977).

1.6 SAFETY ARRANGEMENTS FOR MARKETING PESTICIDES IN BRITAIN

Enlightened self-interest, as well as legal and moral responsibility, requires that no reputable manufacturer should release a new material on to the market before it has been adequately tested. Voluntary or compulsory schemes of notification and 'clearance' of new pesticides, and of new formulations of old ones, operate in many countries. Even where, as in Britain, notification is non-statutory, most manufacturers willingly cooperate because a great sales advantage accrues to a product that has official sanction in consequence of having been cleared by an independent body. In Britain, almost all major firms marketing pesticides belong to the British Agrochemicals Association (BAA), and it is a condition of membership that participants should conform with the requirements of the Pesticides Safety Precautions Scheme (see below). As was seen earlier, launching a new pesticide is extremely expensive, and successful marketing requires worldwide sales, including sales to countries that have compulsory notification and registration. For this reason it is often somewhat academic whether notification is statutory or non-statutory for sales on the home market; research leading to the acquisition of appropriate data is in any case necessary.

The nature of the data needed by government representatives varies somewhat from country to country. A requirement normally exists for information about the properties of the pesticide, including its persistence, metabolism and probable mode of action. Details of reliable and sensitive methods of analysis are also normally required. Data are required on acute and chronic toxic effects on laboratory animals when the substance is applied by a variety of routes of application, including breathing the vapour, ingestion and absorption through the skin. Toxicity data are also required on wildlife mammals, birds and fish. Hazards to bees, to soil organisms and to man may also need to be investigated. The notifier must also indicate for what purpose the pesticide formulation is to be marketed, and what risks might arise from its use. The basic and further toxicological studies expected for pesticides notified under the British scheme are usefully summarised in Appendix 2 of *Pesticides in Agriculture*, a publication of the British Agrochemicals Association (BAA, 1977). Such a list is probably representative of the requirements of schemes, both non-statutory and compulsory, in most other countries. Appendix 4 of the same publication provides information on the types of studies expected on degradation, soil leaching and bio-accumulation of a notified substance. It also lists the types of data that are required on the effect of the substance on non-target organisms, including soil organisms, beneficial insects and fish. More detailed guidelines on the data requirements for the UK Pesticides Safety Precautions Scheme have been published by the Ministry of Agriculture, Fisheries and Food. Revised guidelines were agreed in 1979 (MAFF, 1979).

It is appropriate that an author with no industrial ties of any sort should stress that manufacturers are keenly aware of the importance of the problems discussed in section 1.5. Of course, it is inevitable, short of clairvoyance, that on rare occasions the passage of time uncovers new problems. When this occurs, it becomes necessary by 'positive feedback' to modify or to extend the pre-marketing testing procedure for all future notifications, but at any one time the tests performed by all reputable manufacturers are those demanded by the whole fund of experience and foresight available at that time.

The non-statutory scheme for notification of pesticides in Britain is organised in the following manner. In 1957, the Trade Associations representing manufacturers, importers and servicing companies agreed with the Government that they would not market new pesticides unless the formulations had been cleared for safety. The agreement included not only new active ingredients but also new formulations (or new uses) of old formulations (BAA, 1978). The scheme was revised in 1964, and re-named the **Pesticides Safety Precautions Scheme (PSPS)**. The Ministry of Agriculture, Fisheries and Food (MAFF) is advised on the operation of the scheme by an Advisory Committee on Pesticides, composed of experts in toxicology, medical science, agriculture and other disciplines; the pesticide industry has itself no representatives on this Committee.

The Scheme (PSPS) has a Secretariat comprising Government Scientists. For simple cases of notification, such as a minor modification of an existing formulation, this Secretariat and its consultant advisers can recommend clearance to the Ministry of Agriculture, Fisheries and Food. For all new active ingredients, and often also for old ones when used or formulated differently, the toxicological and other data provided by the notifier are sent to a Scientific Subcommittee. The Subcommittee can hear evidence from the notifier and can call upon the expertise of several specialist panels covering such subjects as medical toxicology, wildlife and residues in food. The recommendations of the Subcommittee are sent to the Advisory Committee, which in turn advises the Minister. If the substance is cleared, the recommendations are printed, and circulated under the title *Chemical compounds used in agriculture and food storage in Great Britain – user and consumer safety – advice of Government Departments*. The salient features of these recommendations must eventually appear on the label of the new product.

Now follows the proverbial tail that wags the dog. When, but only when, clearance is given under the Pesticides Safety Precautions Scheme, the product can be passed to the Agricultural Chemicals Approval Organisation. This body is concerned with the field efficiency of the product and its task is to ascertain that the product fulfils the claims made on the label by the manufacturer and that its efficacy is such as to merit its appearance on the market. The Organisation issues a certificate of Approval, renewable each year, after it has considered the evidence submitted by the manufacturer, backed up, if need be, with independent work done by Agricultural Advisory Services or Research Stations. The Organisation operates under the **Agricultural Chemicals Approval Scheme** and its work safeguards the purchaser in a financial sense. It should be noted that Approval is granted on a specific basis – not for an active ingredient, but for a particular formulation to be used for a specific purpose under United Kingdom conditions.

For each product for which a Certificate of Approval has been issued, the label of the container carries the official mark of the Agricultural Chemicals Approval Scheme. This mark comprises an A with a crown above it, and is sufficiently prestigious to ensure that most manufacturers wish to participate in the Scheme. A **List of Approved Products** is issued each year by the Ministry of Agriculture, Fisheries and Food and is published by Her Majesty's Stationery Office. It not only lists the Approved Products but is a valuable compendium of their uses and limitations under British conditions. It also connects the proprietary or trade names of products marketed in Britain with the internationally-accepted common name for the active ingredient present in each formulation. The latest brochure (MAFF, 1980) lists some 770 proprietary products under the headings of nearly 200 active ingredients.

More recently, the British Agrochemicals Association, in collaboration with other Organisations, has introduced a scheme to extend the principles of safety embodied in the PSPS to the *distribution* of pesticide products. The scheme is the **British Agrochemical Supply Industry Scheme** (BASIS, 1978). The object of this scheme is to ensure that staff in the distribution and supply industries are competent in matters of safety, efficient usage and law. If need be, this will involve the sponsoring of occasional training schemes. By such means it is intended that a high standard of safety in distribution should be maintained and that competent staff are available to help users to buy the right products and to use them efficiently.

Although the three Schemes described earlier in this section do not possess the force of law, the law is, in practice, never far from those who make, distribute or use pesticides. Clearly, if accidents occur and negligence on the part of manufacturer or distributor were to be suspected, civil prosecutions of various kinds could ensue. In addition, numerous types of more general legislation are applicable to the manufacturers, distributors and suppliers of pesticides as they are to those who work in other industries. The British Agrochemicals Association (BAA, 1978) list over twenty pieces of legislation relevant to their industry. Amongst them are the Health and Safety at Work, etc. Act 1974; the Agriculture (Poisonous Substances) Regulations 1975; the Deposit of Poisonous Wastes Act 1972; various Protection of Birds and Animals Acts; the Control of Pollution Act

1974; the Trade Description Act 1972 and the Consumer Safety Act 1978. In addition, as has already been mentioned, British exporters normally have so to order their businesses that their products conform with the statutory requirements of the countries to which they hope to sell large quantities of their products.

1.7 SAFETY ARRANGEMENTS IN OTHER COUNTRIES

Despite its non-statutory nature, the requirements of the PSPS scheme in Britain are not greatly different from requirements elsewhere, but the absence of rigid legal definitions make the operation of the Scheme very different. According to the British Agrochemical Association, an important advantage of the British scheme is that discussions between the company and the authorities can begin at an early stage in the development of a product. This contrasts with, for example, the obligatory system in Germany, for which the completed results are handed over to the authorities who then repeat the tests in their own laboratories at the company's expense (BAA, 1977).

A second difference between the British regulations and those of some other countries (including the USA) is that the regulations of these other countries include maximal limits for residues on harvested crops. An authoritative list of tolerances has been published jointly by the World Health Organisation and the Food and Agriculture Organisation of the United Nations (WHO, 1973, updated yearly). The British regulations do not rely directly on tolerance limits, but probably achieve a similar result indirectly, because the Safety Precautions Scheme is such that the authorities satisfy themselves before clearance that under the conditions of that clearance, harmful residues will not be left at the time of harvesting if the product is used as recommended. In practice residue limit legislation is very difficult to enforce since only a small proportion of marketed produce is likely to be sampled for analysis.

The European Economic Community (EEC) has so far concerned itself with two major directives on pesticides. One of these will prohibit, as from 1st January, 1981, the marketing of specified organochlorine and organomercury pesticides. The other has established tolerances for residues on fruit and vegetables; these maximal residue figures must not be exceeded by national governments when they themselves lay down residue limits. A similar directive for residues on cereals is being discussed. In addition, a draft directive, still under consideration, is designed to harmonise pesticide registration schemes in the nine member states. One consequence of adoption of this proposal would be to enable a company to seek EEC clearance in one country which, if granted, would allow free circulation of the product in all countries of the Community.

In the USA, a major aim of legislation is to control the level of residues on or in a crop at the time of harvesting. Public Law 518 – the Miller Amendment, was enacted in 1954. Briefly, the Act requires that any person registering a pesticide under the **Federal Insecticide, Fungicide and Rodenticide Act** should present a petition for the establishment of a **tolerance** for the new compound. The petition is lodged with the Secretary of Health, Education and Welfare, and, for it to succeed, information has to be provided about the composition and proposed use of the substance, together with data on the acute and chronic toxicity of the compound and the proposed analytical procedure whereby the magnitude of any residue on the crop could be determined (Gunther and Jeppson, 1960). Should the United States Food and Drugs Administration conclude that the substance is safe, it is exempt from a tolerance limit. Otherwise, a tolerance is ascribed which may vary with the manner in which the substance is used and with the crop to which it is applied. The tolerance, expressed as ppm fresh weight, represents a residue limit which must not be exceeded in the harvested food material. **A Federal Register of Official Tolerances** comprises several thousand entries; a compilation from this list was reprinted annually until 1970 by the National Agricultural Chemical Association (NACA, 1970).

The factors taken into consideration by those establishing such tolerances are similar to those discussed in sections 1.5 and 1.6. A dose is determined which just produces some adverse effect in an experimental animal, and it is then divided by an arbitrary safety factor (often 100) to make it applicable to human safety requirements. In addition, tolerance limits may reflect the probable contribution of a particular foodstuff to the diet of the average citizen. For example, a residue is more likely to be harmful to most people if present in milk or maize at 1 ppm than in crabs or caviare at 10 ppm.

The Environmental Protection Agency was established in 1970, and one of its functions has been to take over from the United States Food and Drugs Administration the licensing and monitoring of pesticides. It now has the authority to register pesticides and to assign tolerance limits, but the **Food and Drugs Administration** retains the power to sample foods to determine the levels of pesticide residues.

In addition to these two organisations, various agencies contribute to other aspects of (pesticide) safety in the USA. The National Research Council is concerned with the promulgation of information about the safe use of pesticides. One consequence of this approach is to minimise the improper use of pesticides on food crops. The Occupation Safety and Health Act helps to protect workers who handle poisonous chemicals on the farm. In addition, much spraying is done by air and this aspect of safety is controlled by the *Federal Aviation Administration*. On the ground, transport and storage are regulated by several Acts, two important ones being the Hazardous Materials Transportation Control Act and the Federal Environmental Pesticide Control Act. A detailed consideration of pesticide safety measures and legislation is to be found in the Pesticide Handbook prepared by the Entomological Society of America (ENTOMA, 1977).

REFERENCES

BAA (1977). *British Agrochemicals Association. Pesticides in Agriculture: Submission of Evidence to the Royal Commission on Environmental Pollution*, Alembic House, 93 Albert Embankment, London, SE1 7TU

BAA (1978). *British Agrochemicals Association: Directory*, Alembic House, 93 Albert Embankment, London SE1 7TU

BAA (1980). *British Agrochemicals Association, Annual Report 1979/1980* of the Year ended 31st March, 1980, Alembic House, 93 Albert Embankment, London SE1 7TU

Benitz, K.-F. (1970). *Methods in Toxicology* (ed. G. E. Paget), Blackwell, Oxford p. 82

BASIS (1978). *British Agrochemicals Supply Industry Scheme: A Registration Scheme for Distributors of Crop Protection Chemicals*, 3 Whitehall Court, London SW1A 2EQ

Braunholtz, J. T. (1979). *Chemy. Ind.*, No. 22, Nov. 17, 1979

Brown, V. K. (1980). *Acute Toxicity in Theory and Practice*, Wiley, Chichester

Bryce-Smith, D. and Waldron, H. A. (1974). *Chemy. Br.*, **10**, 202

Du Bois, K. P. and Geiling, E. M. K. (1959). *Textbook of Toxicology*, Oxford University Press

Edson, E. F., Sanderson, D. M. and Noakes, D. N. (1966). *Wld Rev. Pest Control*, 5, 143

ENTOMA (1977). *Entomological Society of America: Pesticide Handbook. 27th edn for the year 1977–78*, 4603 Calvert Road, Box AJ, College Park, Maryland 20740, USA

Fairchild, E. J. (1977). *Agricultural Chemicals and Pesticides: A Sub-file of the Registry of Toxic Effects of Chemical Substances.* US Dept. of Health, Education and Welfare, Public Health Service, Center for Disease Control, National Institute for Occupational Safety and Health, Washington, D.C.

Finney, D. J. (1952). *Probit Analysis*, 2nd edn, Cambridge University Press

Green, M. B., Hartley, G. S. and West, T. F. (1977). *Chemicals for Crop Protection and Pest Control*, Pergamon Press, Oxford

Gunther, F. A. and Jeppson, L. R. (1960). *Modern Insecticides and World Food Production*, Chapman and Hall, London

Hassall, K. A. (1955). *Ann. appl. Biol.*, **43**, 615 (Fig. 1 in the present book shows the data from which the log LD_{50} for *sec*-amyl iodide was calculated)

Headley, J. C. (1972). *Pest Control Strategies for the Future*, Nat. Acad. Sciences, Washington, D.C., p. 100. Available from the National Academy of Sciences, 2101, Constitution Avenue, Washington, D. C., 20418

Horsfall, J. G. (1956). *Principles of Fungicidal Action*, Chronica Botanica Co., Waltham, Mass., USA

Klingman, G. C. and Ashton, F. M. (1975). *Weed Science: Principles and Practices*, Wiley, New York

MAFF. (1979). *Ministry of Agriculture, Fisheries and Food. Pesticides Safety Precautions Scheme Booklet, Revised 1979.* (A loose-leaf dossier comprising *inter alia*, five Appendices A-E. It is obtainable from the MAFF, Great Westminster House, Horseferry Road, London SW1P 2AE)

MAFF (1980). *Ministry of Agriculture, Fisheries and Food (UK), Agricultural Chemicals Approval Scheme: List of Approved Products for Farmers and Growers*, HMSO, London

NACA (1970). *Natn agric. Chem. Ass. News*, **28**, (3). (A list of tolerances compiled from the official tolerance orders as published in the Federal Register, Food and Drugs Administration). NACA News ceased publication in 1970

Ordish, G. (1976). *The Constant Pest*, Peter Davies, London

Sharratt, M. (1977). *The Evaluation of Toxicological Data for the Protection of Public Health*, published for the Commission of the European Communities by Pergamon Press, Oxford, p. 105

USDA (1978). *United States Department of Agriculture: Agric. Statist.* See Table 668 (p. 470) and Table 670 (p. 471). For 1971 production figures, see Table on p. 470 of the 1975 volume

Vettorazzi, G. (1977). *The Evaluation of Toxicological Data for the Protection of Public Health*, Published for the Commission of the European Communities by Pergamon Press, Oxford, p. 207

World Health Organisation (1973). *Pesticide Residues in Food. Tech. Rep. Ser. Wld Hlth Org., No. 525, Annex 1.* Revised and amended ADIs and recommended maximal residue limits are given in Annex 1 of each of the yearly Tech. Rep. No. 545 (1974), No. 574 (1975), No. 592 (1976) and No. 612 (1977)

2 Physicochemical Aspects of Pesticide Formulation and Application

2.1 TARGET SURFACES FOR PESTICIDE APPLICATIONS

The cuticle of plant leaves is frequently treated with pesticidal sprays or dusts. It may be a target by proxy, as when the purpose of spraying is to hit insects or fungi which live on or in the crop plants. Alternatively, leaf cuticle may be a direct target, an *intermediate* barrier which must be penetrated before a herbicide can reach its ultimate objective such as weed chloroplasts or meristems. Similarly, the cuticle of an insect represents an intermediate target for an insecticide, a barrier which must be reached and breached before the active ingredient can penetrate to its site of action which is frequently some part of the nervous system. Fungal hyphal walls and spore walls are similarly intermediate targets for fungicides.

Soil is structurally a very different intermediate target. Rich in silicates, a typical soil in some ways resembles an ion exchange column with fixed negative charges. By successive adsorption and elution, many substances move downwards at different rates. Damp soil contains a continuous aqueous phase and persistence in soil of a pesticide of low aqueous solubility is often (but not always) greater than that of one which is highly soluble. The situation is, however, usually more complex than this. For example, soil contains air spaces and the volatility of the active ingredient may be a factor of importance. The chemical stability of the pesticide in the presence of water and its biochemical stability in the presence of soil micro-organisms must also be taken into consideration. In other cases, the compound is strongly bound within the latticework of clay minerals or is adsorbed on to soil organic matter.

The physicochemical characteristics of waxy cuticular surfaces contrast with those of soils. Consequently, the conditions which favour deposition and persistence of a particular pesticide on a leaf may be different from those which are most favourable for its use in soil. This is one reason why a pesticide manufacturer may **formulate** the same active ingredient in a variety of ways (section 2.4).

In this chapter, the structure of plant leaf cuticle and insect cuticle will be considered; factors associated with soil structure are more conveniently discussed in chapter 11. Later sections of the present chapter deal with application methods in relation to the physical state of the application system. To ensure that the best use is made of an active ingredient, it is often necessary to add supplementary substances to mixtures containing such ingredients. The closing sections of the chapter describe the main groups of supplementary substances and their mode of action.

2.2 PLANT LEAF CUTICLE

An important physical characteristic of the surface of foliage is the existence of a more or less substantial waxy, and therefore hydrophobic, barrier, which diminishes water loss and simultaneously restricts access of aqueous solutions (figure 2.1). This protective layer is deposited outside the cellulose walls of epidermal cells and merges into them. It varies in thickness and composition in different plant species and in different parts of the same plant.

Young parenchymatous cells are initially separated by a middle lamella comprising salts of pectic acid. The primary cell wall is thin and largely consists of cellulose and pectin (a methyl ester of pectic acid). Consequently, few intermolecular ionic forces exist and the young cell wall is elastic, with macromolecular chains able to glide over one another to accommodate extension growth. The thicker secondary cell wall becomes increasingly important as the cell ages, providing strength at the expense of flexibility. Although variable in composition, it contains cellulose, pectins, pectates and hemicelluloses. Unless 'waterproofed' in some way, even secondary cell walls are permeable to water and probably to most substances dissolved in it. Inside the cell wall lies the delicate plasmalemma, a lipoprotein membrane which normally prevents cell solutes from escaping outwards.

Figure 2.1 shows diagrammatically the structure of plant leaf cuticle. Cuticle is not an amorphous structure but comprises several components organised into

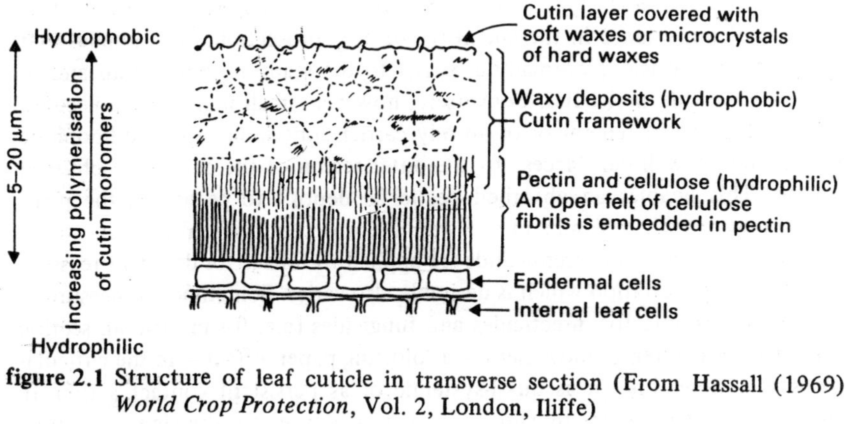

figure 2.1 Structure of leaf cuticle in transverse section (From Hassall (1969) *World Crop Protection*, Vol. 2, London, Iliffe)

a latticework of considerable complexity, and usually between 5 and 20 μm in thickness. A major component of the non-polar region is cutin; this forms a sponge-like lattice within the compartments of which platelets of waxy material are deposited. Cutin is a cross-linked polymer of various hydroxy- and carboxylic acids, insoluble in most solvents. The degree of polymerisation varies, being highest towards the outside of the leaf and least where the cuticle merges into the cellulose cell wall. Permeability to water is inversely correlated with the degree of polymerisation, so the internal cutin is more permeable to aqueous solutions than is the cutin towards the outside (Martin and Juniper, 1970).

Cuticular wax is variable in composition, but in most plants is predominantly composed of straight-chain alkanes or alkanes carrying terminal hydroxyl or carboxyl groups. The hydrocarbons normally contain an odd number of carbon atoms, 21–35 altogether, whereas the acids and alcohols contain even numbers of carbon atoms, the commonest numbers being 20–24. The outermost layer of the cuticle is often a very thin deposit of soft wax, of microcrystals, or of both; this layer, when present, is readily damaged by external solvents. Most of the cuticular wax is, however, embedded in the interstices of the outer region of the cuticle and is not readily removed by immersing the leaf in organic solvents.

Stomata, trichomes (hairs) and hydathode channels are intrusive structures in the cuticles of most leaves. Hydathodes provide channels through the cuticle along which outward-moving water can pass, but they probably do not play an important role in the uptake of pesticidal solutions. Hairs are of some importance in relation to the availability of pesticides, in that they can, when present at appropriate density, form a mesh which prevents spray droplets from running off the leaf, and so allows a toxicant to be deposited in larger amounts than it would be on glabrous leaves. Stomata do not provide entry points for aqueous solutions, or for solutions containing only small quantities of detergents. Some oily liquids can, however, enter the stomata, probably via the sensitive guard cells, and cause extensive damage to the cells lining the stomatal cavity.

The preceding paragraphs and figure 2.1 show that leaf cuticle is a system in which hydrophilic and hydrophobic properties are elegantly balanced. This balance is of importance in the chemistry of crop protection, for it implies that the undamaged cuticle is normally a barrier to the ready ingress of ions and of most water-soluble molecules. Conversely, however, small non-polar molecules, whether of active ingredient or of an oily carrier containing dissolved lipophilic toxicant, might well gain access to the outer region of the cuticle by physical sorption. This enables an absorbed pesticidal deposit to persist on the plant under conditions where wind and rain would remove a superficial deposit.

Since pesticides of different kinds are applied to foliage for different reasons, the degree of penetration which is desirable also varies. Sometimes, a measure of sorption is desirable for insecticides and fungicides (e.g. for malathion, section 4.6). Such local internal movement – a 'blotting-paper' effect – in the cuticular wax is sometimes known as **quasi-systemic** or as persistent contact action. In contrast, some substances remain as superficial deposits possessing an entirely local **contact** action, whilst others, with a **systemic** action, penetrate right through the cuticle and are transported around the plant.

Leaf cuticle represents a macroscopic barrier to the free movement of many pesticides, but there are ubiquitous, if less obvious, microbarriers to be surmounted. These are the lipoprotein membranes which surround all living cells. They are sometimes regarded as consisting of lipoprotein double layers (protein, lipid, lipid, protein) although more recent evidence favours a more indefinite fluid membrane. The fluid membrane is visualised as being composed of lipid penetrated in places by proteinaceous 'plugs' as well as being, in part, coated with protein. Such membranes are dynamically semi-permeable, their properties varying with the physiological state of the cell. They probably act as barriers to the free entry of many types of poisons, although selective permeability is lost if they are damaged by intruding molecules.

2.3 THE INSECT INTEGUMENT

The outermost living layer of the insect body is called the **epidermis**. It secretes a non-living, but structured, protective **cuticle** on its outer surface. The cuticle and epidermis, together with any modified epidermal structures within them, are known as the insect integument (figure 2.2).

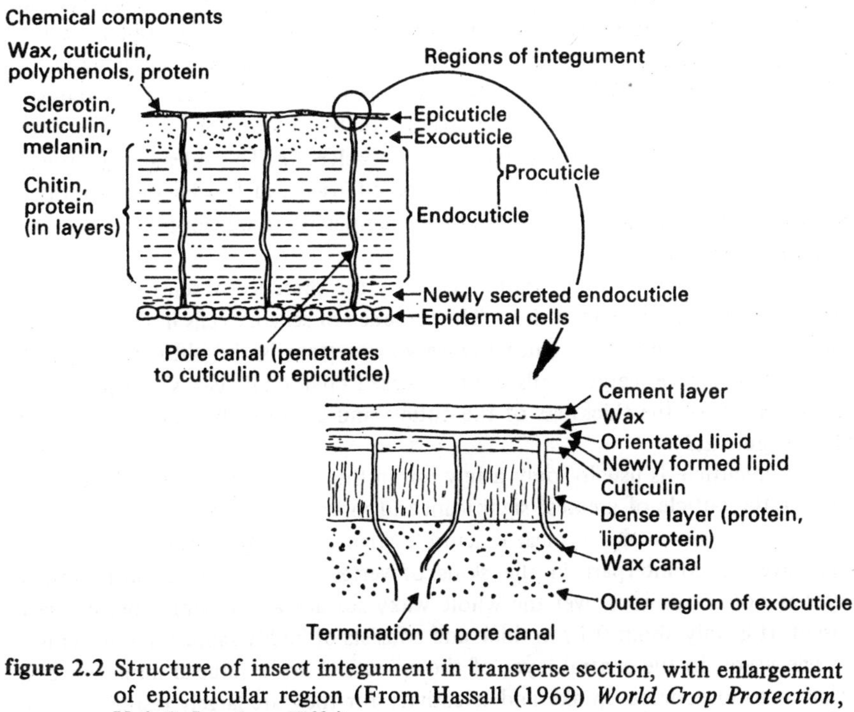

figure 2.2 Structure of insect integument in transverse section, with enlargement of epicuticular region (From Hassall (1969) *World Crop Protection*, Vol. 2, London, Iliffe)

The insect cuticle covers the whole exterior of the insect body and also lines the gut and the cavities of the spiracles, although its thickness varies greatly from insect to insect and from one part of an insect to another. It is a secretion, rich in proteins, and in most insects comprises two main regions. These are the **procuticle**, a thick inner layer consisting largely of glycoproteins, chitin and tanning agents and the **epicuticle**, a thin outer layer which is rich in lipids and lipoproteins.

The procuticle is secreted in distinct layers by the epidermal cells and often represents 90 per cent or more of the total thickness of the cuticle. **Chitin** is its most distinctive component and may account for 60 per cent of the total cuticular weight. It is a stable and water-insoluble polymer of several hundred residues of N-acetyl glucosamine. It contains β-linkages similar to those that render molecules of cellulose resistant to attack by most enzymes (exceptions are some bacterial hydrolases). There are at least three kinds of chitin in cuticle, and these are associated with proteins to form molecular complexes (Rudall, 1963). Chitin helps to provide strength while preserving a measure of flexibility, but it is not responsible for the impermeability of cuticle to many types of water-soluble compounds. Such impermeability is largely a consequence of the presence of wax in the extremely thin cuticle (figure 2.2).

The procuticle can conveniently be considered as comprising a more hydrophilic inner region of **endocuticle** and a more lipophilic outer region of **exocuticle**, although it is, in fact, a single entity, consisting of merging layers which vary in structure and thickness from one insect species to another. The outer, darker exocuticle contains a pigment, **melanin**, which is a complex condensate of various polyphenolic substances. The same polyphenol oxidase that catalyses melanin production is probably responsible for the interaction of proteins with dihydric phenols to form the derived protein, **sclerotin**. This substance is largely responsible for cuticular rigidity in hard-coated insects such as beetles. Although tough, the exocuticular region is probably permeable to aqueous solutions and is saturated with body fluid.

Some parts of the cuticle are modified by the presence of sensory cells, wax canals and various other inclusions. The presence of sensory cells in the exocuticular region is noteworthy, since the nervous system is the ultimate target of many pesticides. Receptor cells are specialised epidermal cells which penetrate through most of the thickness of the cuticle and are directly attached to nerve fibres (figure 2.3).

The epicuticle is usually only a few μm in thickness but it is nevertheless the part of the cuticle of greatest significance in insecticide chemistry. Two of the three main regions of a typical epicuticle (figure 2.2) are extremely lipophilic and have no counterpart in the skins of vertebrates. The outermost cement layer is not continuous over the whole waxy surface and is sometimes entirely absent. It is only about 0.1 μm thick and consists of highly tanned lipoproteins.

The waxes in the central layer of the epicuticle are of crucial importance in relation to insecticide uptake and selectivity for they are largely responsible for the water-impervious nature of the total cuticle. Waxes of different chemical

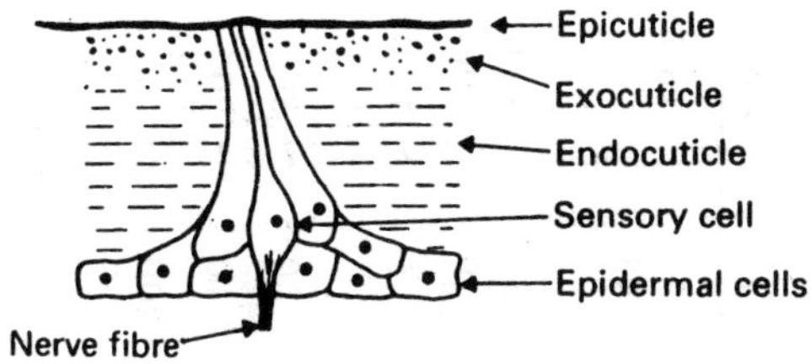

figure 2.3 Modification of insect cuticle in the region of a sensory cell (From Hassall (1969) *World Crop Protection*, Vol. 2, London, Iliffe)

composition, and therefore possessing different physicochemical properties, are present in different organisms. Such differences could well determine how readily an insecticide penetrates the cuticle of a particular sort of insect at some stated temperature. The wax of the cockroach epicuticle, for example, melts at 30°C, while that of the blood-sucking bug, *Rhodnius prolixus*, melts above 50°C. The ability of a lipid-soluble substance to penetrate into these two species of insect might therefore be expected to differ considerably at, say, 40°C.

The inner layer of the epicuticle is penetrated by wax canals which, via the wider pore canals, penetrate right through the procuticle down to the epidermal cells. It is through these that the wax originating from the (living) epidermal cells is eventually deposited in the wax layer of the epicuticle. This inner layer consists mainly of protein but it is also rich in polyphenols and cuticulin. The latter is a lipoprotein tanned with quinones and polyphenols.

It will be evident from the foregoing account that the insect cuticle in some ways resembles plant cuticle. In both cases, the inner layer of the protective barrier is less waxy, less rich in lipoprotein and less cross-polymerised than the outer region and is more or less freely permeable to water and to solutes dissolved in it. The lipid, hydrophobic nature of the outermost layers of both leaf and insect cuticles is of importance in plant protection chemistry, for many pesticides, when applied to leaves, are suspended or dissolved in water. A 'water-off-a-duck's-back' problem can clearly arise in such circumstances and is overcome by formulating in the presence of surfactants (section 2.9).

There can be no doubt about the ease with which certain insecticides in suitable formulation can gain access to the bodies of many insects. As well as the remarkable and nearly instantaneous knock-down effect of natural pyrethroids (section 7.4) a comparison of the LD_{50}s of insecticides applied topically and by injection, first to insects and then to mammals, shows that, with a few exceptions, insect cuticle gives little protection against oil-soluble substances. Thus the values for the ratio

$$\frac{\text{Cuticular (or Dermal) } LD_{50}\text{, ppm body weight of insect (or mammal)}}{LD_{50}\text{ by Injection, ppm body weight of insect (or mammal)}} = \frac{C(\text{or } D)}{I}$$

are often small for insects (C/I) and should be contrasted with the larger values (D/I) often observed for mammals (table 2.1).

Table 2.1 LD_{50} of insecticides applied by different routes (ppm body weight)

Substance	Insect	LD_{50} (ppm)			Mammal	LD_{50} (ppm)		
		To Cuticle (*C*)	Injected (*I*)	*C/I* ratio		To skin (Dermal, *D*)	Injected (*I*)	*D/I* ratio
DDT	Cockroach	10	5–8	1.5	Rat	2500	15	16.7
	Japanese beetle	93	162	0.6	Rabbit	–	45	–
	Bee	114	0.2	570.0				
Lindane	Cockroach	5	4	1.2	Rat	500	35	14.0
					Rabbit	180+	6	30.0
Parathion	Cockroach	1.2	0.95	1.3	Rabbit	50	~ 5	~ 10.0
	Japanese beetle	3.3	0.45	7.3				
Malathion	Cockroach	–	8.4	–	Rat	4000+	50	~ 80
DNOC	Cockroach	20	14	1.4	Rat	200+	10	~ 20
Pyrethrins	Cockroach	6.5	6	1.1	Rat	1800+	200	~ 9
	Japanese beetle	40	40	1.0	Dog	–	7	–
Rotenone	Cockroach	>2000	10	~200	Rat	~1000	6	~167
	Japanese beetle	40	25	1.6				

Most of the figures for Mammals are quoted from Edson *et al.* (1966) or Negherbon (1959) and most of the figures for insects are from Metcalf (1955). Mammalian injection figures are for intravenous, or occasionally for intraperitoneal or subcutaneous, injections. The *C/I* and *D/I* ratios for any one substance are a rough index of the extent to which the cuticle of insects and the skin of mammals can act as barriers to penetration.

2.4 PESTICIDE FORMULATION AND STATE OF APPLICATION

With the exception of many substances used as fumigants, few active ingredients can be conveniently and effectively used without subjecting them to some grinding, mixing or dissolving process. The preparation of the active material in a form which is convenient for its intended use is known as formulation. Most active ingredients are very biologically active and it would be difficult to spread them evenly over a field at a rate of less than about a kilogram per hectare if they were not diluted in some way. In addition, many are greasy, organic substances insoluble in water, so simply mixing such substances with water in the spray tank would be very inefficient. A common maxim is that no active ingredient can perform more adequately in the field than its method of formulation will allow.

In later sections, some of the principal methods of formulation will be described. Briefly, however, an active ingredient may be applied in the dry state, in which case it may be formulated and applied as a **dust** or as **granules**. Alternatively, it may be applied as, or in the presence of, a liquid. The three major formulations for this purpose are the **aqueous concentrate**, the **wettable powder** and the **emulsifiable concentrate**, each of which is diluted by addition of water in the spray tank. Miscellaneous formulations of occasional importance are **smoke generators, aerosols, emulsified liquids, poison baits** and **microcapsules.**

It should be observed that the effective state of an active ingredient at the time of application has little to do with whether it is a solid or a liquid in the pure state. For example, a liquid active ingredient such as malathion can be adsorbed on to a solid and applied as a dust. Conversely, a solid active ingredient such as DDT can be so formulated that it can be applied so that there is an aqueous continuous phase, the DDT being dissolved in oil which in turn is emulsified in water. It should also be noted that the state in which an active ingredient is applied may be different from its state at the moment of uptake by the target organism. For example, aldrin is frequently applied to soil as a solid but it often kills soil-living insects by uptake as a vapour.

2.5 FUMIGATION

When an active ingredient is applied as a vapour, the process is termed fumigation. Within moderate limits, the concentration (c) required for a given level of response is inversely related to the time (t) of exposure; therefore,

$$c \times t = \text{constant} \qquad (2.1)$$

The advantage of using a toxicant in the form of a vapour is that it can penetrate to every crevice of a system that constitutes a continuous gaseous phase. However, the successful use of fumigants also implies the requirement for an enclosed space; once that enclosed space is opened up, the toxic vapour rapidly disperses. This is not necessarily a disadvantage, for a notable use of fumigants is to protect food during storage and transportation. Holds of ships and freight wagons can be readily sealed, and since such vehicles often contain foodstuffs on their way to market, the use of a non-persistent active ingredient (e.g. methyl bromide) is an advantage in that it presents a low residue hazard to the consumer.

The need for an enclosed space limits the usefulness of fumigation in the protection of growing crops, but fumigation of greenhouses is of considerable value to horticulturists. As well as using true gases or vapours, it is possible to place volatile solids upon the steam pipes in greenhouses. The solid then volatilises to form a vapour which usually condenses to a suspension of colloidal solid particles; this has almost the same penetrative power as a true vapour. The 'false fumigant' condition can also be achieved by smoke generators; these are commercially available for applying several insecticides and acaricides. Similarly, aerosol preparations produce a suspension of tiny droplets of liquid and this mist also simulates the true gaseous state. For the fumigation of grain, 'fumigation tablets' can sometimes be used. An example is aluminium phosphide, which slowly emits phosphine (PH_3) on contact with the moist atmosphere in a sealed grain store.

Soil restricts the diffusion of volatile toxicants and it is possible to fumigate soil by the use of a special injection device which can be plunged into the soil at regular intervals of distance and which delivers a pre-determined dose of a volatile liquid to a specified depth. Partial soil sterilants such as 'D-D' mixture (dichloropropene plus dichloropropane) and dibromochloropropane can be administered to soil in this way, to control some nematodes and soil fungi.

Some compounds of relatively low volatility are so toxic that they are able to vaporise in the soil air space to an extent that enables them to reach a concentration great enough to kill soil pests by vapour contact. Both aldrin and lindane (chapter 5) probably control wireworms and leatherjackets by a predominantly fumigant action. Triallate may similarly kill wild oats by fumigant action (chapter 13).

Aerosols that contain pesticides are commonly employed by home gardeners. Pyrethroid aerosols are used in the home against flies and mosquitoes, and malathion and pirimicarb aerosols are also used for small-scale application purposes. In such formulations the active ingredient (1–10 per cent w/v) is dissolved in a high boiling point solvent such as kerosene (1–10 per cent w/v) and the solution is mixed with a liquid which is gaseous at atmospheric pressure (80–98 per cent w/v). The latter is known as the **propellant**. The mixture is enclosed in a container under pressure exerted by the vapour of the low boiling point liquid and the device works on the soda-water siphon principle. Dichlorofluoromethane is frequently used as the propellant and exerts a pressure of 90 psi when liquid at room temperature. This solvent rapidly evaporates leaving a mist of fine droplets comprising the active ingredient dissolved in the high boiling point liquid.

2.6 APPLICATION OF SOLIDS AS DUSTS OR AS GRANULES

The preparation of pesticidal **dusts or powders** has been described by Polon (1973). Usually, a suitable sorptive material is impregnated with a concentrated solution of the pesticide. The solvent is then evaporated and the powder diluted with a large bulk of the same sorptive substance. To minimise transportation costs, the process may be done in two stages, the primary concentrated dust being shipped to a local depot prior to secondary dilution and packaging. An alternative method of preparation is to mill 1–10 per cent of the active ingredient with the sorptive diluent.

The minerals whose particles are impregnated with pesticides are of two kinds, The **clays** include kaolinite and attapulgite (fuller's earth) and the **silicas** include talc, diatomite and pumice. Such absorptive diluents are often referred to as **fillers** or **carriers**. Most dusts are so prepared that almost all the particles are less than 50 μm diameter and the majority normally lie within the range of 3 to 30 μm. Although small particles are often more efficient than larger ones they are also a greater inhalation hazard, and suitable face masks should be worn by all who handle dry powders (mineral particles may be abrasively damaging to lung epithelia, even if the pesticide is innocuous).

The application of dusts to foliage, in contrast to their use for soil or seed treatment, is not often done on a large scale. The reason for this is that not more than 10–20 per cent of the applied material normally adheres to dry foliage, although retention is much improved if the dust is applied when the plant leaves are damp with dew. Other problems relate to the tendency for dusts to be blown to neighbouring crops in all but the calmest weather, and to the tendency for the particles to cake under humid conditions. In addition, deposited dusts are often vulnerable to wind and rain occurring shortly after application, although small particles ($<$ 10 μm) adhere much more tenaciously than large ones.

In countries where water is scarce and the weather predictable, dusting may be the preferred method of foliage application. Dusting appliances are usually turbine blowers which are relatively light, cheap and manoeuvrable compared to medium-volume spray machines and the process of dusting is usually more rapid than that of spraying a similar acreage. However, fine particles tend to clump together when blasted out of a dusting machine, and technical difficulties exist in designing dusting machines that dispense powder evenly.

The method of formulation may greatly influence the effectiveness of the dust, for, within reasonable limits, the number of particles per unit area of foliage is more important than the weight of active ingredient per unit area. Consequently, a finely ground mixture of toxicant with a suitable diluent operates more efficiently than one which is less finely milled or less well mixed. On the other hand, particles smaller than about 20 μm may not settle out on warm days because of the effect of convection currents, and electrical forces may influence, favourably or adversely, the foliar deposition and retention of particles.

Pesticidal dusts are often applied to soils. Particle size may also be significant for dusts used in this way since the rate of volatilisation of a substance with an appreciable vapour pressure increases with the degree of dispersion. As was seen above, some dusts applied to soils function, at least in part, by evaporation of the active ingredient into the soil air spaces, and for all substances, dispersion is improved by decrease in particle size. Many insecticides and soil-acting herbicides can be applied in this way. In some cases the dust is sprinkled in trenches before placing tubers (e.g. potatoes) or in holes before transplanting seedlings (e.g. brassicas). In others, the dust is scattered around row crops (e.g. for carrot root fly control). Some insecticides and herbicides are dispersed as dusts on the surface of the soil and then tined into the soil to an appropriate depth. Dusts may well be more convenient than surface-applied sprays for many of these purposes, and most of the disadvantages associated with foliage application of dusts are not applicable.

Granular formulations are occasionally used for treatments above ground level, an important example being their lodgement in the leaf funnels of maize as an ideal means for the control of the stem borer. Nevertheless, they are most commonly employed for ground level application, especially for application through a standing crop. This minimises wastage and avoids possible phytotoxicity. Granules are safer to handle and more convenient to apply than liquids and are consequently often sold in sprinkler tins for domestic use in Britain.

In seed furrows, it is easier to sow seeds and apply granules simultaneously than it is to sow seeds and to apply a spray liquid at the same time. Their application is less affected by wind and air currents than is the application of dusts and of some sprays and they can be placed rapidly but accurately on the top of the ground around individual plants. This procedure is particularly useful for the control of insects which lay eggs near the base of the stem (e.g. carrot root fly) and for weed control near fruit bushes and similar established crops. Rather similarly, granules are useful for application to cultivated soil before a crop is sown, the application often being followed by tining, rotavation or watering. Dusts and sprays in these circumstances can lead to clogging of machinery (Hartley, 1976). Granular formulations sometimes make it possible for root-systemic action to be exploited for substances which are poorly systemic when applied to foliage. Moreover, for appropriate types of pest control, granules are often the preferred formulation for application by aircraft. An example is to bounce granules through foliage to swampy ground to control mosquito larvae.

In addition to the advantages mentioned above, granules are a highly convenient formulation, especially for the small-scale user who tends to prefer solid products to liquids. Moreover, granular formulations are considerably safer than dusts or sprays for such hazardous substances as aldicarb (section 5.7) and paraquat/diquat (section 12.4). In the latter case, for example, small sachets of granules purchased by the home gardener can be emptied directly into water in a watering can, a procedure that avoids contact with a concentrated liquid or with an inhalable dust. A further advantage is that the toxic substance is usually released slowly from the granule, so that a granular application can often replace two or more spraying operations. A current disadvantage is that granular formulations tend to be expensive. In addition, granules are seldom used for direct foliage application, although as in the case of paraquat/diquat, they may be indirectly involved in this process for active ingredients that are water-soluble.

Sometimes the inert mineral carrier of a granular formulation is a porous reticulum impregnated with a solution of the active ingredient, In other cases the inert support is bead-like, with the active ingredient deposited on its surface. Inert supports include crushed and sieved silicate minerals (including attapulgite, vermiculite and kaolin), brick dust, limestone chips and walnut shell. Less frequently, granules are made by aggregating dusts by adding a solution containing an adhesive to the dust enclosed in a rotating drum. Elvy (1976) has described laboratory tests on potential carriers for granules.

Factors that can influence the choice of carrier for a granular formulation have been discussed by Polon (1973), by Hartley (1976) and by Matthews (1979). A particular carrier may be chosen for its sorptive capacity, its hardness and for the speed with which it disintegrates in water. For example, attapulgite clay granules will disintegrate in water unless they are strongly heated prior to use. A granular carrier for a water-soluble but hazardous substance has done its job by the time the material has been mixed with water and hence an unfired attapulgite might well be used, whereas a granular formulation for soil application would normally contain fired carrier particles. A granular formulation often only contains 5–20 per cent of active ingredient (and is thus costly to transport) and the rate of release of this substance from the granule will vary with the choice of carrier. If, when other parameters are satisfactory, the release of the active ingredient would still be too fast for a particular purpose, this can often be rectified by providing the already-treated granules with a plastic coating.

Many granular formulations contain particles that are mostly within the range 0.3–1.0 mm diameter (the apposite British Regulations require that not more than 4 per cent by weight should be able to pass through a 0.25 mm sieve). This means that there may well be many millions of granules per kilogram. However, as would be expected, any granular formulation will contain a range of particles of different sizes. A granular formulation is thus described by a double mesh number; one of these typifies a mesh with pores so large that all the particles pass through, the other a mesh with pores so small that all the particles are retained. For example, a '20/60 mesh' granular formulation will have granules of such a size that all will pass through a 20 mesh sieve and none will pass through a 60 mesh sieve. The relationship between sieve mesh, pore size and the approximate number of particles per kilogram is shown in table 2.2.

When granules are applied to soil they are not necessarily randomly distributed for, as was seen earlier, placement techniques are often to be preferred to broadcast treatments. The rate at which the pesticide escapes from a given formulation is largely determined by the extent of leaching by rainwater. Leaching, in its turn, is influenced not only by the duration and intensity of the rain but by the

Table 2.2 Relationship between sieve mesh, particle size and approximate number of particles per kilogram of a granular formulation

Mesh size of sieve	Size of particle which just passes through (mm)	Approximate number* of particles per kg (millions)
8	2.36	0.2
15	1.08	1
20	0.83	4
30	0.54	12
40	0.40	40
60	0.25	100
80	0.20	300

*The actual number may vary several-fold according to the precise nature of the carrier, its degree of hydration, etc.

temperature. The rate of release will normally decrease with depletion and thus with the passage of time. The degree of control over effective release from granules also depends upon the extent of adsorption of the active ingredient on soil particles and the rate at which it decomposes once it reaches the soil.

The theory of granule leaching has been discussed by Furmidge and coworkers (1966, 1967). The amount of pesticide leached from granules per unit area of land is expressed approximately by the equation

$$\text{Amount leached per unit area} = \frac{kR\alpha}{Dp\rho} \tag{2.2}$$

where R is the volume of rain per unit area, D is the diameter of the granules, p is the percentage of active ingredient in the formulation at any moment, α is the amount of toxicant applied per unit area and ρ is the density of the granules.

Hartley and Graham-Bryce (1980) have provided a detailed account of the physical principles which determine pesticide behaviour when granules (*inter alia*) are applied to soils. Each granule or droplet is a point source from which the pesticide is released, the hemispheres of diffusion from adjacent point sources overlapping to give patterns of distribution which can be calculated for stated idealised systems. Further complications arise in most soil situations since simple diffusion is often greatly modified by the superimposed effect of leaching while, simultaneously, progressive decomposition of many active ingredients occurs. The same authors (Hartley and Graham-Bryce, (1980 p. 763) present mathematical models describing the release of active ingredient from the matrix of slow-release granules and the effect of the factors which diminish evaporative loss from granules applied to the soil surface.

2.7 APPLICATION AS LIQUIDS

The most frequently used method for applying a pesticide is to spray it as an aqueous solution or as a suspension in water. Less commonly, an active ingredient may be applied dissolved only in a small volume of an oily organic solvent and, in a few special cases, a liquid active ingredient may be applied undiluted.

For the vast majority of uses the manufacturer supplies a concentrated pesticidal formulation which is diluted by the user by addition of an appropriate volume of water. Sometimes the substance (e.g. aminotriazole) is soluble in water and the concentrate is a true solution containing the active ingredient and a spreading agent (section 2.8). More usually, the pesticide is an organic substance insoluble in water. For such substances the concentrated formulation may consist of the active ingredient dissolved in a suitable organic liquid, and accompanied by one or more spray supplements (see below). Alternatively, it may consist of a **wettable powder** comprising the active ingredient, an inert solid diluent and various spray supplements. Oily preparations will usually tend to soak into the leaf, providing a more resistant and permanent deposit than is obtained with a wettable powder. An advantage of wettable powders, on the other hand, is that packaging and transportation tend to be cheaper.

Formulations containing the active ingredient dissolved in an organic liquid are of two kinds. In the most important of these, the **emulsifiable oil**, the solubility characteristics of the active ingredient are such that it is possible to form a moderately concentrated solution of it in a liquid mixture comprising an oil and an oil-soluble emulsifier. The oil may be an aliphatic or aromatic hydrocarbon, cyclohexanone, or some similar substance. In contrast to dusts and granules, the concentration of active ingredient is usually between 15 and 40 per cent w/v. This is advantageous in that it reduces the cost of transportation per unit weight of active ingredient.

An emulsifiable oil is a single-phase mixture, and when it is stirred into water it forms an emulsion, the particles of which are usually 8 μm or less in diameter. The emulsion is stabilised by the use of emulsifying agents, and in practice an emulsifiable concentrate normally contains a mixture of a non-ionic detergent of the polyethylene oxide type with an anionic detergent. It has been found that such a mixture gives better results than either detergent employed separately. One possible explanation for this is that anionic detergents increase in solubility as the temperature is raised whereas non-ionic detergents in such circumstances becomes less soluble. The stability of emulsions also depends upon the choice of emulsifiers possessing the optimal hydrophilic–lipophilic balance (HLB) for the particular active ingredient and organic liquid carrier (section 2.9).

In the second kind of liquid formulation, the manufacturer's concentrate is a two-phase system, and consists of the active ingredient dissolved in oil, the oily solution then being emulsified in water. This **stock emulsion** has a mayonnaise consistency and is diluted down by the purchaser by adding it to more water in the spray tank. These emulsified (as distinct from emulsifiable) concentrates were widely used in the days before the development of non-ionic detergents of the polyethylene oxide variety, for earlier detergents were mostly too insoluble in oils of appropriately low cost, low flammability and low phytotoxicity. Now that this problem has been solved, emulsified concentrates are not greatly used, principally because of their erratic storage qualities and because the presence of water causes many active ingredients to hydrolyse.

Wettable powders are solids so formulated that they disperse when stirred into water in the spray tank. They are thus quite different from dusts which are applied dry. Wettable powders consist of an active ingredient which may be solid or liquid, one or more spreading and dispersing agents, and a solid, inert diluent or filler. In manufacture, the active ingredient is often coarsely ground and then converted to very fine particles by subjecting the coarse particles to air blasts of such high velocity that they are literally blown into small pieces. The dispersers and spreaders ensure that the active ingredient does not separate out quickly in the spray tank and that the water droplets are able to flatten on a leaf surface. The filler is usually a porous mineral such as a clay or a diatomaceous earth, but may be a more alkaline substance such as calcium or magnesium carbonate.

Diluents or fillers are almost always needed because organic substances tend to be greasy, and this causes the particles to adhere together. On the other hand, the amount of filler should be kept to a minimum because it is both abrasive to equipment and adds to the cost of transportation. A typical wettable powder contains about 50 per cent active ingredient, 47 per cent filler and 3 per cent detergents. The active ingredient normally remains strongly adsorbed on the filler when the wettable powder is suspended in water, thus ensuring that it, too, is well dispersed. The particles of the majority of wettable powders are mostly within the range of 1–10 μm diameter. A liquid active ingredient can be as readily adsorbed on to a suitable filler as can a solid toxicant; this often provides a preparation which is not only safer to handle than a liquid formulation but also possesses good keeping qualities, low volatility and low corrosion.

Although water is almost always present when liquid applications are made, there has been a tendency in recent years for lower spray volumes to be used. **High volume** spraying involves applying some 1000 litres/hectare (90 gal/acre). In such circumstances the water functions as a carrier, i.e. the mass and the velocity of the large droplets which leave the spray tank are able to overcome air resistance and carry the spray to its target. **Medium** and **low volume** spraying usually require the use of high-powered fans, the droplets being impelled from the jet by a blast of air. Thus the principles underlying the use of high volume and low volume spraying are different – it is not just a question of the volume of water employed (table 2.3). Moreover, when high air pressure is used to provide droplet momentum, the average droplet size is lower than is the average for droplets from most high-volume spray machines. Since the concentration of the mixture in the spray tank is normally high for low volume application, the area covered per tank filling is roughly the same for low and high volume spraying.

Table 2.3 Application volumes and common methods of impulsion of spray droplets

	l/ha	Imperial gal/acre	Carried or impelled by:
Ultra low volume (ULV)	2–30	<3	air or centrifugal energy
Very low volume	30	3	air
Low volume	150	13	air
Medium volume	350	31	water/air
High volume	>900	>80	mass of water

Ultra low volume (ULV) sprays occasionally comprise a pure liquid pesticide, but more usually the ULV spray comprises a solid or liquid active ingredient dissolved in oil. Aqueous formulations of herbicides have also been applied at a rate of only a few litres per hectare.

Two methods of impulsion are commonly employed to project ULV spray droplets to their target surface. The use of air as a carrier is the older method. In equipment utilising this principle a fan (of which there are several varieties) accelerates air; the spray liquid is fed into the air stream at the nozzle so that the liquid is shattered into small droplets. The blast of air then carries the droplets to their target. The second method relies upon the centrifugal energy of a rotating disc or cup to provide the impulsion for the droplet.

The development of spinning disc technology for spray application has proved increasingly popular as the use of lower application volumes became more practicable. The particular technique known as **controlled droplet size application,** CDA, depends on liquid that has been placed near the centre of a spinning disc being thrust towards its edges, the speed of disc rotation determining the size of the droplets. The CDA technique when used with appropriate fluids allows the correct size of droplet to be applied for a given task and, importantly, provides droplets of a much more uniform size than is normally obtainable using hydraulic nozzles. In consequence, losses caused on the one hand by small droplets blowing away, and on the other by large droplets running off the foliage, can be minimised. This in turn enables a lower dosage of active ingredient to be applied per hectare than would be needed for equally effective control using conventional spraying techniques. Similarly, CDA can be a useful technique in tropical areas where water supply is limited.

The spray volume which needs to be applied by CDA is a function of both droplet size and of the density of deposited droplets. It can be shown (Matthews, 1979) that the same density of droplets (one droplet per mm^2 of sprayed surface) is achieved by using 9.8 litres per hectare with spray droplets of 70 μm diameter or by using 42 litres per hectare with droplets of 200 μm diameter. A well-known design of hand-carried CDA applicator has a rotating disc driven by electrical batteries located in the handle.

A detailed consideration of the structure of spray machinery and of the factors affecting spray patterns is beyond the scope of a book on pesticide chemistry. For general information on higher volume equipment, the Bulletin issued by the Ministry of Agriculture, Fisheries and Food (MAFF, 1961) should be consulted. For more technical data, and for more recently introduced machinery, the textbook by Matthews (1979) is a valuable source of information.

2.8 TYPES OF SPRAY SUPPLEMENTS

Water is present in greater or lesser quantity at the time of application for most agricultural spraying operations. For reasons of availability, cheapness and lack of phytotoxicity, water is an excellent, even an essential, spray component. In most other ways, unfortunately, it is almost the worst possible choice for a material, the primary function of which is to deposit a pesticide on a waxy surface. Water has low affinity for such surfaces and so tends to run off, with wastage of the active ingredient. Its high surface tension (78 dynes per cm) is a further cause of difficulty, for small droplets of water thus tend to be spherical and to bounce off a lipophilic surface. To minimise these difficulties, various **spray supplements** must be present in the majority of spray mixtures to enable the active ingredient to be utilised as effectively as possible.

Two types of supplements, the **emulsifiers** and the **dispersing agents** or **stabilisers**, operate at the spray tank level, enabling heterogeneous mixtures to be kept thoroughly mixed (a process often assisted by mechanical stirring and circulation). Another type of supplement, the **spreading agent** or **wetting agent**, operates at the moment of impact; it reduces surface tension and so increases the contact between the spray droplets and the sprayed surface. These three types of supplements all have surfactant properties (section 2.9).

The retention of a droplet by a lipid surface is influenced by several interacting physicochemical factors, including droplet size, surface tension, volatility of carrier and, if relevant, the presence of an acquired electrostatic charge. In the absence of a spreader, droplets larger than about 250 μm but smaller than about 2.5 mm bounce off leaves and almost no residue is left. However, the percentage retention of water droplets by waxy surfaces increases suddenly as the size of the droplet falls below about 100 μm. Retention is also affected by the surface tension of the aqueous carrier. If too little spreader is present, the contact between an aqueous spray and a lipid surface is so poor that most of the droplets bounce off; if too much spreader is present, the droplets (especially when application involves medium to high volumes of liquid) may tend to flatten and coalesce, so that much is lost by run-off. Another possibility, favoured by high impact velocity, is that a large droplet may explode into many smaller droplets upon impact, and many of the latter then bounce off. The surface tension of aqueous sprays should probably not exceed 50 dynes per cm if adequate retention is to be assured.

Leaves of different types of plants retain or reflect aqueous droplets to different extents. In the case of herbicides this difference can sometimes be exploited to achieve a greater selectivity of action. This is particularly true for weed control in some legume and brassica crops, although, naturally, some types of weeds will also have highly reflective leaves. If, therefore, the amount of wetting agent is sufficiently low to ensure reflection from leaves of crop plants but little reflection from those of most kinds of weeds, any selectivity possessed by an active ingredient (e.g. dinoseb) can be accentuated. In this respect it should be noted that it is possible to select dispersing agents which do not reduce surface tension greatly and thus wettable powders as well as water-soluble concentrates can be used to exploit differences in reflection by different species of plants (Hartley, 1976).

From what has been said above, it is evident that to maximise deposition on leaves requires the presence of the optimal amount of spreading agent. Except at high application volumes (500–1200 litres per hectare) droplets do not overlap and coalesce. Thus for lower application volumes it is possible to use much spreading agent so that the droplets spread out on the leaves, almost to the point of incipient run-off, and then dry *in situ*. This ensures that any oily residue is sorbed into the cuticle.

After deposition emulsions will break. At higher volumes of application the rate at which this occurs may well determine how much active ingredient remains on the leaf. If breaking occurs rapidly enough, the oily disperse phase (and any dissolved pesticide it contains) will be left behind when the water runs off the leaf.

The **sticking agent** is another type of spray supplement of occasional importance. Stickers are substances that improve the adhesion of deposited residues; they are of particular importance for use with active ingredients that are deposited upon, rather than sorbed within, the wax of the leaf surface. Amongst the numerous materials that have been used as stickers are casein, flour, oils, gelatin, gums, resins and synthetic polymers.

Humectants are sometimes added to a spray mixture to increase the length of time between the application of the droplets and their evaporation to dryness. This interval of time is occasionally of considerable importance, especially when penetration to some target is a prerequisite for success. An example is the penetration of crevices of bark so that the active ingredient is able to reach overwintering eggs; another is the need to penetrate the 'wool' produced by woolly aphids in order to kill the insects underneath. Fungal mycelium can regenerate from the smallest strands, so in this case also, total wetting prior to the last of the water evaporating may greatly improve performance. Water-soluble organic liquids such as propylene glycol are often employed as humectants because the fugacity of water molecules (i.e. their tendency to escape) becomes less as the solution becomes more concentrated with respect to the propylene glycol.

Synergists and activators enable a fixed amount of certain toxicants to be more poisonous than they would be if applied alone under otherwise identical

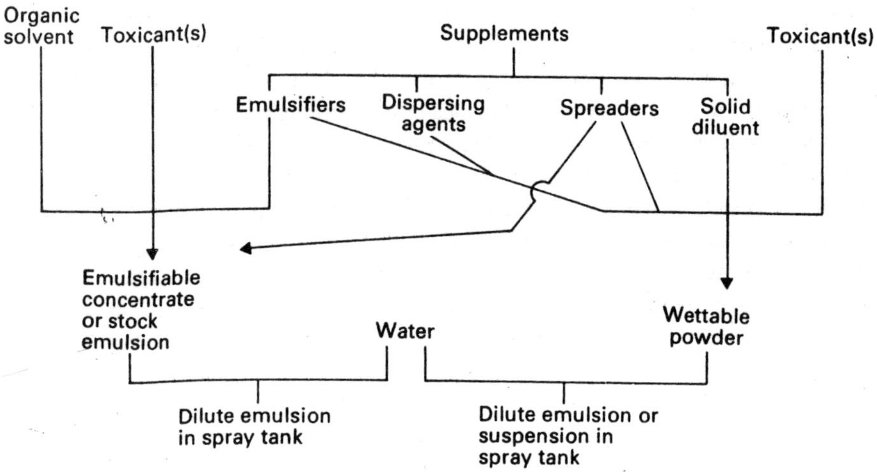

figure 2.4 Possible components of spray mixtures

conditions. **Synergists** act at the biochemical level, often by inhibiting the microsomal oxidases (section 3.4) which inactivate many, but not all, toxic substances. If the enzymes are inactivated, the toxic substances persist longer at the site of action, and can therefore do greater damage. **Activators** often alter the physical state of the applied spray mixture, allowing better access of the active ingredient to its target. One example is the use of a bisulphate to alter the degree of ionisation of active ingredients that are weak acids. Figure 2.4 depicts a scheme showing components that may be present in spray mixtures, with some of the ways in which they interact with one another.

2.9 CHEMISTRY OF SURFACTANTS

It was seen in the previous section that emulsifiers, stabilisers and spreading agents increase the availability of an active ingredient before and at the moment of impact. These three types of spray supplements share an important physicochemical attribute, namely, their molecules are able to separate preferentially at lipophilic–hydrophilic phase boundaries. For electrical reasons, surface activity results in the stabilisation of emulsions and suspensions. In addition, the presence of surfactant molecules at phase boundaries pushes apart close-packed water molecules and so causes a marked reduction of aqueous surface tension. In order to justify and illustrate these observations, it is necessary to review briefly the chemical structure of surface active substances.

Surfactants (soaps and detergents) are divisible into three main groups. The first of these, the **anionic** surfactants, is the group to which soap belongs. Compounds of this type are characterised by the possession of a long non-polar portion to the molecule with a compact hydrophilic, ionised group at the end:

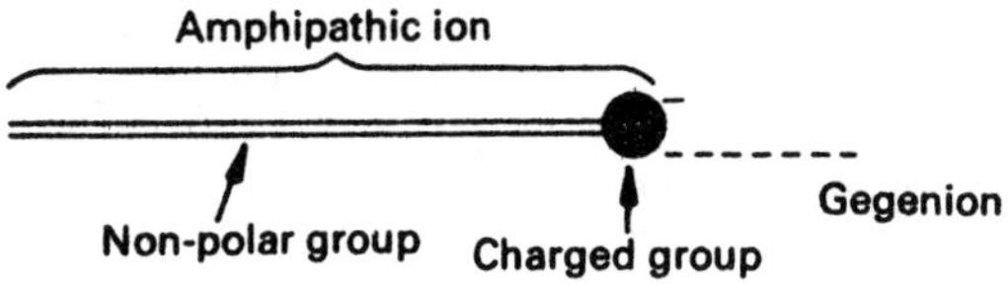

(2.3)

These compounds are termed anionic surfactants because the ion which separates at surfaces, the **amphipathic** ion, carries a negative charge (compare Cl^-, SO_4^{2-}). The balancing, or gegenion (German 'opposing') carries a positive charge.

Anionic surfactants are of several kinds. Amongst them are the sulphated alcohols and sulphonated hydrocarbons:

$C_{12}H_{25}$—O—SO_2O^-............Na^+ sulphated alcohol (e.g. Dreft®)

R_2CH—O—SO_2O^-...............Na^+ sulphated alcohol (e.g. Teepol®) (2.4)

$C_{12}H_{25}$—C_6H_4—SO_2O^-......Na^+ sulphonated hydrocarbon (e.g. Dispersols®)

An advantage possessed by many anionic surfactants is that they can be readily synthesised from natural oils or from petroleum products and are therefore competitively priced. They do, however, suffer two major disadvantages. The first of these is that the amphipathic ions can react with metal ions to form insoluble compounds – solubility is a prerequisite for surface activity. Interfering metal ions may be present in hard water or be a part of certain active ingredients. Similarly, amphipathic ions may react with hydrogen ions with the formation of acids which in some cases (e.g. soap) are only slightly ionised and are in consequence poor surfactants. Secondly, most anionic surfactants are of low solubility in hydrocarbon oils. Consequently, emulsifiable concentrates do not usually contain this type of surfactant (see non-ionic detergents below).

The second group comprises substances the amphiphatic ion of which carries a compact positively charged group;

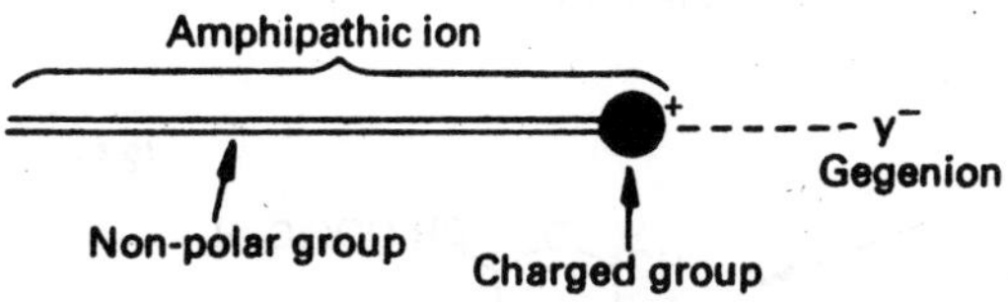

(2.5)

Compounds with molecules of this sort are known as **cationic** surfactants. Representative examples are quaternary ammonium compounds and quaternary heterocyclic compounds such as pyridinium derivatives:

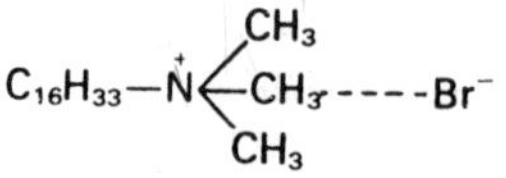

A tetraalkylammonium salt (cetyl trimethyl-ammonium bromide)

(2.6)

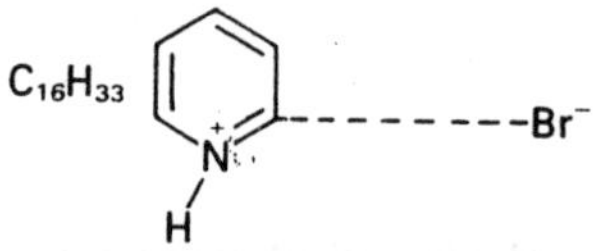

An alkylpyridinium salt (cetyl pyridinium bromide)

The group $C_{15}H_{31}CH_2-$ is a cetyl group; the long carbon chain confers hydrophobic qualities which counterbalance the hydrophilic charged nitrogen atom and lead to separation of the ion at water/non-water phase boundaries. Cetyl alcohol is often made by the reduction of palmitic acid, $C_{15}H_{31}$.COOH, which itself is obtained by hydrolysis of natural fats and oils.

Cationic surfactants are sometimes expensive but they have the advantage that, being positively charged, the amphipathic ion cannot interact with cations in water or with cationic active ingredients. In practice, many anionic surfactants also work quite well in hard water because their heavy metal salts are fairly soluble. It is essential to avoid mixing a formulation containing an anionic detergent with one containing a cationic detergent, for the two oppositely-charged amphipathic ions interact; both are then precipitated as a slimy oil and both the formulations are ruined.

The third group of surfactants contains **non-ionic** substances. A major reason for the increasing popularity of some of these compounds in recent years is that they are more soluble in hydrocarbon and similar oily liquids than are most ionic detergents. As was seen in section 2.7 this property has greatly simplified formulation of liquids by making it possible to make single-phase emulsifiable concentrates to replace the earlier two-phase stock emulsions which often possessed low storage potential.

Many non-ionic surfactants are **polyethylene oxide** derivatives formally derived as follows:

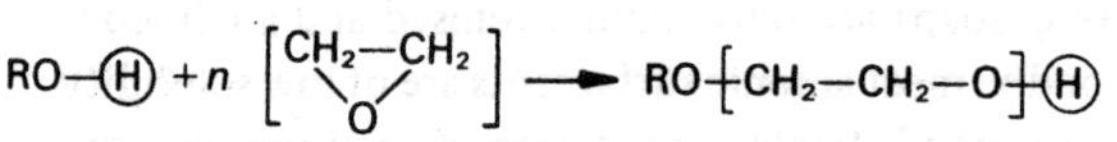

(2.7)

Early members of the group included the Tween® series of compounds introduced by the Atlas Chemical Company where R in equation (2.7) is an ester containing sorbitol anhydride united with various fatty acids. For instance, Tween 40 contains isomers of polyoxyethylene sorbitan monopalmitate:

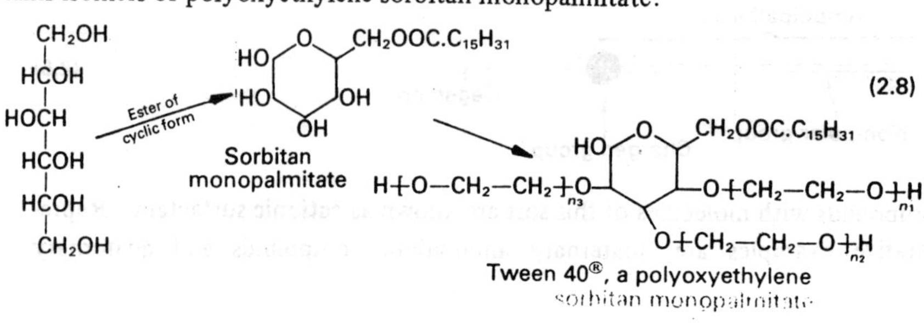

(2.8)

Tween 40®, a polyoxyethylene sorbitan monopalmitate

Numerous other series exist with different groups, R, attached to polyoxyethylene chains. A particularly valuable series for pesticide formulation is a condensate containing a phenol substituted with an alkyl group containing a long carbon chain:

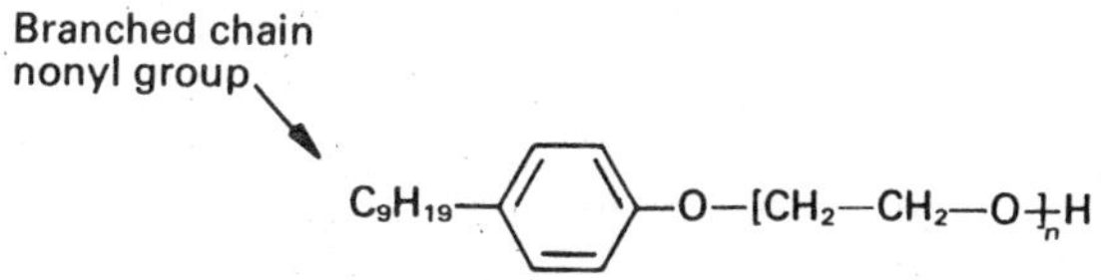

(2.9)

Polyoxyethylene nonyl phenolate

The mean value of *n* varies considerably in different marketed products, and gives to each product its distinctive properties and uses. Mixing together two such non-ionic surfactants, for which the average value of *n* differs widely, often results in a mixture with a field performance superior to that achieved using either separately. Their surface active properties depend on a balance of hydrophilic properties (a property of the regularly occurring oxygen atoms) and hydrophobic properties (a function of the $C_9H_{19}-$ group and of the regularly occurring $-C_2H_4-$ groups). Other pesticide formulations often contain a mixture of a branched chain anionic surfactant and a non-ionic surfactant, for again the mixture often proves superior to either used by itself.

On adding an anionic surfactant to a suspension of solid particles or of oil droplets in water, the amphipathic ions orientate themselves in the way shown in figure 2.5a. Since each suspended particle becomes surrounded, in effect, by a sphere of negative electricity, the tendency of the suspended particles to separate out under the influence of gravity is opposed by the electrical charges upon them (figure 2.5b). Cationic surfactants act in a similar manner, except that the suspended particles are surrounded by spheres of positive electricity.

By this means, surfactants help to prevent non-uniform application of pesticides, a situation which would be economically detrimental if it were to occur. This chemical dispersion is often aided by mechanical agitation of spray tank contents. It should be noted that the presence of the **stabiliser** or **emulsifier** confers upon a suspension or emulsion electrical properties characteristic of the colloidal state, although the particles are much larger than colloidal particles. In addition, of course, the electrical charges are artificially induced whereas those of colloids are inherent in the colloidal state itself.

Spreaders are surfactants that operate at the moment of impact, and they function for the following reason. The compact stereochemistry of water molecules allows them to approach one another very closely, with the result that the forces of cohesion between them are very great (compare the inverse square law). The magnitude of these forces is manifest at the surface as a high surface tension. The latter can be regarded as an elastic film which, by contracting, causes a water droplet to assume as nearly spherical a shape as its size and environment will allow.

When a spreading agent is added to water, the molecules separate preferentially

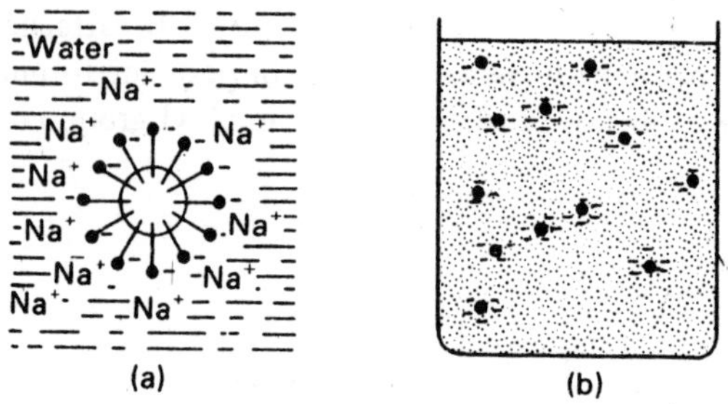

figure 2.5 How a surfactant stabilises an emulsion. (a) Amphipathic ions of anionic detergent separate at the surface of an oil droplet and water. (b) Each droplet thus acts as though surrounded by a sphere of negative electricity, so droplets repel one another

at the surface, thrusting the close-packed water molecules apart. The high forces of cohesion of water molecules are thereby partly replaced by the lower forces of adhesion between water molecules and surfactant molecules. The consequence of the presence of a spreader in an aqueous spray droplet is, therefore, to make the droplet 'sag' more than it would have done in the absence of surfactant (figure 2.6). This increases the area of contact between spray and sprayed surface; that is to say, it increases wetting and spreading. As was seen in section 2.8, an optimal quantity of a particular surfactant gives maximal retention in any particular set of circumstances. The physical chemistry of coverage and of spreading, including mathematical aspects of the contact angle and of the rate of spreading, have been discussed by Hartley and Graham-Bryce (1980, p. 402).

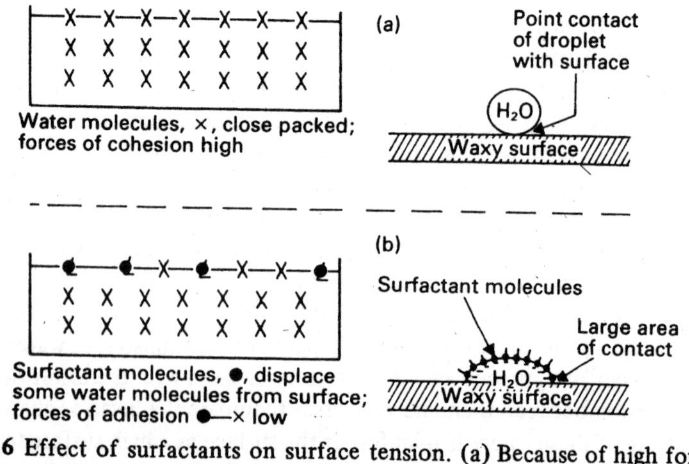

figure 2.6 Effect of surfactants on surface tension. (a) Because of high forces of cohesion, there is little contact between droplet and surface in the absence of surfactants. (b) The area of contact increases when lower forces of adhesion replace the high forces of cohesion

The choice of an appropriate surfactant, or mixture of surfactants, for a particular purpose is still a rather empirical procedure, for optimal amounts have to be found by trial and error. However, a system has been developed for classifying surfactants that allows some level of selection of detergents for particular work. The system is called the **Hydrophilic–Lipophilic Balance**, or HLB, system.

All surfactants contain hydrophilic (water-loving) and lipophilic (oil-loving) groups and the balance between their opposing effects can be measured. The lower the HLB value, the more lipophilic (or hydrophobic) the product is. Typical values range from 2 to 18, those of low HLB tending to be oil-soluble while those of high HLB tend to be water-soluble. In the Span® series, for example, Span 60 (sorbitan monostearate) has a value of 4.7 and Span 20 (sorbitan monolaurate) a value of 8.6. Similarly, in the Tween® series, Tween 85 (polyoxyethylene sorbitan trioleate) has a HLB value of 11 and Tween 20 (polyoxyethylene sorbitan monolaurate) a value of 16.7. Once a body of data has been built up it is possible, within certain limits, to predict the right mixture of surfactants to provide an appropriate combination of emulsifying (or stabilising) and spreading properties to optimise the availability of active ingredient for any desired use in the field.

REFERENCES

Edson, E. F., Sanderson, D. M. and Noakes, D. N. (1966). *Wld Rev. Pest Control*, 5, 143

Elvy, J. H. (1976). *Br. Crop Prot. Council Monogr.*, 18, 102

Furmidge, C. G. L., Hill. A. C. and Osgerby, J. M. (1966). *J. Sci. Fd Agric.*, **17**, 518

Furmidge, C. G. L. and Osgerby, J. M. (1967). *J. Sci. Fd Agric.*, **18**, 269

Hartley, G. S. (1976). *Drug Design*, **6**, Academic Press, London, chapter 6

Hartley, G. S. and Graham-Bryce, I. J. (1980). *Physical Principles of Pesticide Behaviour*, 2 vols, Academic Press, London

MAFF. (1961). *Ministry of Agriculture, Fisheries and Food, Bulletin No. 182*, HMSO, London

Martin, J. T. and Juniper, B. E. (1970). *The Cuticles of Plants*, Arnold, London

Matthews, G. A. (1979). *Pesticide Application Methods*, Longmans, London

Metcalf, R. L. (1955). *Organic Insecticides*, Interscience, New York

Negherbon, W. O. (1959). *Handbook of Toxicology, vol. 3, Insecticides*, W. B. Saunders, Philadelphia

Polon, J. A. (1973). *Pesticide Formulations* (ed. W. Van Valkenburg), Marcel Dekker, New York, p. 143

Rudall, K. M. (1963). *Adv. Insect Physiol.*, **1**, 257

3 Principles of Pesticide Metabolism

3.1 DEFENCE SYSTEMS AND THEIR IMPORTANCE

All living organisms possess defence mechanisms intended to protect them from the deleterious effects of small quantities of many sorts of foreign compounds, including pesticides. If a toxic substance enters an organism more quickly than it can be eliminated (with or without prior metabolism) it will accumulate until a toxic concentration is reached at the site of action. Numerous anatomical, physiological and biochemical factors interact to determine how much of a particular substance penetrates in unit time, how it is distributed in the body, how rapidly and by what route it is metabolised and by what mechanism it is excreted. The importance of metabolic changes described in this chapter must be considered in this overall context, and particularly in relation to differences in the nature of such intermediate barriers as insect cuticle, vertebrate skin and plant waxes (chapter 2).

Enzymes that metabolise foreign compounds serve two related functions. In the first place the metabolic changes so alter the molecular structure that usually (but not invariably) the product is less harmful than the original substance. Secondly, these same changes render many substances more soluble in water and more polar, thereby increasing the rate of elimination from the body.

Most substances used as pesticides are not very soluble in water and their oxidation or hydrolysis (or both) leads to the **insertion** or the **uncovering** of polar groups. Such reactions are **primary** metabolic reactions. Often, but not always, the products of these primary reactions undergo **secondary** changes whereby they are conjugated with endogenous substances to form molecules that are more readily excreted. In some measure, the nature of the conjugating substance is characteristic of the organism or even of the Phylum (section 3.6). In higher animals, these changes facilitate the transportation of metabolites of pesticides in blood, which in turn leads to their eventual excretion. Despite obvious differences in detail, analogous processes occur in invertebrates and in plants.

The preceding paragraphs explain the rationale underlying the near-universality of foreign compound metabolism in organisms as diverse as moulds and man. However, the need for a pre-existing defence mechanism which is versatile enough to cope with any one of an almost infinite range of foreign compounds requires that the enzymes concerned should, in general, be of very low substrate specificity. It will be seen later that this flexibility sometimes causes the system to misfire so that it is not unknown for the organism to become its own executioner. Indeed, this accidental self-destructiveness has occasionally been exploited in the design of specifically insecticidal or herbicidal chemicals.

The majority of the enzymes responsible for the primary metabolism of foreign compounds are hydrolases and oxygenases. The primitive function of these enzymes is presumably to catalyse steps in the normal metabolism of various endogenous compounds. In the case of the oxygenases, exogenous lipophilic substances also fit their surfaces, perhaps because lipophilic properties and van der Waals bonding are more important than precise stereochemical fitment. Whether or not this is so, it is unlikely that those interested in evolutionary philosophy will contemplate the possibility of 'pre-cognitive evolution' whereby enzymic defence mechanisms against specific poisons came into being millions of years before those poisons were an environmental hazard. There must be some more rational explanation.

Before considering some of the enzymes involved in pesticide metabolism, it is necessary to indicate why this subject is of fundamental importance to all concerned with chemical pest control. In the first instance, metabolism plays an important (but not an exclusive) role in relation to **selectivity** of action of pesticides, and thus plays a part in determining the safety of these compounds to man and his farm animals. If evolution had taken a different turn an intelligent insect could certainly have devised selective mammalicides. Secondly, the level of metabolism is a factor determining the **persistence** of pesticides in soil, in plants and in animal bodies and therefore contributes significantly to the reduction of environmental pollution. Once pesticides have undergone primary metabolic changes, most of them can be completely degraded by micro-organisms of various kinds. In some cases, bacteria are known which can use the carbon skeletons as a source of food. Finally, and often detrimentally from a farmer's point of view, metabolic factors are frequently implicated in the mechanism leading to the development of pest **resistance**. If organisms – especially fast-breeding ones such as insects – are exposed to doses of a pesticide insufficient to wipe out the whole population, the surviving organisms, by interbreeding, can give rise to a more resistant population than the original one. It is sometimes found that this greater resistance can be attributed at least in part to increased levels, or increased efficiency, of an inactivating enzyme compared with what was, on average, present in individuals of the original population. In some cases, the location of a gene coding for an enzyme associated with resistance has been identified on a specific chromosome.

3.2 ENZYMES RESPONSIBLE FOR THE METABOLISM OF PESTICIDES

Hydrolytic enzymes are widely distributed in various tissues of plants and animals and in different parts of individual cells. In vertebrates, hydrolases present in blood plasma are capable of attacking a wide range of extraneous esters although the natural substrates of these esterases remain uncertain. In addition, membrane-bound hydrolases exist in the **microsomal** fraction obtained on differential centrifugation of a tissue homogenate and soluble **hydrolases** are present in the **cytosol**. A schematic representation of how subcellular particles can be separated by centrifugation of a crude homogenate is shown in figure 3.1; the microsomal fraction represents the remnants of the original endoplasmic reticulum (see below).

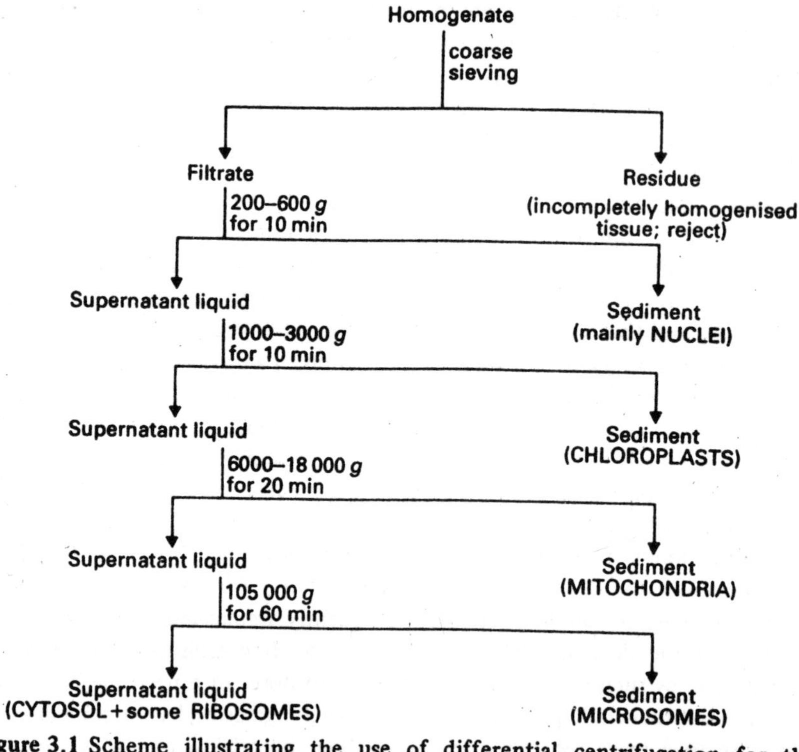

figure 3.1 Scheme illustrating the use of differential centrifugation for the separation of subcellular particles (from Edwards and Hassall (1980) by permission of McGraw-Hill)

The various kinds of hydrolases and their roles in pesticide detoxication are considered in section 3.3.

Enzymes capable of **oxidising** foreign compounds occur in various parts of individual cells of most, perhaps all, organisms. Within the cells of higher animals, those associated with the endoplasmic reticulum of liver cells often play a major role in the metabolism of hydrophobic foreign compounds, including many pesticides. They can be studied *in vitro* by preparing a 105 000 g sediment of ice-cold liver homogenate along the lines shown in figure 3.1. Microsomal enzymes are, in fact, of several kinds, those responsible for pesticide oxidation being termed microsomal **mono-oxygenases** or mixed function oxidases. They are considered in section 3.4. In addition to these membrane-bound enzymes, there are enzymes in the cytosol that are able to oxidise certain alcohols and aldehydes.

Less is known about the oxidative systems in plants and in invertebrates than is known about those in warm-blooded vertebrates, but systems capable of performing in a similar manner to vertebrate mono-oxygenases are undoubtedly present. The level of mono-oxygenase activity, and indeed the activity of other enzymes that metabolise foreign compounds, varies considerably between species in both the plant and the animal kingdom, but, with some exceptions, the types of primary metabolites produced under aerobic conditions are remarkably constant from Phylum to Phylum.

Reduction of certain types of pesticides can also be effected by microsomal and other enzymes. Reduction reactions are, however, more evident in a reducing milieu, such as occurs in the gut of vertebrates, in the mud beneath ponds and in certain types of soils. In each of these cases micro-organisms can reduce suitable pesticides; for example, nitro groups can be reduced to amino groups.

Conjugation processes usually require cooperation between membrane-bound enzymes in the microsomes and other enzymes and cofactors present in the cytosol. For example, **glucuronic acid** conjugates of metabolites are common in higher animals (section 3.6) and the enzyme that facilitates the conjugation is a membrane-bound microsomal enzyme. However, before it can operate, the glucuronic acid must first be activated, and this is brought about in two steps each of which involves an enzyme in the cytosol. In addition, the endogenous conjugating substances (in the example above a glucuronic acid derivative) are water-soluble compounds.

3.3 HYDROLASES

Many pesticides contain ester, amide or phosphate linkages which are more or less readily attacked by hydrolases. Examples are the organophosphorus, pyrethroid and carbamate insecticides, the dithiocarbamate and dinitrophenol fungicides, and the urea and carbamate herbicides. In contrast to the oxidases (section 3.4) and the transferases (sections 3.5, 3.6) the hydrolases do not require any coenzymes although they occasionally need cations to activate them.

Hydrolases are present in all living organisms and play an important role in pesticide metabolism. Nevertheless, the situation is not so straightforward as was originally thought. Hydrolysable linkages are not necessarily the locus of initial metabolic attack and, if they are attacked, it does not necessarily follow that the enzymes which attack them are hydrolases. As an example, many organophosphorus insecticides contain alkoxy groups attached to the phosphorus atom (R–O–P) and these linkages look likely candidates for hydrolytic attack; in fact, while hydrolase action is possible, the end-product (H–O–P) is often formed by the action of a quite different enzyme system (section 3.6).

A second difficulty relates to the naming and substrate specificity of hydrolases. The latter act *inter alia* upon R–O–P linkages ('phosphatases'), R–COOR′ linkages (carboxylesterases) and R–CONHR linkages (carboxylamidases). However, many of the enzymes responsible for these reactions have not been purified and it is often uncertain whether a particular carboxylesterase can also hydrolyse an amide linkage, or whether more than one enzyme is present.

Esterases can be divided into three groups with respect to their response to organophosphorus compounds – the latter became involved historically because they inhibit acetylcholinesterase and some other esterases. A-type esterases can hydrolyse organophosphates whereas B-type esterases are inhibited by true organophosphates (although they may split certain thiophosphates). Most (carboxyl) esterases are specifically inhibited by EPN (Walker, 1975):

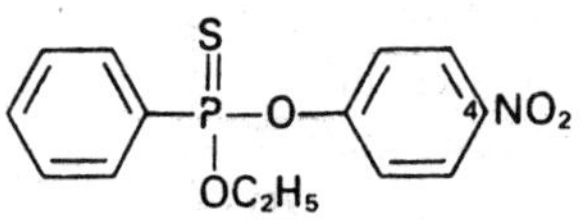

(3.1)

EPN, O-ethyl O-4-nitrophenyl phenylphosphonothionate

Up to 7 per cent of all microsomal proteins are membrane-bound esterases, and these are predominantly of the B-type. The different behaviour of A- and B-type esterases in the presence of organophosphates is explained by the relative magnitudes of the rate constants governing the reactions leading to the binding of the organophosphate to the enzyme and the subsequent dephosphorylation process (section 4.4). This leads to a paradox, for these esterases are often called **carboxylesterases**, yet those of type A can evidently exhibit phosphatase activity in certain circumstances and with certain substrates. C-type esterases are less well defined but they are neither inhibited by organophosphates nor do they degrade them. They may act preferentially on esters of acetic acid (Junge and Krisch, 1975).

The following examples illustrate some of the hydrolytic processes referred to in later chapters. Malathion (figure 4.5) offers a straightforward instance in which the action of carboxylesterase leads to detoxication. In this example, the esterase attacks its 'proper' substrate, the carboxylethyl linkage:

$$R{-}COO{-}C_2H_5 \xrightarrow[H_2O]{\text{carboxylesterase}} R{-}COOH + C_2H_5OH \qquad (3.2)$$

The efficiency of this process in some higher animals helps to explain why malathion is of low toxicity to them (section 4.6).

A second important reaction that is mediated by carboxylesterase is the hydrolysis within plants of the esters of 2,4-D (section 12.1). These esters readily penetrate into weeds and are then hydrolysed to release the biologically active dichlorophenoxyacetic acid. The esterase is inhibited by 2×10^{-4}M copper ions (Morre and Rogers, 1960).

Carbamates have a high affinity for B-type esterases. Just as normal substrates acylate enzymes of this group (section 4.4), so carbamates form carbamoyl derivatives. Some of these hydrolyse slowly, others do so more rapidly. Where hydrolysis is sufficiently fast, the substrate is destroyed and the enzyme regenerated, so that such carbamates can be regarded as substrates for the enzyme. Metcalf (1967) observed that the LD_{50} of carbaryl (section 5.5) to house flies was reduced to 3 per cent of its former value when flies were pre-treated with tri-orthocresol phosphate, another substance that is regarded as a specific inhibitor of esterases.

It is probable that hydrolysis plays an important part in the metabolism of the oxime sub-group of carbamates after an initial oxidative step has weakened the molecule (compare figure 6.5). The sulphide is first converted to a sulphoxide or sulphone and both of the latter then readily hydrolyse to the non-insecticidal free oximes and a carbamic acid derivative.

Many pesticides contain alkoxy groups such as $-O-CH_3$ and $-O-C_2H_5$. The alkyl parts of these groups are often removed during metabolic degradation but the mechanism of removal is not always clear. For some relatively unstable compounds (e.g. dichlorvos, figure 4.11), hydrolytic removal may be of quantitative importance, but there is increasing evidence that small alkoxy groups in many compounds are removed by a *glutathione transferase* system (section 3.5). This system is, for example, probably responsible for a major proportion of the O-dealkylation of phorate (figure 4.9).

The **amide** linkage is somewhat similar to that in esters, and it is attacked by **amidases**. Dimethoate is an important organophosphorus insecticide which is hydrolysed in this way:

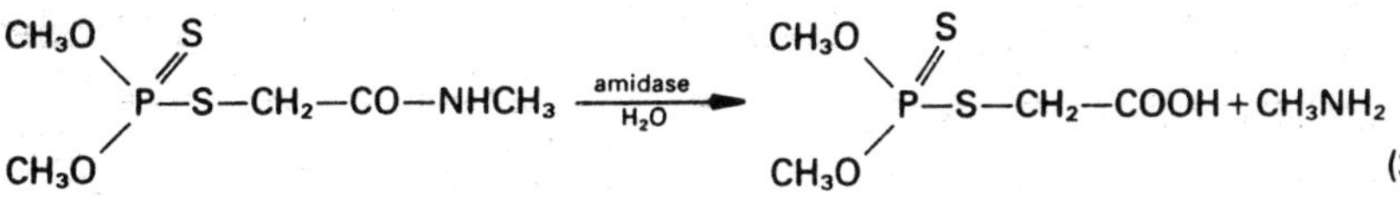

(3.3)

Dimethoate — Dimethoate acid

Hacskaylo and Bull (1963) found that about half the radioactivity liberated from ^{32}P-labelled dimethoate was in dimethoate acid after the insecticide had been applied to excised cotton leaves.

Yih *et al.* (1968) postulated the existence in rice plants of an aryl carboxylamidase which was capable of hydrolysing propanil (section 12.10), a herbicide of importance for weed control in this crop:

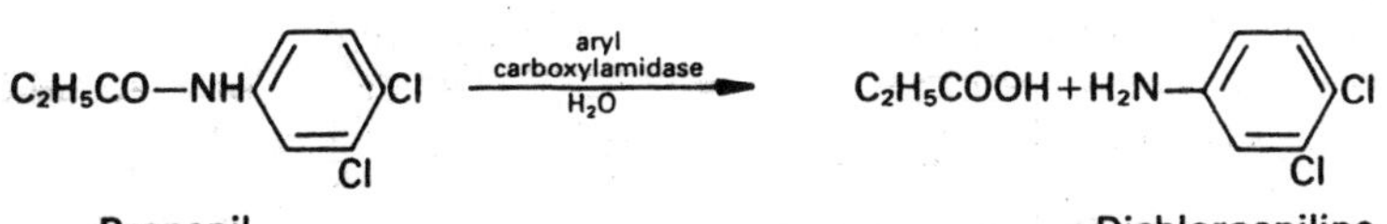

(3.4)

Propanil — Dichloroaniline

As was seen earlier, it is possible that some carboxylamidases are ordinary carboxylesterases capable of selecting amides as substrates. This possibility is of more than academic importance because the problems of cross-resistance between such compounds as malathion (an ester) and dimethoate (an amide) are likely to be more severe if the same enzyme attacks both molecular groupings. There is no doubt that crude carboxylesterases sometimes attack amide linkages. This could be due to the presence of two or more similar enzymes, but the esterase and amidase activities cannot usually be separated by purification. However, the ratio of esterase activity to amidase activity has been found to be different in four forms of pig liver esterase, an observation that could be interpreted as implying the possibility of partial separation (Junge and Krisch, 1975).

Another complication of considerable importance relates to the possible role of hydrolases in insect resistance. Resistance to some insecticides is known to be associated with high levels of carboxylesterase. The assumption in such cases has usually been that the insecticide is inactivated by hydrolysis. However, instances have been reported where resistance appears to be correlated with a lower than usual carboxylesterase activity. The matter is, as yet, unresolved, but it seems possible that a mutant form of carboxylesterase exists which, while less effective than the non-mutant form against carboxylic esters, is a more potent attacker of some other types of hydrolysable linkage (Plapp, 1970).

Epoxide hydrolase (or **hydratase**, or **hydrase**) is a different kind of hydrolytic enzyme which plays an important role in the metabolism of certain foreign compounds. This enzyme is present in the microsomal fraction of liver and other cells and hydrolyses epoxides to dihydrodiols. A well-known example is the (slow) hydrolysis of dieldrin to aldrin trans-dihydrodiol (figure 6.10). Dieldrin is unusual in being a very stable epoxide. However, epoxides of a more transient nature are almost certainly intermediate oxidation products in the biodegradation of many aromatic compounds. Epoxide hydrolase therefore probably plays a vital role in the sequential breakdown of a wide range of pesticides and of other foreign compounds. One such example is carbaryl (section 5.5). This insecticide can be metabolised by several routes; in one of these it is oxidised by microsomal oxidases to give carbaryl epoxide, and this is then converted to a trans-dihydrodiol by the action of epoxide hydrolase.

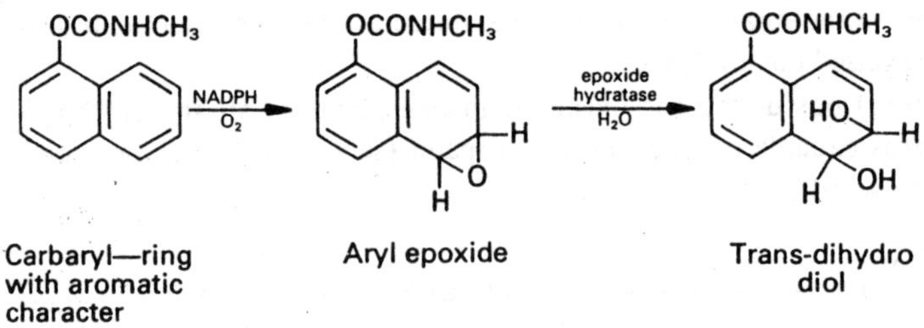

(3.5)

3.4 MICROSOMAL MONO-OXYGENASES

Tightly bound to the smooth endoplasmic reticulum, especially of hepatic cells in higher animals, are enzymes which are capable of inserting one of the two oxygen atoms from an oxygen molecule into an appropriate substrate, R–H. The other oxygen atom eventually forms a part of a water molecule:

$$R\text{–}H + O_2 + [2H] \longrightarrow R\text{–}OH + H_2O \qquad (3.6)$$

The **mono-oxygenase** system (also known as the microsomal oxidase or mixed function oxidase system) gets its name from the way the atoms from the oxygen molecule are separated from one another and end up in different substances. A highly characteristic feature of the system is that, indirectly, the reducing power (designated [2H] in equation (3.6)), originates from NADPH. A second feature is that the final electron carrier of the microsomal electron transport system is a haem protein called **cytochrome P_{450}**. The system has been most extensively investigated in liver cells of vertebrates but it also operates in cells of several other vertebrate organs, including those of the kidneys, lungs and intestine. Similar but not necessarily identical systems appear to exist in organisms as diverse as insects and plants.

The unique haem protein, cytochrome P_{450}, is so named because of the position of a characteristic spectral line formed when the cytochrome is poisoned with carbon monoxide. This cytochrome behaves as 'active oxygen' (compare acetyl coenzyme A, biotin carboxylic acid and ATP as 'active acetate', 'active CO_2' and 'active phosphate' respectively). Cytochrome P_{450} is remarkable in that it is functionally a composite between mitochondrial cytochromes, for which the iron can assume Fe^{2+} and Fe^{3+} forms, and haemoglobin, in which the iron remains in the ferrous state but can coordinate with one oxygen molecule per porphyrin group to form oxyhaemoglobin.

The exact mechanism of oxygen activation by cytochome P_{450} is still uncertain. It is a difficult substance to study because it is tightly bound to microsomal membranes but it has recently been solubilised. Literature on this subject has been reviewed by Hutson (1977) and by Bentley and Oesch (1979). Cytochrome P_{450} was originally thought to be one cytochrome of extremely low substrate specificity but it now seems likely that several closely related enzymes of a more discriminating nature exist (Thomas *et al.*, 1976).

The main features of the mono-oxygenase system are summarised in figure 3.2, a scheme essentially similar to one proposed by Estabrook *et al.* (1971). The substrate, XH, probably becomes attached to the ferric form of cytochrome P_{450}. An electron from NADPH vicariously reduces the complex to the ferrous form. This incidentally, explains the paradox that mono-oxygenase action requires the obligatory presence of a reducing agent. The oxidising agent, molecular oxygen, then attaches itself to the ferrous form of a cytochrome P_{450}-substrate complex. The bond between the two oxygen atoms is then weakened, possibly by addition of a second electron. An intramolecular electron shift then leads to the formation of a [peroxide-XH-ferric P_{450}] complex. This apparently loses an oxide ion to protons in the medium, leaving behind a highly active oxygen atom which reacts with the XH attached nearby to form the metabolite, X–OH. The ferric form of cytochrome P_{450} is then ready to accept more substrate.

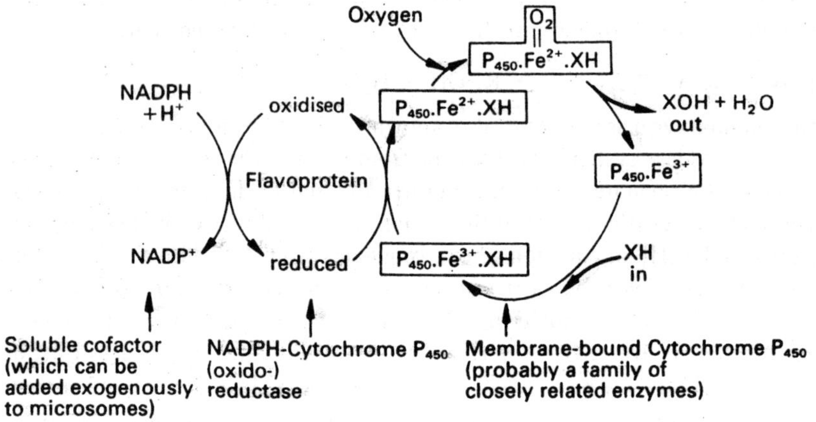

figure 3.2 Proposed scheme for microsomal oxidation of a foreign compound, XH

The enzyme which passes electrons from NADPH to cytochrome P_{450} is a flavoprotein. Thus the microsomal electron transport system comprises at least three components reminiscent of those in the mitochondrion, namely, a reduced pyridine nucleotide, a flavoprotein and a cytochrome. A cytochrome b_5 is also present in microsomes and may participate in this system, as may a non-haem iron protein.

The mono-oxygenase system is rather specifically inhibited by a group of substances called **methylene dioxyphenols**, of which **piperonyl butoxide** is an important example. The significance of these substances to pesticide science was first evident in relation to pyrethroids, under which heading they are further discussed (section 7.4). Later, they were found to delay oxidative metabolism of carbamates (section 5.5) and of certain other pesticides.

Mono-oxygenase reactions of importance in pesticide metabolism are summarised in table 3.1. The reactions are of five sorts. In the first, a C–H bond is broken and an oxygen atom is interpolated (aliphatic and aromatic hydroxylation). Reactions of the second sort are analogous to the first but involve alkyl groups attached to oxygen or nitrogen atoms. The hydroxylation often, but not always, leads to unstable products which split up to yield an aldehyde together with an alcohol (O-dealkylation) or an amine (N-dealkylation). Thirdly, oxygen can be attached to an olefinic bond to form an epoxide. This leads to the formation of the three-membered heterocyclic **oxirane** ring. Fourthly, an element (sulphur) in Group 6 of the Periodic Table can be replaced by the more electronegative element oxygen. Finally, oxygen atoms can coordinate with sulphide sulphur atoms and with certain nitrogen atoms to give often unstable and sometimes biologically more active derivatives. These reactions will now be illustrated by reference to specific examples involving crop protection chemicals.

An example of **aliphatic hydroxylation** is the formation of an hydroxymethyl

Table 3.1 Oxidation reactions involving NADPH-dependent microsomal oxidations

	Type of reaction	General representation	Name of product
1(a)	Aliphatic hydroxylation	$R{-}CH_2{-}H \xrightarrow{[O]} R{-}CH_2{-}OH$	alcohol
(b)	Aromatic hydroxylation	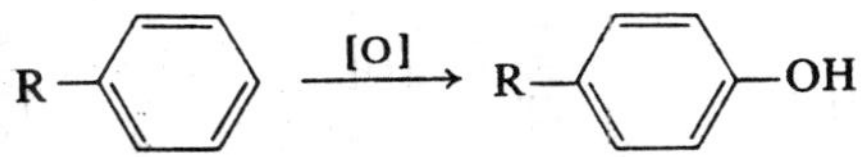	phenol
2(a)	O-Dealkylation	$R{-}O{-}CH_3 \xrightarrow{[O]} R{-}O{-}CH_2O{-}H \xrightarrow{[O]} R{-}OH$	alcohol
(b)	N-Dealkylation	$R{-}NH{-}CH_3 \xrightarrow{[O]} R{-}NH{-}CH_2O{-}H \longrightarrow R{-}NH_2$	primary amine
3	Epoxidation	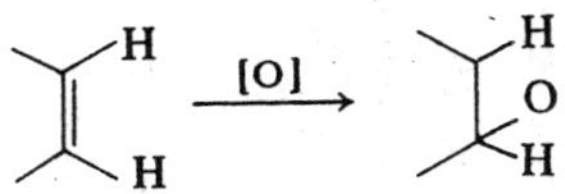	epoxide
4	Substitution of oxygen for sulphur	$>P{=}S \xrightarrow{[O]} >P{=}O$	'oxon' derivative
5(a)	Formation of sulphoxides and sulphones	$R{-}CH_2{-}S{-}CH_3$ $\quad R{-}CH_2{-}S(=O){-}CH_3$	sulphoxide
		$R{-}CH_2{-}S(=O)_2{-}CH_3$	sulphone
(b)	Formation of amine oxides	$R{-}N(CH_3)_2 \xrightarrow{[O]} R{-}N(=O)(CH_3)_2$	amine oxide

group from a part of the isobutenyl group of pyrethrin (section 7.4).

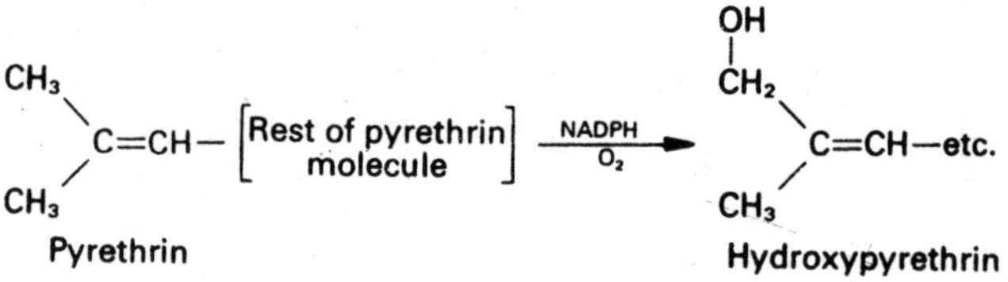

(3.7)

Numerous crop protection chemicals are aromatic or heterocyclic and NADPH-dependent **ring hydroxylation** frequently occurs. Thus carbaryl (section 5.5) hydroxylates to give a mixture of 4-hydroxy and 5-hydroxy carbaryl.

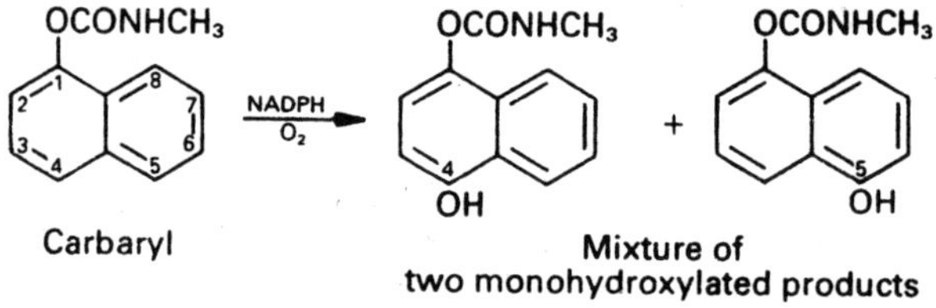

(3.8)

Many pesticides contain the alkoxy group. Dealkylation can occur in more than one way, but oxidative **O-dealkylation** probably occurs in the case of methoxychlor (Kapoor *et al.*, 1970). This compound is a minor organochlorine insecticide and differs from DDT in that it has p,p′-dimethoxy groups instead of chlorine atoms:

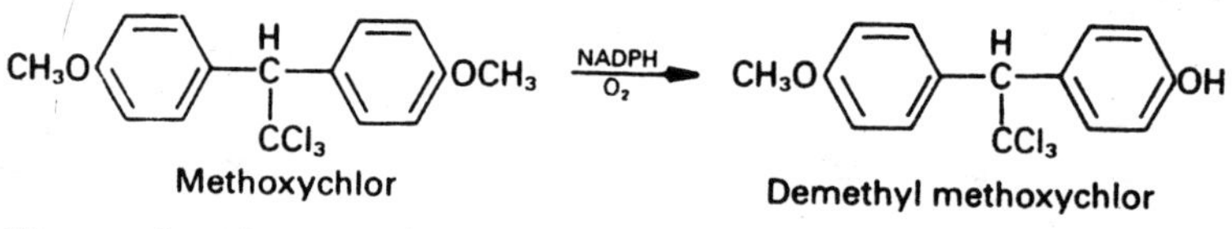

(3.9)

The product is not only more polar but, because it possesses a free hydroxyl group, can undergo secondary conjugation, a reaction which further facilitates its removal from animal bodies*.

N-dealkylation reactions are also common in pesticide metabolism since many pesticides are substituted amines or amides. The carbamate carbofuran (section 5.6) is an example. For this compound the intermediate hydroxymethyl compound is fairly stable:

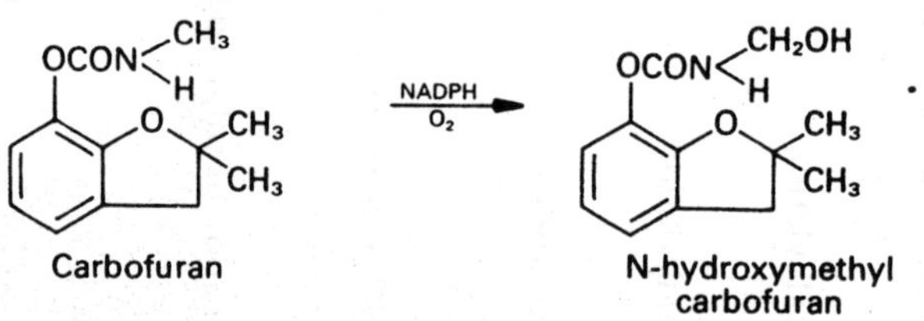

(3.10)

* The late Dr. Barnes, an eminent toxicologist, long ago said to the author. 'if methoxychlor had been marketed instead of the slightly cheaper DDT, the persistence of organochlorine insecticides may never have been regarded as a factor of major ecological importance'.

In the case of the herbicide diuron the reaction is similar but, since the product is unstable, the hydroxymethylation is followed by the elimination of the elements of formaldehyde (Ernst and Bohme, 1965):

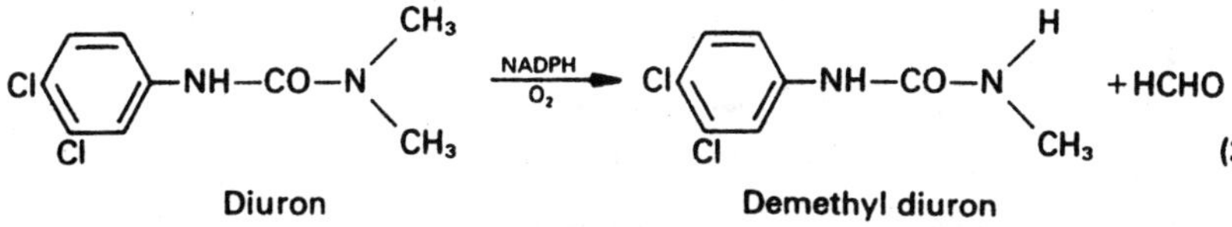

(3.11)

Alkyl groups other than methyl are usually removed in a similar manner; for example, trifluralin (section 13.8) undergoes N-depropylation.

A somewhat different type of NADPH-dependent reaction results in the formation of epoxides. As was mentioned in the previous section, epoxides are probably transitory intermediates in the metabolism of numerous unsaturated compounds. In the case of the organochlorine compound, aldrin, however, the epoxide is a stable product which not only persists in the fat of organisms but is more toxic than the parent olefine:

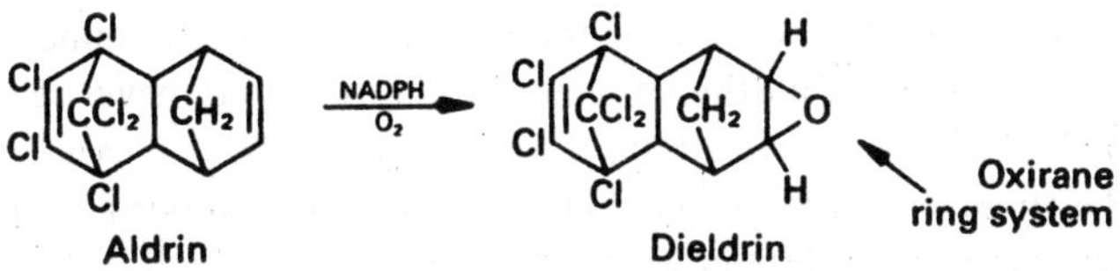

(3.12)

The transient epoxide of carbaryl was referred to earlier (equation (3.5)).

Replacement of a sulphur atom by oxygen is an important step in the metabolism of many thionphosphates. Almost invariably the oxidation results in a molecule which is a better cholinesterase inhibitor than was the original substance, and in consequence when this reaction is a step in the metabolism, it actually results in biological **activation**. When this happens in insects, **lethal metabolism** occurs – the insects are executed by their own 'detoxifying' enzyme systems. A well known example is the activation of parathion.

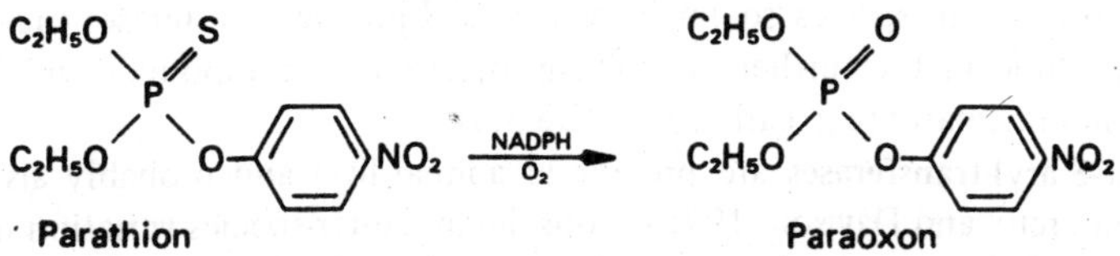

(3.13)

Examples of NADPH-dependent oxidation of sulphides to sulphoxides and sulphones are to be found amongst the organophosphorus and carbamate insecticides. In the following example, the carbamate aldicarb (section 5.7) is oxidised in two steps to its sulphone:

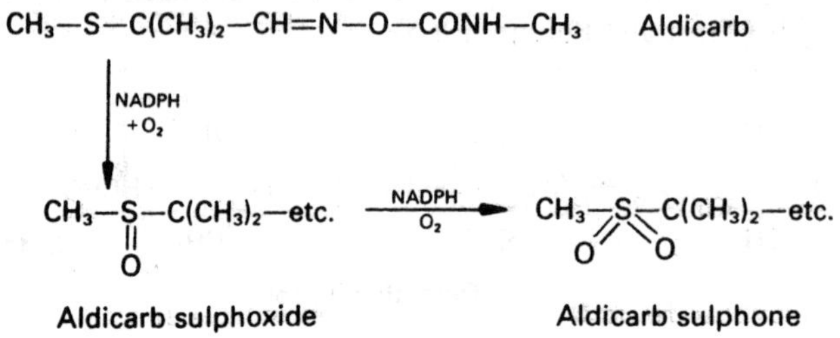

(3.14)

3.5 ROLE OF GLUTATHIONE IN PESTICIDE DEGRADATION

Glutathione (GSH) is a tripeptide comprising residues of glycine, cysteine and glutamic acid. It is important in relation to the degradation of pesticides for a major metabolic pathway involves **glutathione-S-transferase** reactions. In addition, however, glutathione can participate in reactions not involving these enzymes.

Glutathione often forms a conjugate with the intruding foreign compound, whereas most other conjugating endogenous substances (section 3.6) take part in secondary reactions which only occur after the original substance has undergone oxidation or hydrolysis. Glutathione conjugation reactions have been reviewed by Boyland (1971). Several groups of glutathione-S-transferases are known, of which the three shown in table 3.2 are of considerable importance in pesticide detoxication. There transferases are present in the soluble cell fraction of mammalian liver and have molecular weights of the order of 45 000. It is of interest that the true function of these enzymes, if it is other than to detoxify foreign compounds, remains unknown.

Glutathione-S-epoxide transferases open the epoxide ring and insert glutathione by an addition reaction. They may be important in the destruction of carcinogenic oxides (Boyland, 1974). Benzene epoxides are not readily attacked unless the ring contains chlorine atoms, but naphthalene and phenanthrene epoxides are satisfactory substrates for these enzymes.

Amongst pesticidal compounds, dieldrin, endrin and heptachlor epoxide are all organochlorine epoxides, but their glutathione derivatives are only minor products in most of the species so far investigated. Epoxide transferases may, however, play a role in the further metabolism of aromatic compounds which form epoxide intermediates (e.g. carbaryl, figure 5.3).

Glutathione-S-aryl-transferases are present in animal liver and probably also in plants (Lamoureux and Davison, 1975). Considerable inter-species variation in the distribution of these enzymes exists, the activity in the female dog being, for example, some 40 times greater than the activity in adult humans (Walker, 1975). These transferases probably contain active thiol groups because they are blocked by alkylating agents and by mercuribenzoate. Typically, they unite with

Table 3.2 Three groups of glutathione-S-transferases of importance in pesticide degradation (GSH = Glutathione)

Type	Name	Reaction examples
1	Glutathione-S-epoxide transferase	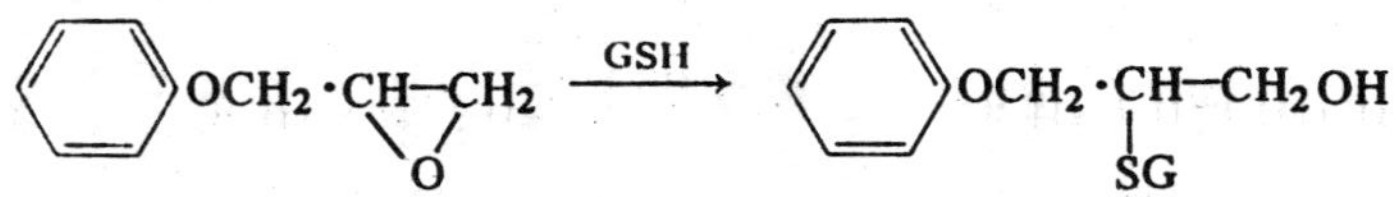
2	Glutathione-S-aryl transferase	(a) 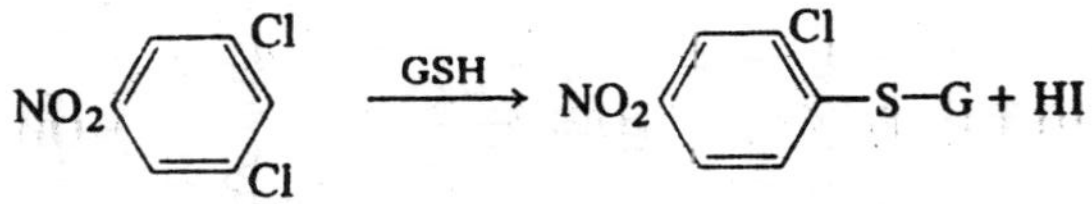 3,5-dichloronitrobenzene
		(b) 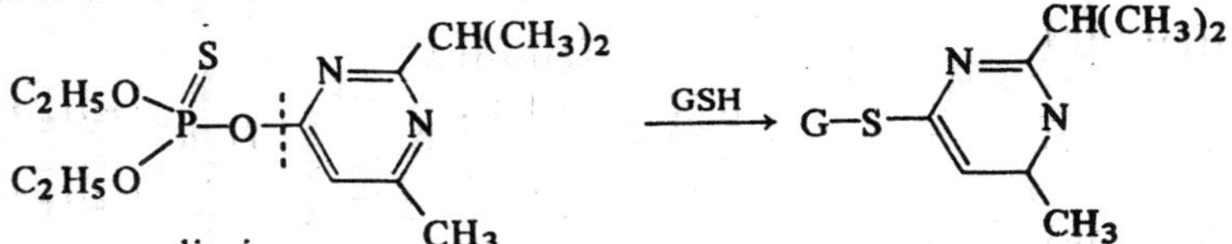 diazinon $+ (C_2H_5O)_2 \cdot PS.OH$
3	Glutathione-S-alkyl transferase	(a) $CH_3I \xrightarrow{GSH} CH_3{-}S{-}G + HI$
		(b) 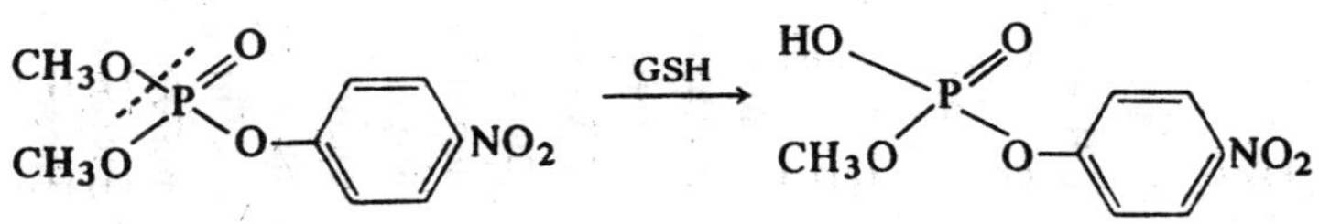 methyl parathion $+ CH_3{-}S{-}G$

a substrate in such a way that hydrogen halide is eliminated (e.g. dichloronitrobenzene in table 3.2). A similar reaction occurs in plants to give rise to glutathione conjugates of atrazine and some other members of the chlorinated triazine sub-family of herbicides (section 13.4). It is of interest that maize and other crops that are resistant to atrazine are well endowed with this enzyme system.

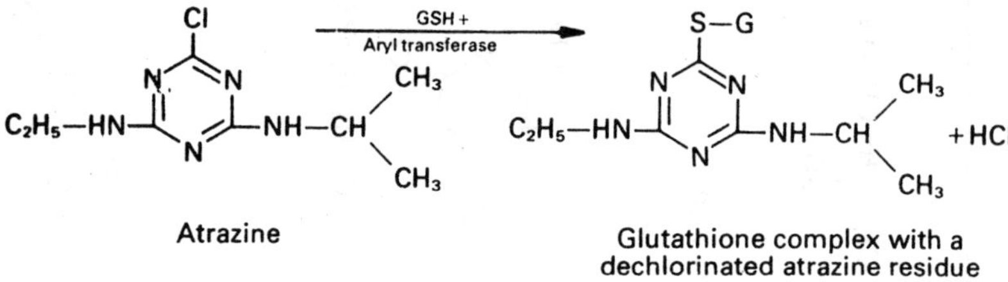

(3.15)

Another important reaction usually attributed to enzymes of this group is the combination of glutathione with the 'leaving group' of certain organophosphorus insecticides (section 4.4). The enzymes responsible are still ill-defined, but it is now evident that some of the reactions leading to detoxication of organophosphorus compounds originally ascribed to hydrolases are, in fact, glutathione-dependent. Thus Shishido *et al.* (1972) found that diazinon was cleaved in the way shown in table 3.2. It should be noted that enzymes of this class lead to the union of glutathione with an aryl or heterocyclic group which constitutes a sizeable part of the original molecule of the pesticide.

The **glutathione-S-alkyl transferases** typically react with alkyl halides in the manner shown in example 3a of table 3.2. More important in relation to pesticide metabolism, however, these or similar enzymes can remove methyl groups from those organophosphorus insecticides that contain the CH_3–O–P group. In contrast to the aryl transferases above, the present group of enzymes therefore catalyse reactions in which the glutathione becomes attached to what was a small part of the original molecule. Since the major part of the molecule remains unconjugated, the role of glutathione was overlooked in much of the earlier work and many such reactions were mistakenly attributed to hydrolytic cleavage. The glutathione-S-transferase reaction is particularly effective in the removal of methyl groups; ethyl and larger alkyl groups are removed much less readily, at least from organophosphorus compounds.

Methyl parathion (table 3.2) and fenitrothion were shown by Hollingworth (1969) to be demethylated, in part, by a glutathione-dependent mechanism. Morello *et al.* (1968) and Miyata and Matsumura (1971) observed similar glutathione-dependent demethylation of mevinphos and dichlorvos respectively (chapter 4).

From what has been said in this and other sections of this chapter, it is evident that the removal of alkyl groups attached to oxygen or nitrogen atoms can be achieved by three distinct mechanisms. The first of these is a simple hydrolytic action, catalysed by an esterase, amidase or phosphatase (section 3.3), the second is oxidative dealkylation (section 3.4) and the third involves the removal of an alkyl group by a glutathione-S-alkyl transferase (table 3.2, example 3b). The relative importance of each of these mechanisms in particular cases is not always known but it varies both with species and for different substrates.

As was mentioned at the beginning of the present section, glutathione also participates in a number of detoxication reactions where enzymes other than S-transferases are involved. In one important instance, the dehydrochlorination of DDT, a transitory glutathione complex is probably formed, but the glutathione is eventually recovered unchanged (figure 3.3). The reaction is almost certainly catalysed by glutathione (Lipke and Chalkley, 1962) but the complex has not been isolated.

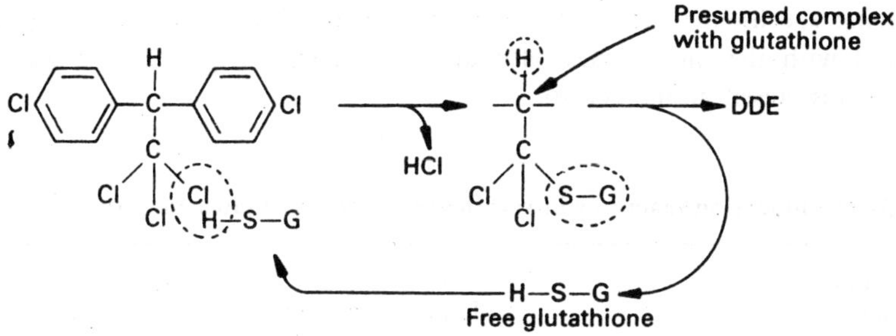

figure 3.3 Proposed catalytic role of glutathione (H–S–G) in the dehydrochlorination of DDT

Glutathione can also react non-enzymatically with certain electrophilic molecules. Amongst pesticidal substances this probably happens during the degradation of propachlor (section 13.9).

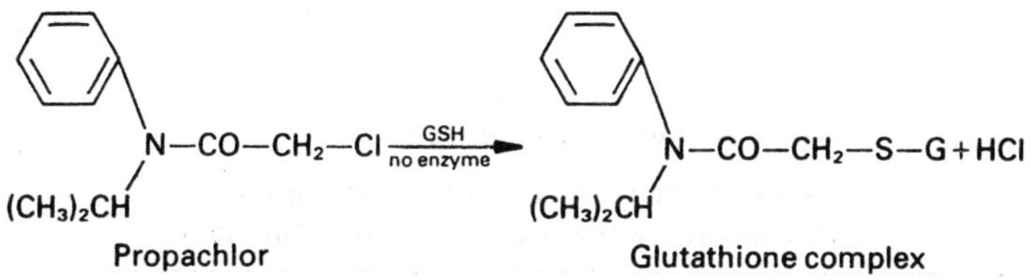

(3.16)

The glutathione complex later decomposes to give mercapturic acids (section 3.6).

During the breakdown of the fungicides folpet and captan (section 9.4) the trichloromethylthio moiety, $-SCCl_3$, is detached in the form of thiophosgene, $CSCl_2$. Glutathione can unite with this to give rise to a heterocyclic compound.

3.6 OTHER CONJUGATION REACTIONS

Primary oxidative, hydrolytic or reductive changes usually uncover or insert active groups such as hydroxyl or carboxyl which make the molecules more polar. These polar groups may then undergo secondary **conjugation** with endogenous molecules, a process which further increases polarity and, in animals, facilitates excretion in urine or faeces. In plants, these more soluble substances are usually stored in the cell vacuoles and deciduous plants get rid of them when leaves fall to the ground.

Conjugation reactions are classified according to the nature of the conjugating agent (table 3.3). **Glucuronides** are the principal conjugates in most terrestrial vertebrates. There are some exceptions – for example, the cat cannot synthesise O-glucuronides, although it can apparently make certain N- or S-glucuronides (Walker, 1975). The donor of the glucuronic acid is uridine diphosphate glucuronic acid, UDPGA, and thus the reaction involves the transfer of glucuronic acid from one organic molecule to another. The enzymes, of which several exist with overlapping specificities, are termed **glucuronyl transferases**. They are largely associated with the microsomes, although the UDPGA (figure 3.4) is water-soluble and is present in the cytosol.

Table 3.3 Conjugation reactions in vertebrates, in invertebrates and in plants

Endogenous conjugant	Type of organism[1]
Glucuronic acid	Mammals, birds, reptiles, terrestrial amphibians, some fish
Glucose	Higher plants, insects, some molluscs
Arginine	Arachnids, myriapods
Glutamic acid or glutamine	Arachnids, some myriapods
Glycine	Fish, crustaceans
Sulphate	Amphibians, echinoderms, some crustaceans, some molluscs, some insects
Others	Acetylation (fish, amphibia, a few insects) Phosphate (a few arachnids, some dipterans)

[1] Found in many investigated members – not necessarily true of every member of the group. Some organisms produce more than one type of conjugation product.

Many insecticides, fungicides and herbicides are converted to glucuronic acid conjugates after primary metabolism by animal liver enzymes. In rats, for example, carbaryl is converted to 4-hydroxycarbaryl and 1-naphthol, and both of these are excreted mainly as O-glucuronides (figure 3.5). Dichlorvos (section 4.8) and resmethrin (section 7.5) are two other insecticides that can be ultimately converted to O-glucuronides.

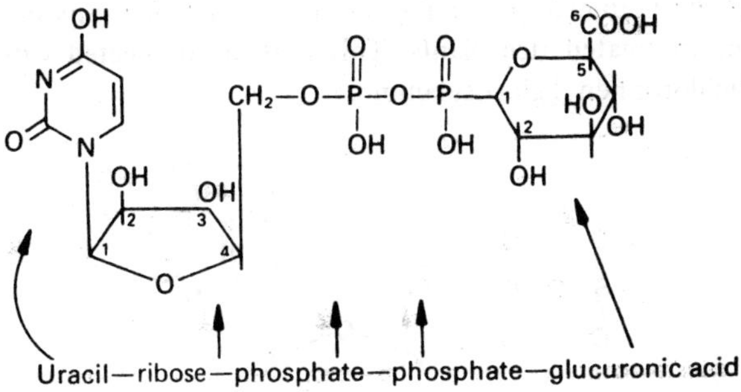

figure 3.4 Uridine diphosphate glucuronic acid (UDPGA)

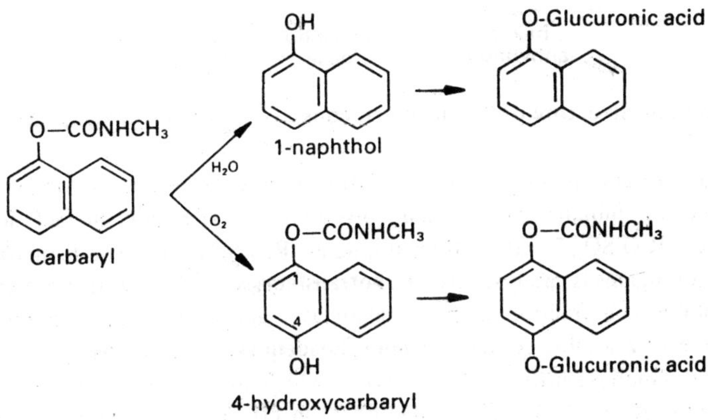

figure 3.5 O-Glucuronide formation from hydrolytic and oxidative products of carbaryl

The fungicide ferbam (section 9.2) provides an example of a substance that is converted, in part, to an S-glucuronide in some animals (Hodgson *et al.*, 1975).

$$\left[(CH_3)_2N-\underset{\underset{S}{\|}}{C}-S-\right]_3 Fe \xrightarrow[\text{UDPGA transferase}]{\text{UDPGA +}} 3\ (CH_3)_2N-\underset{\underset{S}{\|}}{C}-S-\text{glucuronic acid} \qquad (3.17)$$

In contrast to most terrestrial vertebrates, insect and plants usually produce glucose conjugates (table 3.3). The mechanism resembles that for glucuronide formation but uridine diphosphate glucose, UDPG, replaces UDPGA as donor. The glucose becomes attached to alcoholic, thiol or amino groups by a β-glucoside linkage, the enzyme responsible being a glucose transferase. Glucosides, like glucuronic acid derivatives, are more polar and water soluble than the parent aglycones (after the sugar of a glycoside is removed by hydrolysis, the non-sugar residue is termed an aglycone). A specific example of a substance that forms an

N-glycoside is the herbicide propanil (section 12.12) which is hydrolysed to 3,4-dichloroaniline in treated rice plants. This is then conjugated with glucose to form N-3,4-dichlorophenyl glucosylamine.

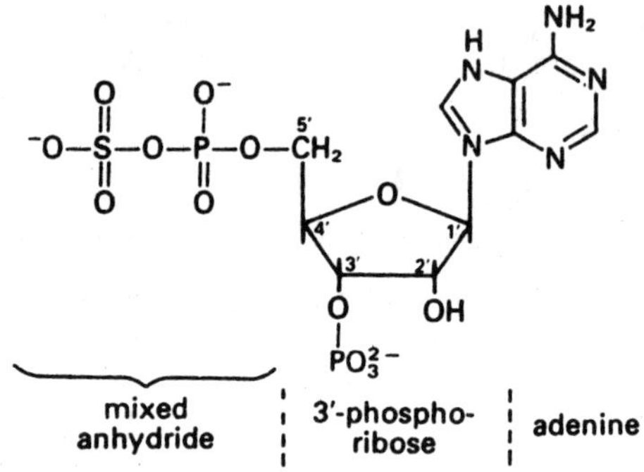

figure 3.6 Structure of 3′-phosphoadenosyl-5′-phosphosulphate, PAPS

Sulphate esters are important in relation to the excretion of foreign compounds by amphibians and by many invertebrates. They are, in fact, ionised bisulphates, $R.O.SO_3^-$, rather than sulphates, $R.O.SO_2.OR'$. The formation of sulphate conjugates is catalysed by **sulphotransferases**. These enzymes are present in the soluble cell fraction of mammalian liver and kidney. They require the cooperation of a soluble cofactor, **3′phosphoadenosyl 5′-phosphosulphate**, PAPS (figure 3.6) which is synthesised by enzymes which are also located in the soluble cell fraction (Roy, 1971). This cofactor is, in effect, 'active sulphate' and its synthesis occurs in steps which can be summarised by the overall equation:

$$2\ ATP + SO_4{}^{2-} \xrightarrow[+\ enzymes]{Mg^{2+}} PAPS + ATP + pyrophosphate \qquad (3.18)$$

Sulphate esters of phenols formed by primary metabolism of pesticides occur in many higher animals, but usually as a minor pathway compared to glucuronide formation. For example, small amounts of sulphate conjugates of carbofuran derivatives are eliminated by several mammals. In insects, however, the route may sometimes be as important as glucoside production. Conjugation with sulphate is rare in higher plants.

Glycine conjugates occur mostly in vertebrates and are formed especially from aromatic pesticides which, after primary metabolism, give compounds containing carboxyl groups. Thus chlorthiamid (section 13.9) breaks down to give dichlobenil, which, in turn, gives rise to 2,6-dichlorobenzoic acid or its hydroxylated derivatives. The latter then conjugate with glycine (figure 3.7). An interesting characteristic of those arachnids so far investigated (including ticks, spiders, and scorpions) is that three other amino acids – arginine, glutamic acid and glutamine – are major conjugating agents (table 3.3).

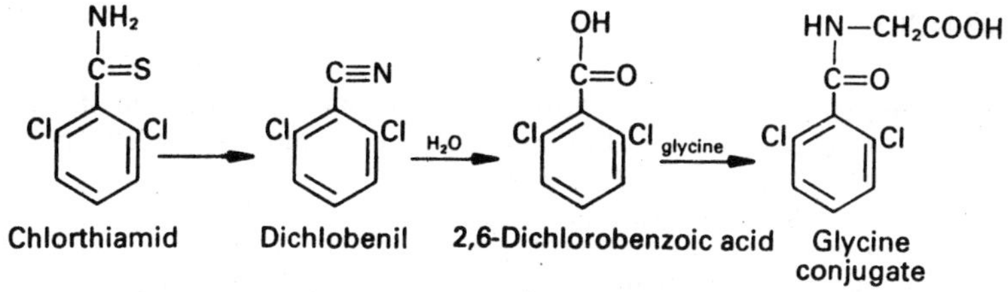

figure 3.7 Glycine conjugates of chlorthiamid derivatives

Finally, it will be recalled that glutathione occupies a somewhat anomalous position in that it usually reacts with the initial foreign compound and not with a metabolite of that compound (section 3.5). However, glutathione conjugates can themselves undergo secondary changes, for excreted products in many organisms are not glutathione conjugates as such, but derivatives called **mercapturic acids.**

Glutathione is a tripeptide and it conjugates with foreign compounds to give complexes of the sort shown in figure 3.8. If the glycine and glutamic acid are hydrolytically removed, a cysteine derivative of the foreign compound remains (figure 3.8b). Acetylation of the amino group of the cysteine then frequently occurs so that the excreted material in many higher animals is an N-acetyl cysteine derivative. This is called a mercapturic acid. In insects, glutathione conjugates tend either to be excreted unchanged or converted to cysteine rather than to acetylcysteine derivatives (Walker, 1975).

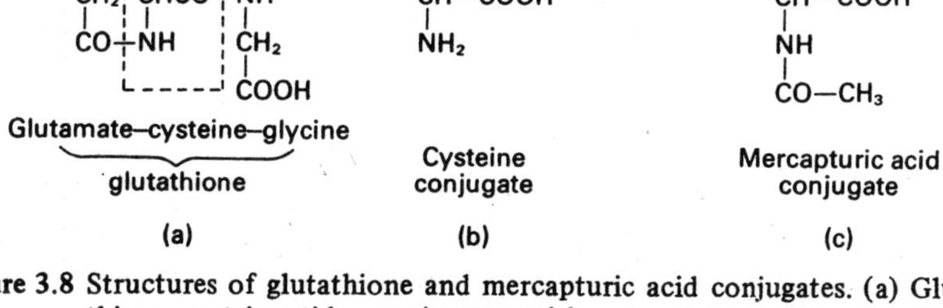

figure 3.8 Structures of glutathione and mercapturic acid conjugates. (a) Glutathione, a tripeptide, conjugates with a pesticide residue, R. (b) A cysteine conjugate is formed on hydrolysis of (a). (c) N-acetylation of (b) gives a mercapturic acid derivative

REFERENCES

Bentley, P. and Oesch, F. (1979). *Foreign Compound Metabolism in Mammals*, Vol. 5, The Chemical Society, Burlington House, London, W1V 0BN, p. 89

Boyland, E. (1971) *Handb. exp. Pharmak.*, **27**, 584

Boyland, E. (1974) *Biochem. Soc. Trans.*, 2, 167

Ernst, W. and Bohme, C. (1965). *Fd Cosmet. Toxic.*, **3**, 789, 795

Estabrook, R. W., Baron, J., Peterson, J. and Ishimura, Y. (1971). *Biochem. J.*, **125**, 3[illegible]

Hacskaylo, J. and Bull, D. L. (1963). *J. Agric. Fd Chem.*, **11**, 464

Hodgson, J. R., Hoch, J. C., Castles, T. R., Helton, D. O. and Lee, C.-C. (1975). *Toxic. appl. Pharmac.*, **33**, 505

Hollingworth, R. M. (1969). *J. Agric. Fd Chem.*, **17**, 987

Hutson, D. H. (1977). *Foreign Compound Metabolism in Mammals*, Vol. 4, The Chemical Society, Burlington House, London, W1V 0BN, p. 259

Junge, W. and Krisch, K. (1975). *Critical Reviews in Food Science and Nutrition*, p. 371

Kapoor, I. P., Metcalf, R. L., Nystrom, R. F. and Sangha, G. K. (1970). *J. Agric. Fd Chem.*, **18**, 1145

Lamoureux, G. L. and Davison, K. L. (1975). *Pestic. Biochem. Physiol.*, 5, 497

Lipke, H. and Chalkley, J. (1962). *Biochemistry, N.Y.*, 85, 109

Metcalf, R. L. (1967). *Ann. Rev. Ent.*, **12**, 229

Miyata, T. and Matsumura, F. (1971). *Pestic. Biochem. Physiol.*, **1**, 267

Morello, A., Vardanis, A. and Spencer, E. Y. (1968). *Can. J. Biochem.*, **46**, 885

Morre, D. J. and Rogers, B. J. (1960). *Weeds*, **8**, 436

Plapp, F. W. (1970). *Biochemical Toxicology of Insecticides* (ed. R. D. O'Brien and I. Yamamoto), Academic Press, New York, p. 179

Roy, A. B. (1971). *Handb. exptl. Biochem. Pharmac.*, Part 2, p. 536

Shishido, T., Ulsui, K. and Fukami, J. (1972). *Pestic. Biochem. Physiol.*, 2, 27

Thomas, P. E., Lu, A. Y. H., Ryan, D., West, S. B., Kawalek, J. and Levin, W. (1976). *Molec. Pharmac.*, **12**, 746

Walker, C. H. (1975). *Organochlorine Insecticides: Persistent Organic Pollutants* (ed. F. Moriarty), Academic Press, New York, p. 73.

Yih, R. Y., McRae, D. H. and Wilson, H. F. (1968). *Pl. Physiol., Lancaster*, **43**, 1291

4 Organophosphorus Insecticides

4.1 INTRODUCTION

During the Second World War, German scientists under the direction of Gerhard Schrader were engaged on the task of developing highly toxic nerve gases for potential use in warfare. As a spin-off, several organophosphorus compounds structurally related to the nerve gases tabun and sarin were found to be effective insecticides. This association of organophosphorus insecticides with war gases was hardly an auspicious beginning for a group of compounds which was destined to become one of the main weapons in our armoury against insect pests of importance in agriculture, public hygiene and medicine. The earliest members of the group, marketed in the immediate post-war years, did little to allay these fears for most of them were very toxic to vertebrates.

Altogether some 100 000 organophosphorus compounds have been screened for their possible insecticidal action and over 100 have been marketed for this purpose. An important feature of the group is that different members possess very different physicochemical properties; in particular, they have different vapour pressures at room temperature and different solubilities in water. They also vary considerably in their chemical stability and their toxicity to mammals (table 4.1).

This wide spectrum of physicochemical and of biological properties enables appropriate substances to possess a correspondingly wide range of uses in agriculture and in animal hygiene. Some are used as fumigants, others as contact poisons and yet others as systemic compounds. Some are used for crop protection early in the growing season whereas some others are particularly valuable near harvest time. Those used early in the season must be sufficiently persistent to avoid the necessity of multiple application with its attendant increase in cost and effort. On the other hand, those used late in the season are specifically chosen because they do *not* persist very long; they must disappear sufficiently rapidly to ensure that negligible residues remain when the crop is harvested.

4.2 STRUCTURAL DIVERSITY OF ORGANOPHOSPHORUS INSECTICIDES

Most organophosphorus insecticides have the following general structure

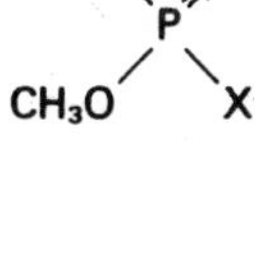

(4.1)

Table 4.1 Physicochemical and biological characteristics of some organophosphorus insecticides

Common name	Water solubility 20°–25°C	Vapour pressure (mm Hg at 20°C)	Persistence	Oral LD_{50} mg/kg male rats	Systemic properties (by foliage application)
Mevinphos	miscible	not given (5)	low	7[2], 3[5]	transient
Dichlorvos	10 000 ppm[1]	1.2×10^{-2}[4]	low	25[3], 56[5]	No
Dimethoate	25 000 ppm[1]	8.5×10^{-6}[5]	moderate	320[5]	Yes
Demeton-S-methyl	3300 ppm[1]	3.6×10^{-4}[5]	moderate	57[5]	Yes
Menazon	250 ppm[1]	1×10^{-6}[5]	moderate	1950[5]	slight
Malathion	145 ppm[5]	(1.2×10^{-4}[4]	low/moderate	2800	No (QS)
Pirimiphos methyl	5 ppm[6]	1×10^{-4}	moderate	2000	No (QS)
Fenitrothion	almost insoluble	{ 6×10^{-6}[5] 5.4×10^{-5}[4]	moderate	250+[4][5]	No
Bromophos	40 ppm[5]	1.3×10^{-4}[5]	moderate	3750+[5]	No
Phorate	50 ppm	8.4×10^{-4}[5]	moderate	2[5]	Yes
Vamidothion	33 ppm	v. low	long	64[5]	No (QS)
Azinphos-methyl	30 ppm[4]	2.2×10^{-7}[5]	long	16[4]	No (QS)

QS: quasi-systemic.

Refs. (1) Mitchell, Smale & Metcalf (1960).
(2) Price Jones & Edgar (1961).
(3) Edson, Sanderson & Noakes (1966).
(4) Matsumura (1975).
(5) British Crop Protection Council (1979).
(6) ICI Technical Data Sheet, May 1970.

The two R groups are usually methyl or ethyl and are the same in any one molecule, while X is frequently a rather complex aliphatic, homocyclic or heterocyclic group. For reasons explained in section 4.4, X is often referred to as the **leaving group**. Sometimes the leaving group is such that a carbon atom in it is joined directly to the phosphorus atom but more often it is joined via an ester or thioester linkage, viz. P–O–X or P–S–X.

Some of the more important variants of the basic molecular structure, a chemical naming system and some examples of insecticides possessing each structure are shown in table 4.2. Phosphates have four oxygen atoms arranged around the phosphorus. If one of these oxygen atoms is replaced by sulphur the compounds are thiophosphates and if two are replaced they are dithiophosphates. If the sulphur is in the double bond part of the molecule, the structure is that of a thionphosphate; the term *thion* (compare ket*one*) distinguishes the substance from one with a structure in which the sulphur atom is in the *thiol* position (compare alcoh*ol*). If a carbon atom of the leaving group, X, is attached directly to phosphorus, the compound is designated a phosphonate. The latter has thionphosphonate analogues. There are, in addition, a few instances where R–O–P is replaced by R–P or by R_2N–P.

The structural variability of organophosphorus compounds is reflected not only in their range of physical properties (table 4.1) but also in a considerable diversity of mechanisms by which they can be attacked by enzymes. This has two important consequences. Firstly, species selectivity is sometimes achieved because the amounts or the activities of different enzymes vary from species to species. Secondly, a multiplicity of possible types and positions of attack by enzymes minimises the risk of a uniform development of tolerance to all organophosphorus compounds (section 4.11). The major types of enzymes which can attack various linkages in organophosphorus compounds are shown in figure 4.1.

4.3 PRACTICAL USES OF ORGANOPHOSPHORUS INSECTICIDES

Organophosphorus compounds have overtaken organochlorine compounds as the most used insecticides, whether the criterion of usage is tonnage produced or wholesale value (table 1.1). In 1971, the last year for which a reliable breakdown of American production statistics is available, they were two or three times as dear as organochlorine compounds, but oil-price inflation has probably helped to close the gap since then. They have tended to replace organochlorine compounds as concern about the persistence and polluting effect of such compounds as DDT has increased. However, for the very reason that organophosphorus compounds are, in general, less persistent than organochlorine compounds, a greater number of applications to a crop may be necessary during the course of the growing season. Thus a price is exacted of the farmer, and hence indirectly of the consumer, for the privilege of growing organochlorine-free produce.

The majority of organophosphorus compounds are effective in the control of

Table 4.2 Main chemical groups of organophosphorus insecticides

Group	Structure	Examples
1. Orthophosphates	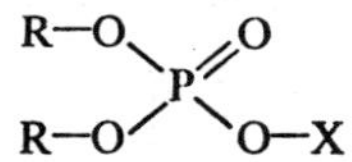	e.g. Chlorfenvinphos, Dichlorvos, Mevinphos, Phosphamidon
2. Thionphosphates (phosphoro*thion*ates)	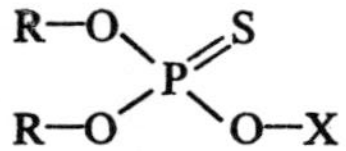	e.g. Bromophos, Diazinon, Fenitrothion, Parathion, Pirimiphos (methyl and ethyl)
3. Thiolphosphates (phosphoro*thiol*ates)	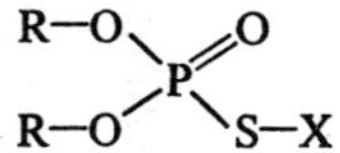	e.g. Demeton-S-methyl, Oxydemeton-methyl Vamidothion
4. Dithiophosphates (phosphoro*thiolothion*ates)	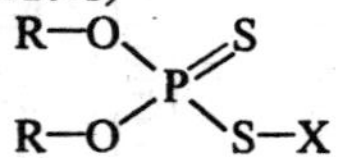	e.g. Azinphos-methyl, Dimethoate, Disulfoton, Malathion, Menazon, Phorate
5. Phosphonates	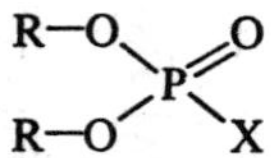 (isomeric phosphonates exist, as do related thiophosphonates)	e.g. Trichlorphon, Butonate
6. Pyrophosphoramides	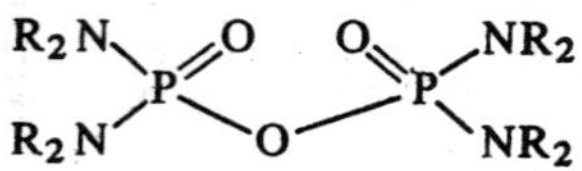	e.g. Schradan

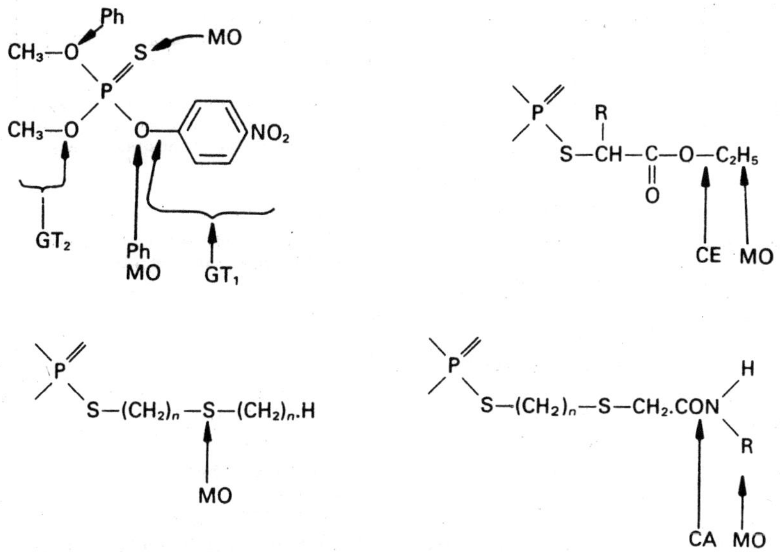

figure 4.1 Enzymes which attack organophosphorus compounds. (Ph, phosphatase/type A esterase; MO, microsomal mono-oxygenase; CE, carboxylesterase; CA, carboxylamidase; GT_1, glutathione-S-aryltransferase; GT_2, glutathione-S-alkyltransferase

aphids and similar soft-bodied small insects, but many of the newer compounds have a wider range of uses. Some of these uses are listed in table 4.3, but only experience can determine the best substance to use for a particular purpose in a particular locality. In most countries, this information is available from governmental agricultural services as well as from commercial literature and sales representatives. Some of the compounds marketed for killing the more refractory insects possess certain disadvantages, such as somewhat high toxicity to vertebrates, or high persistence, or high cost. Killing pests can in some ways be compared to trying to keep law and order in a humane society – the golden rule in the choice of a pesticide is never to use a powerful, undiscriminating weapon when a small selective one will do. Applying this rule, almost all organophosphorus compounds will control aphids and so, if the latter are the only problem, there is usually little need to employ one of the less safe or more costly members of the group.

It is possible to classify organophosphorus insecticides according to their practical uses and such a classification is of more value in the present context than one based primarily on chemical constitution. Each sub-group in the following scheme is discussed in more detail in later sections of the chapter.

Sub-group 1. Compounds of low chemical stability soluble in water but more or less rapidly hydrolysed by it. They are used as **contact** insecticides. E.g. mevinphos, tetraethyl pyrophosphate, tetrachlorvinphos.

Table 4.3 Use of Organophosphorus compounds against common pests of some representative crops

Crop	Pest	Chemicals
Apples, pears some other top fruit	Aphids	demeton-S-methyl, dichlorvos, dimethoate, fenitrothion, formothion, vamidothion
	Codling moth	azinphos-methyl, fenitrothion, phosphamidon
Avocadoes	Soft brown scale; thrips	malathion
Bananas	Aphids, thrips, red spider	malathion
Beans (several types)	Aphids	malathion, demeton-S-methyl, dimethoate, disulfoton, menazon, mevinphos, formothion
	Looper caterpillar	dichlorvos
Cereals	Leatherjackets	fenitrothion
Citrus	Mealy bug	malathion
	Mediterranean fruit fly	
	Red scale	parathion
Coffee	Antestis bug	parathion, trichlorphon
	Leaf miner, stem borer	parathion
Maize	Aphids	demeton-S-methyl
	Fruit fly	phorate
	Stalk borer	trichlorphon
Pea	Aphid, pea moth, thrips, weevils	azinphos-methyl, fenitrothion
Potatoes	Aphids	as for beans; also oxydemeton-methyl
Stored grain	Weevils	dichlorvos, malathion, pirimiphos-methyl
Sugar beet	Aphids (and virus)	as for potatoes; also mevinphos, pirimicarb
Sweet potatoes	Leaf miners	trichlorphon
Tobacco	Aphids	dimethoate, malathion
	Leaf miners	phosphamidon
Tomatoes	American bollworm	mevinphos
	Aphids	demeton-S-methyl, disulfoton, malathion, oxydemeton-methyl
	Red spider mite	formothion, oxydemeton-methyl
Vegetables (various, e.g. beetroot, carrot, onion, parsnip)	Mangold fly	dimethoate, formothion
	Aphids	demeton-S-methyl, oxydemeton-methyl, disulfoton
	Carrot fly	chlorfenvinphos, diazinon, phorate disulfoton
	Celery fly	malathion
	Onion thrips	malathion

Sub-group 2. Compounds of moderate to high chemical stability, usually of low solubility in water but soluble in oil. These **persistent contact** or **quasi-systemic** compounds soak into leaves but do not travel around the plant. The sprayed substance is activated before it reaches its site of action in insects. E.g. malathion, methyl parathion, diazinon, fenitrothion, trichlorphon.

Sub-group 3. Compounds of moderate to high chemical stability. Their oil/water partition coefficients are such as to enable them both to enter plants and to be translocated within them. These **systemic** compounds are activated before reaching their site of action in insects. E.g. dimethoate, demeton-methyl, phorate, formothion, disulfoton.

Sub-group 4. Compounds with sufficiently high vapour pressure and low chemical stability to enable them to be used as **fumigants**. E.g. dichlorvos, sulfotep.

Sub-group 5. Substances suitable for formulation as **granules** for soil application. E.g. chlorfenvinphos, phorate, diazinon, bromophos.

Sub-group 6. Substances suitable for surface application to, or systemic action in, animals. E.g. fenchlorphos, coumaphos, dimethoate.

4.4 MODE OF ACTION AND TOXIC EFFECTS OF ORGANOPHOSPHORUS INSECTICIDES

The primary site of action of organophosphorus compounds is the enzyme (acetyl) **cholinesterase** which is present in the nervous system. Differences undoubtedly exist between the nervous systems of vertebrates and insects but the following simplified account based on what happens in the vertebrate nervous system probably reflects the essential features of the situation in insects also (Dowson, 1977). When a nerve impulse reaches the end of an axon, a very small gap exists before the next nerve cell. This is called a synapse. A similar gap exists between the last nerve cell and the effector cell (e.g. a muscle). A chemical transmitter carries the signal across the gap, and must be destroyed immediately it has done its job or else the consequence would be the same as if the second nerve cell (or the effector cell) were to receive a continuous impulse. An analogy can be drawn with a Morse key, which only transmits a meaningful message if it is released between dots or dashes.

An important chemical transmitter is **acetylcholine**:

$$CH_3—\overset{O}{\overset{/\!/}{C}}—O—CH_2—CH_2—\overset{+}{N}(CH_3)_3 \qquad (4.2)$$

Acetylcholine cation

It is present at many, but not all, of the junctions mentioned above. The task of cholinesterase is to destroy this substance the moment its task is done. Acetylcholine binds to cholinesterase at two attachment sites. One of these, the **ester-forming site**, probably contains a serine residue in the protein chain while the other, the negative or **anionic site**, may contain a glutamic acid residue (figure 4.2).

The carbon atom of the carbonyl group of the substrate carries a slight positive charge and makes an electrophilic* attack on the hydroxyl group of the serine. This results in **acetylation** of the enzyme and in the splitting and deactivation of the acetylcholine. The free choline readily leaves the enzyme surface (figure 4.2b). The esteratic bond, EB, is weak and is rapidly hydrolysed during the **recovery**

* Electrophilic; 'electron-loving'; an attack by a slightly positively-charged centre in one molecule on a slightly negatively-charged centre in another.

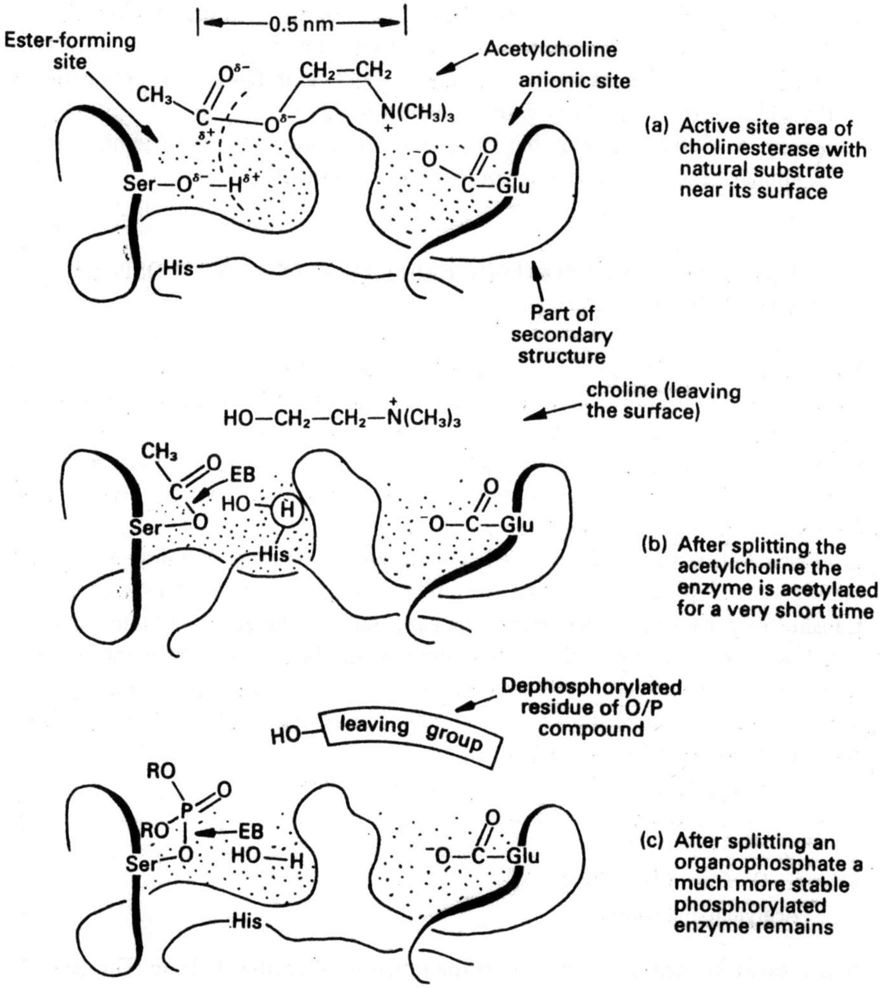

figure 4.2 Cholinesterase: a representation of its secondary structure to illustrate how its active site is believed (a) to align acetylcholine, (b) to become transiently acetylated when it splits acetylcholine and (c) to become phosphorylated when it splits organophosphorus compounds. Ser, serine; Glu, glutamic acid; His, histidine; EB, esteratic bond (see text)

stage of the enzyme, the hydrolysis probably being facilitated by a basic histidine residue nearby (Eto, 1974). The surface of the enzyme is then free to accept another molecule of acetylcholine. It has been estimated that about 300 000 molecules of acetylcholine are destroyed by one molecule of enzyme per minute at 37°C.

Organophosphorus compounds have some structural similarity to the natural substrate, acetylcholine. The phosphate group is attracted to the esteratic site and the rest of the molecule is aligned by interaction with the numerous side groups of amino acids which form the total active area of the enzyme. Most organophosphorus compounds have no positively charged groups to associate with the normal anionic site. Up to the moment of fission, they undergo a series of changes analogous to those described above. However, they **phosphorylate** the enzyme rather than acetylating it and the bond EB is relatively stable to hydrolysis (figure 4.2c). The stability of the bond depends on the nature of the phosphorylating entity, which is frequently either dimethyl phosphate or diethyl phosphate. It takes about 80 minutes for half the enzyme molecules to be dephosphorylated when dimethyl phosphate is attached to it but some six times longer when it is phosphorylated by diethyl phosphate. These times are such that it takes up to a million times longer for a molecule of the enzyme to become operative again than it does when it undergoes the natural process of acetylation.

The ease with which the bond EB is formed (as distinct from being destroyed) depends on the stability of the P–X bond in formula 4.1, and this stability depends on the electron-attracting capability of the group X. This is one reason why the various groups X in the many different organophosphorus insecticides are important. Other reasons are that they are attacked in different ways by enzymes capable of activating or destroying organophosphorus compounds, and that they influence the ease of access of the molecule to its potential site of action in insects and in vertebrates.

A somewhat more rigorous approach to the subject of cholinesterase inhibition and enzyme recovery follows, with the aid of minimal mathematics but utilising simple enzyme kinetics, based on the concept of the intermediate formation of an **enzyme–inhibitor complex**. These considerations apply equally to organophosphorus and carbamate insecticides (chapter 5). Both of these groups of compounds inhibit cholinesterase, but whereas organophosphorus compounds work by phosphorylating the enzyme, the carbamates often seem to compete with acetylcholine for the enzyme surface. The following account shows that the difference is, in fact, one of degree, depending on the relative magnitudes of certain velocity constants.

Classical enzyme–substrate interactions (e.g. succinate dehydrogenase, E, and succinate, S) can be represented by the equation:

$$\mathrm{E+S} \underset{k_2}{\overset{k_1}{\rightleftharpoons}} \mathrm{ES} \xrightarrow{k_3} \mathrm{A+B+E} \tag{4.3}$$

where k_1, k_2, k_3 are velocity constants, ES is the enzyme–substrate complex and A, B are products of the reaction.

Analogously, if an inhibitor, IX, reacts with an enzyme in an initially reversible manner, the reaction can be represented by the equation:

$$E + IX \underset{k_2}{\overset{k_1}{\rightleftharpoons}} E.IX \xrightarrow{k_3} EI \xrightarrow[H_2O]{k_4} E + IH \qquad (4.4)$$
$$\qquad\qquad\qquad\qquad\qquad \searrow X$$

Enzyme–substrate complex (the two are not covalently linked) — E.IX

Inhibited enzyme (inhibitor attached covalently) — EI

Leaving group — X

where IX is the inhibitor and I is that part of the inhibitor molecule which eventually blocks the enzyme. The group X is a leaving group, released from the enzyme–inhibitor complex, E.IX, as the van der Waals forces of attraction are replaced by a covalent bond in a reaction with rate constant, k_3.

The extent of **enzyme blocking** is thus determined by two separate factors, namely, the rate at which the complex E.IX is converted to 'blocked enzyme', EI, and the concentration of E.IX undergoing the change

$$\text{Rate of overall reaction} = k\,[\text{reactant}] = k_3\,[E.IX] \qquad (4.5)$$

where square brackets indicate the molar concentration of the substance inside the bracket, and the overall reaction is that which leads to blocked enzyme, EI.

If either [E.IX] or k_3 is low, much of the enzyme may not be inhibited. Applying the law of mass action, the concentration of enzyme–inhibitor complex [E.IX], is often measured approximately by the relationship

$$K_D = \frac{k_2}{k_1} = \frac{[E]\,[IX]}{[E.IX]} \qquad (4.6)$$

although this assumes that the distorting effect of k_3 is slight. The constant K_D is evidently the dissociation constant of the complex, E.IX; it is often called the affinity constant (of enzyme for inhibitor), a large affinity of E for IX being reflected in a low value of K_D.

EI, the inhibited form of the enzyme, will be formed from a fixed concentration of the enzyme–inhibitor complex, E.IX, most rapidly when k_3 is high (equation (4.4)). An index which takes account of the contributions of both K_D and k_3 is the inhibition constant K_i. It is defined as

$$K_i = \frac{k_3}{K_D} \qquad (4.7)$$

Clearly, the smaller K_D and/or the larger k_3, the larger is K_i; thus a large K_i value implies a high level of enzyme blocking.

Finally, it is necessary to refer to the rate of **unblocking** or **recovery** of the inhibited molecules of enzyme. This is, in the present context, the rate of hydrolysis of phosphorylated or carbamylated cholinesterase, a reaction governed by the rate constant k_4 (equation (4.4)).

It is now possible to compare and contrast the enzyme blocking actions of organophosphates and carbamates. When organophosphates inhibit acetylcholinesterase there is typically a very low concentration of enzyme–substrate complex, E.IX, present in the steady state condition. The reverse reaction governed by k_2 is in consequence negligible. In addition, the compound, EI, is rather stable and the phosphorylated, inactive enzyme only dephosphorylates slowly. Conversely, for carbamates, the rate-limiting step is the carbamylation of the enzyme (i.e. the rate at which E.IX is converted to EI). When the steady state condition has been reached a finite proportion of the enzyme is present in the form of the enzyme–carbamate complex, E.IX. Moreover, the inhibited enzyme, (EI, which here is E-carbamate) is often hydrolysed moderately rapidly back to E, thus freeing enzyme which can then participate in further reactions with natural substrate or with more carbamate. Consequently, in *in vitro* systems, carbamate inhibition can be partly reversed by addition of acetylcholine, i.e. there is a competition between carbamate and acetylcholine for the enzyme surface.

In mammals, symptoms of severe poisoning by organophosphorus insecticides include involuntary muscular twitching, convulsions and contraction of the pupils of the eyes. Atropine is one of several drugs which can ameliorate the condition; it blocks the action of the excess acetylcholine which, in vertebrates, often leads to death by causing a failure of the muscles of the diaphragm. Artificial respiration can be a valuable emergency measure. Other types of drugs accelerate the rate at which dephosphorylation of the phosphorylated enzyme takes place (e.g., pyridine-2-aldoxime methiodide; section 5.4). In insects there is no diaphragm and the precise cause of death in uncertain.

In man, mild poisoning by organophosphorus compounds can lead to a number of behavioural changes. For example, spray machine operators have been known to suffer depression and insomnia, others have lost the power to concentrate on their work or to remember things. Occasionally, affected workers may suffer a slurring of speech or acquire feelings of anxiety or irritability (Levin and Rodnitzky, 1976).

The widespread use of organophosphorus compounds has stimulated research into the possible existence of chronic effects unrelated to their anticholinesterase activity. Teratogenic (embryo-deforming) effects have been observed in avians treated with several organophosphorus compounds, but especially with diazinon. Introduction of malathion into the yolk sac of incubating eggs has similarly resulted in deformed chicks. These effects are apparently much less marked in mammals, although trichlorphon has caused embryonic abnormalities in mice and liver damage in cats. Chromosome abnormalities (mutagenic effects) have been observed to occur with greater than average frequency in human beings acutely poisoned by malathion. This has caused some concern, because mutagenic symptoms sometimes signal carcinogenic (cancer-producing) potential. Again, parathion and methyl parathion can cross the placental barrier of pregnant rats, leading to a larger than usual percentage of still-born offspring. These and other examples have been described in more detail by Fishbein (1976).

4.5 LOW PERSISTENCE CONTACT POISONS (Sub-group 1)

In this and subsequent sections, an account will be given of the chemistry, biochemistry and uses of organophosphorus insecticides, based on the classification outlined in section 4.3.

Mevinphos (figure 4.3) is an important member of this small family of low-persistence contact organophosphates. **Tetraethyl pyrophosphate,** TEPP, one of the earliest commercially-used organophosphates, also belongs to this group, although it is seldom used nowadays. TEPP is a water-soluble compound of low stability, but is of high toxicity to vertebrates – one drop of the concentrate, splashed into the eye, can be fatal. **Tetrachlorvinphos** (figure 4.3) is of relatively low persistence and probably qualifies for inclusion in this group. It is a relatively minor insecticide and acaricide used on certain types of bush fruit and brassicas. **Dichlorvos** could also be regarded as a member of this group were it not for the fact that its high vapour pressure introduces a new feature and leads to it having different practical uses (section 4.8).

The water solubility and low hydrolytic stability of members of this group preclude the possibility of long persistence in a biological environment. In practice, therefore, field action must be almost immediate and this requires that action is primarily by direct contact with the pest. Some of them, including mevinphos, do additionally have a very short-term systemic action.

Mevinphos is one of the more toxic organophosphorus insecticides still in use (table 4.1). It is miscible with water and moderately soluble in lipids. Mevinphos usually only persists for a few days after application, a property which enables it to be used as an end-of-season spray. In Britain, it is recommended

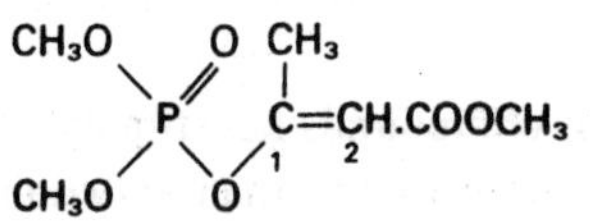

(a) Mevinphos

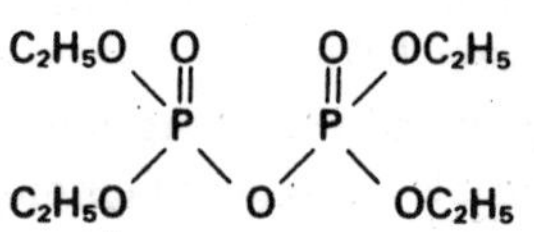

(b) TEPP

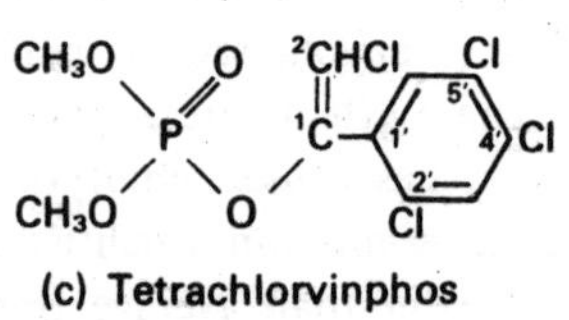

(c) Tetrachlorvinphos

figure 4.3 Contact poisons of low persistence. ((a) *cis-trans* 2-methoxycarbonyl-1-methylvinyl dimethyl phosphate; (b) tetraethyl pyrophosphate; (c) 2-chloro-1-(2′,4′,5′-trichlorophenyl) vinyl dimethyl phosphate) that three days should elapse between the last application and harvesting edible crops. It can also be employed in situations where a rapid kill is necessary, as, for example, when mechanical breakdown or weather conditions delay spraying to such an extent that a harmful insect population becomes established on a crop before action is possible.

Unlike substances in sub-groups 2 and 3, mevinphos is probably the actual toxicant at the site of action; that is to say, enzyme systems in insects or plants do not 'activate' it by converting it to a more potent cholinesterase inhibitor. For this reason it possesses little selectivity between target species and any beneficial insects which happen to be in the vicinity at the time of spraying (contrast compounds in sections 4.6 and 4.7).

Mevinphos exists in *cis* and *trans* isomers (figure 4.4). The *cis* isomer predominates in commercial products and depending on the insect species is 10–100 times as toxic as the *trans* isomer. The *cis* form, which has methyl and carboxymethyl ($COOCH_3$) groups on the same side of the double bond, is destroyed more rapidly than the *trans* isomer, so the less toxic form is the one which persists the longer.

Both forms of mevinphos are broken down in stages with the eventual production of innocuous phosphoric acid and acetoacetate. In the case of the *trans* isomer, the initial breakdown appears to be by straightforward hydrolysis at the P–O–X linkage, catalysed by an **esterase** (figure 4.4). The more insecticidal *cis* form may be similarly hydrolysed to some extent, but the **glutathione-S-alkyl transferase** system (section 3.5) is probably involved in the main catabolic route in most species. It brings about demethylation and the product containing the phosphorus is the same as would be formed if an esterase had removed the methyl group as methanol. The initial products are soon further degraded with the eventual production of phosphate and acetoacetic acid, both of which are natural components of biochemical systems.

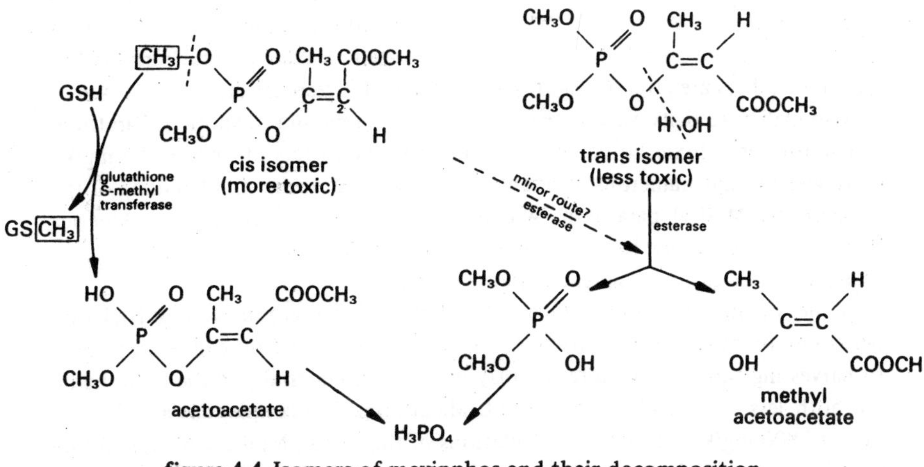

figure 4.4 Isomers of mevinphos and their decomposition

Much work on the degradation of tetrachlorvinphos has been done by Oppenoorth and coworkers (1977, 1979). The principal mechanism is O-demethylation, but whether this is hydrolytic, oxidative or involves a glutathione transferase remains uncertain (section 4.11). Some resistant strains of house fly do, however, possess high levels of glutathione-S-transferase.

4.6 QUASI-SYSTEMIC COMPOUNDS (Sub-group 2)

Substances of this group are of low solubility in water but are soluble in lipids. Their chemical stability varies, but they characteristically persist a few days to a few weeks after application to plant leaves. They are therefore also known as *persistent contact poisons*. Their lipophilic nature enables them to soak into the waxy cuticle of leaves and to diffuse short distances from the point of original contact, sometimes travelling as far as the lower surface of the leaf. This property has practical value for it allows control of insects on low-growing plants (e.g. strawberries) where direct contact spraying would be impracticable. Once absorption has occurred they are not readily removed by rain.

The low aqueous solubility of members of this group precludes the possibility of gross translocation, although radioactive labelling techniques have shown that minute amounts may be transported, especially after application to plant roots. Another characteristic of most, if not all, members of this family is that they undergo activation before they reach their site of action in the nervous system of the insect. Since activation is an oxidative process (see below) this renders the product more polar than the original molecule so that it may possess some systemic action even where the original molecule does not.

The persistent contact or quasi-systemic sub-group of organophosphorus insecticides is rather large. It can be conveniently subdivided into three principal families (as well as several minor ones). The three important sub-sections are here referred to as the malathion family, the parathion family and the family of compounds whose members contain a heterocyclic leaving group. Formulae of typical members are given in figures 4.5, 4.6 and 4.7 respectively.

Mammalian toxicity varies greatly from one compound to another. **Parathion** is a notoriously toxic liquid with an oral toxicity to rats of only 3–10 mg/kg body weight, and numerous accidents have occurred during the handling of the concentrate. **Methyl parathion**, which has been used extensively in America (table 1.1) is somewhat less toxic and its near relative, **fenitrothion**, is even safer, with an oral LD_{50} for rats greater than 250 mg/kg.

Some of the earlier-introduced members of the group were sufficiently persistent to cause residue problems especially when used on edible crops near to harvesting time. Many more recently introduced materials, on the other hand, are both less toxic and less persistent. Malathion for example, has an LD_{50} to rats of 2800 mg/kg. The Recommendations of the British Ministry of Agriculture suggest that it can be used, in appropriate cases, up to one or two days before harvesting, whereas the corresponding period for **diazinon** is two weeks, for azinphos-methyl three weeks, for parathion four weeks and for vamidothion six weeks.

The Malathion Family

Malathion is one of the most generally useful quasi-systemic compounds. As can be seen (figure 4.5) it is a dithiophosphate and the 'leaving group' is a succinic acid ester. It is principally used to control aphids and similar sap-sucking insects, although various weevils, small beetles, scale insects and red spider mites are often also susceptible. The technical material is a brownish liquid with a smell of garlic. Although somewhat expensive, it owes its popularity to the relative safety with which it can be handled, to its low persistence and to the possibility of application as an ultra low volume spray.

Malathion (figure 4.5) provides an elegant example of the way selective action can sometimes be achieved by exploiting differences in the enzymic constitution of different organisms. Almost always, hydrolysis leads to detoxication, so organisms in which a **carboxylesterase** is particularly active are able to protect themselves from malathion by removing one of the ethyl groups from the succinic ester moiety (section 3.3). Fortunately, most vertebrates seem to be well endowed with this enzyme (or at least with an enzyme which has a similar ultimate effect). When, however, the thion sulphur atom is replaced by oxygen, a much more powerful cholinesterase inhibitor, **malaoxon**, is formed. It so happens that most

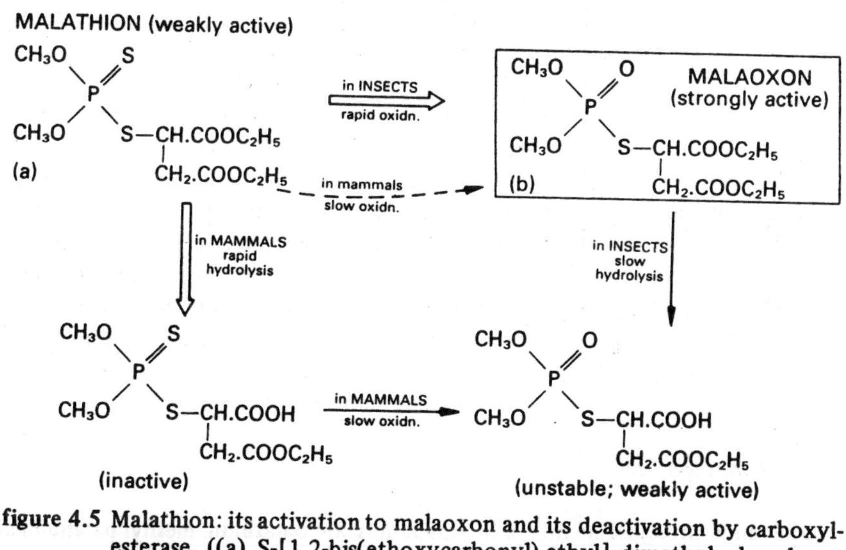

figure 4.5 Malathion: its activation to malaoxon and its deactivation by carboxylesterase. ((a) S-[1,2-bis(ethoxycarbonyl)-ethyl]-dimethyl phosphorothiolothionate; (b) S-[1,2-bis(ethoxycarbonyl)-ethyl]-dimethyl phosphorothiolate

insects seem to possess a very active oxidative enzyme system. This can activate malathion by attacking the P–S linkage, so that the insect's own defence system actually helps to kill it. It is stressed that all insects and all vertebrates probably possess both an esterase and an NADPH-dependent oxidase system; it is the balance of the action of these two systems which varies from one organism to another. The breakdown of malathion is, in practice, more complex than is shown in the figure, for further metabolic changes can remove both ethyl groups from the succinate moiety. There is also some evidence that a glutathione-S-methyl transferase (section 3.5) can remove methyl groups from one or both of the two methoxy groups.

Within a decade of the introduction of malathion, several species of insects had developed resistance to it. For example, Spiller (1961) reported that by that time mosquitoes and house flies had developed a 40–100 fold increase in tolerance under field conditions in both California and Florida. The problem in recent years has become increasingly troublesome in relation to the use of malathion for the protection of stored food products (McCullum-Deighton, 1978). Sometimes, but not always, the development of resistance to malathion is associated with a simultaneous development of resistance to other organophosphorus compounds. This phenomenon is known as **cross-resistance**. On the other hand, strains of insects which have developed resistance to organochlorine insecticides such as DDT and dieldrin are usually initially susceptible to malathion.

Townsend and Busvine (1969) reported that malathion-resistant strains of insects show high carboxylesterase activity, thus providing a clue to the probable cause of this particular type of insect resistance. Dauterman and Matsumura (1962) observed that malathion-resistant strains of the mosquito *Culex tarsalis* showed cross-resistance to analogues of malathion which had propyl or isopropyl groups attached to the succinic acid instead of ethyl groups, but that there was no such cross-resistance to the carboxymethyl analogue. This suggests that carboxylesterase may not readily remove methyl groups from $-COOCH_3$ linkages.

The role of carboxylesterase as a likely cause of selective action and of insect resistance has been described in some detail to illustrate the rationale underlying a systematic approach to the synthesis of new pesticides. If different species have different drug-metabolising enzymes – in type, in amount, or in availability to the substrate – such differences can often be exploited to achieve **selective toxicity**. As malathion illustrates, the differences need not be absolute. Similarly, although resistance to organophosphorus compounds occurs in insects for several reasons (section 4.11) it is often possible to pin-point one enzyme system which is found to be more effective in resistant than in non-resistant strains.

The Parathion Family

The three principal members of this family are **parathion, methyl parathion** and **fenitrothion** (sumithion®). Parathion is more persistent than methyl parathion and much more persistent that fenitrothion. Their mammalian toxicities were

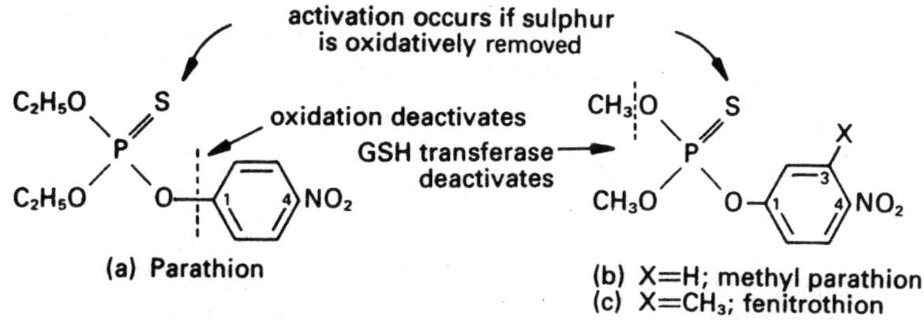

figure 4.6 The parathion family of persistent contact insecticides. ((a) diethyl 4-nitrophenyl phosphorothionate; (b) dimethyl 4-nitrophenyl phosphorothionate; (c) dimethyl 3-methyl-4-nitrophenyl phosphorothionate)

mentioned on p. 80 and their formulae are given in figure 4.6. Their agricultural uses are rather similar and wide ranging – in most localities they control aphids, capsids, leafminers, sawflies and weevils. They also tend to have individual uses – in Britain, parathion is recommended for the control of mealy bugs, scale insects, root-knot nematodes and woodlice whereas the less toxic fenitrothion is used against flour beetles, grain beetles and grain weevils. Desmarchelier (1978) studied the rate of loss of fenitrothion from stored wheat and rice grain and has provided several references to work on this subject. For some substances the problem of food residues can be minimised by spraying the insecticides on to the absorbent walls of the warehouse rather than admixing with the bulk of the food material. Other aspects of the use of organophosphorus compounds to protect food in storage are considered in section 4.8.

The metabolism of parathion and its analogues illustrates how closely field safety is associated with minor variations in chemical structure which, in turn, lead to changes in the balance established between the various pathways of metabolism. All three compounds shown in figure 4.6 are oxidised by mono-oxygenases in animals, insects and plants and are thereby changed to derivatives containing the P=O group and which are more powerful inhibitors of cholinesterase than was the original thion phosphate. Their degradation, however, follows different routes. Parathion is principally degraded by rupture of the P–O–phenyl linkage, whereas the two methyl analogues are not.

The rupture of the P–O–phenyl linkage of parathion is not a hydrolytic reaction but is an oxidative process mediated by an NADPH-dependent oxidase:

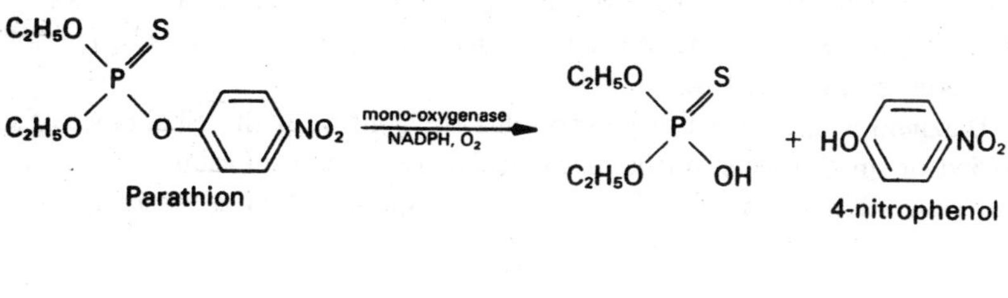

(4.8)

In contrast, methyl parathion and fenitrothion are principally destroyed in most species investigated by rupture of a P–O–CH_3 linkage;

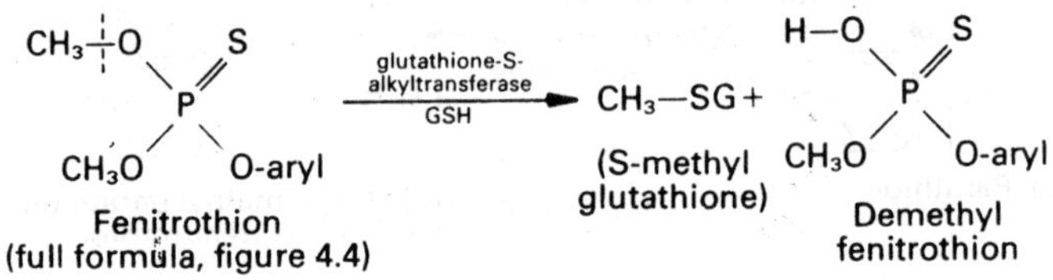

(4.9)

Fukami and Shishido (1966) found that rat liver could carry out this change five times as fast as a comparable insect system. It seems probable that this metabolic difference could be a factor contributing to the lower mammalian toxicity of these methyl analogues compared to parathion itself, for parathion does not readily de-ethylate in this manner. Rather similarly, it will be recalled that methyl parathion is normally more toxic to mammals than is fenitrothion, and Rainsford (1978) attributed this difference to the more rapid metabolism of the latter compound.

Experiments by Baker *et al.* (1979) on whole perfused liver have provided a further insight into the higher animal's total defence mechanism against parathion (and probably many other organophosphorus compounds). In these experiments, perfusion fluid is recycled, the ingoing and exiting liquids being monitored for parathion. They showed that, in a single cycle, absorption of parathion was very high, varying at different parathion concentrations between 69 per cent and 98 per cent of the dose. As the authors point out, the facility with which the liver is able to extract most of the toxicant reaching it is the most immediate and most protective of all mechanisms of defence.

When sulphur is removed from thionphosphates like parathion in *in vitro* systems, it becomes bound to microsomal macromolecules, sometimes to the detriment of microsomal enzyme activity. It is probable that such organic binding is in the form of hydrodisulphides (R–S–S–H). Eventually, in an *in vivo* situation, most of this organically bound sulphur is liberated and appears in the urine as sulphate and thiosulphate. The formation of inorganic sulphate is enhanced by the presence of reduced glutathione (Morelli and Nakatsugawa, 1979).

The Diazinon Family (with Heterocyclic Leaving Groups)

Of the numerous substances in this family of persistent contact poisons, figure 4.7 gives the structures of diazinon and pirimiphos-ethyl and, additionally, the formulae of azinphos-methyl and menazon. Others include chlorpyriphos, phosalone and triazophos.

Diazinon is used as a foliage spray for the control of the usual range of soft-bodied or small insects, but is of particular use to control cabbage root fly and late generation carrot fly. In Britain, **pirimiphos-methyl** is primarily used to

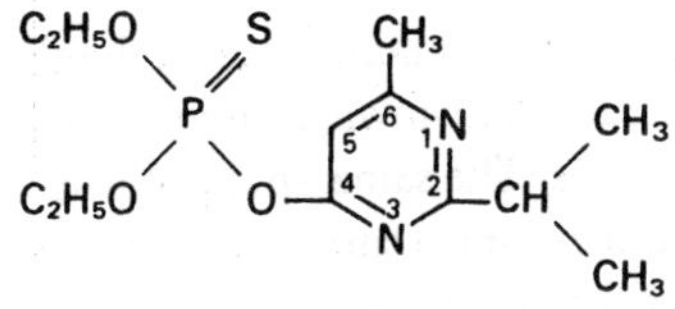

(a) Diazinon

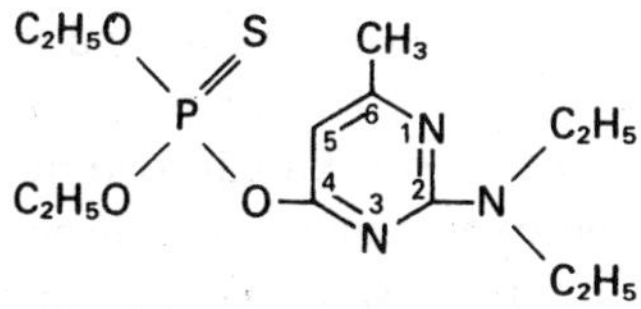

(b) Pirimiphos-ethyl

figure 4.7 Persistent contact poisons with heterocyclic leaving groups. ((a) diethyl 2-isopropyl-6-methylpyrimidin-4-yl phosphorothionate; (b) dimethyl 2-diethylamino-6-methylpyrimidin-4-yl phosphorothionate; others are (c) azinphos-methyl: dimethyl S-(3,4-dihydro-4-ketobenzo-[d]-1,2,3-triazin-3-yl-methyl) phosphorothiolothionate; (d) menazon: dimethyl S-(4,6-diamino-1,3,5-triazin-2-yl-methyl) phosphorothiolo-thionate)

control mites, flour beetles, grain weevils and similar organisms in stored products. Wherever possible, it is advisable to avoid direct spraying of food products, for spraying the walls of storage areas or treating sacks containing produce often works adequately. It has found use for the control of organisms that have developed resistance to malathion (McCullum-Deighton, 1978). **Azinphos-methyl** is particularly used in Britain to control codling, tortrix and winter moths, as well as a variety of other pests of top fruit, bush fruit and cane fruit. Unfortunately, red spider mites are tending to develop resistance to it. **Menazon** is a good aphicide, with slight systemic activity by foliage application but it can enter roots quite readily (compare section 4.7).

Most, if not all, of the heterocyclic compounds are activated by mono-oxygenases with the formation of P=O (oxon) derivatives that are powerful inhibitors of cholinesterase. A common route of degradation for diazinon (and probably for other members of the group) involves the cleavage of the P–O–pyrimidine linkage by the action of an NADPH-dependent oxidase. This is analogous to the reaction undergone by parathion (equation (4.8)). On the other hand, it is becoming increasingly clear that glutathione-S-aryl and -S-alkyl transferases often catalyse contributory, and sometimes major, detoxifying pathways. Thus Usui *et al.* (1977) isolated several glutathione-S-transferases from the fat bodies of female *Periplaneta americana.* These transferases had different but overlapping specificities, two being particularly active on diazinon and three on methyl parathion. As a result of genetic studies done on diazinon-resistant house flies, Motoyama *et al.* (1977) concluded that glutathione-dependent enzymes are not a major biochemical mechanism responsible for housefly resistance to diazinon.

The Phosphonate Family

Amongst the miscellaneous non-systemic poisons not covered by the previous three families, the phosphonates merit brief consideration because some of them exhibit a feature normally absent in other groups, namely, *stereoisomerism.*

EPN is a minor insecticide and acaricide which controls a number of orchard pests (Matsumura, 1975). Its three-dimensional structure, shown in figure 4.8, indicates that it can exist in two optically active forms. The same applies for its oxon. The oxon isomers have been isolated by Nomeir and Dauterman (1979) and their toxicities and rates of metabolism studied individually. It was found that the (+) isomer of EPN-oxon was much more toxic to mice than was the (−) isomer. This was due to a higher affinity of the toxic isomer for cholinesterase rather than a difference in chemical stability; the rate of alkaline hydrolysis was the same for both enantiomers.

A somewhat similar result was obtained by Lee *et al.* (1978) using the experimental phosphonate fonofos and its oxon. They found that one of the enantiomers of fonofos-oxon (S) was between four and 60 times more inhibitory to several esterases than was the other (R). *In vivo* studies showed similarly that the S isomer was, in different tests, from 2 to 12 times as toxic as the R isomer to mice and to house flies. A rather interesting further observation was that the more toxic S isomer of fonofos-oxon was formed in microsomal systems *in vitro* by oxidative desulphuration of the R enantiomer of fonofos.

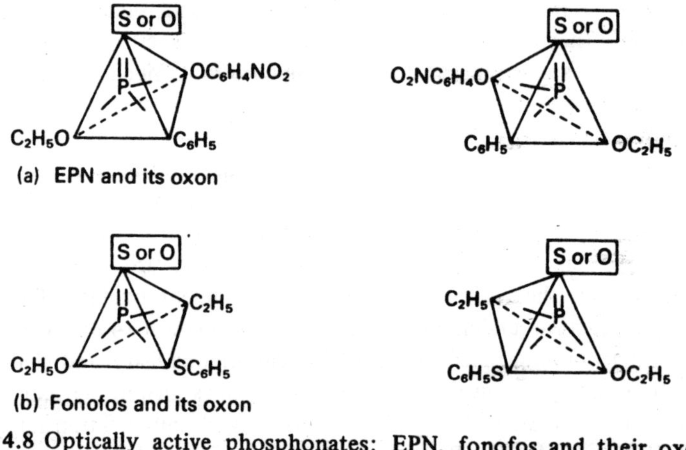

figure 4.8 Optically active phosphonates: EPN, fonofos and their oxons. ((a) ethyl p-nitrophenyl phenylphosphonothionate; (b) ethyl phenylthio ethylphosphonothionate)

4.7 SYSTEMIC INSECTICIDES (Sub-group 3)

These substances are soluble in lipids but they also usually possess a higher water solubility than substances in sub-group 2. Consequently, their partition coefficients are such that they pass through cuticular waxes and lipoprotein membranes to emerge in the phloem or xylem streams of sprayed plants. Most are of moderate stability and persist for a period of a few days to a few weeks within plants.

It should be noted that the word 'systemic' can be used in slightly different ways, for a substance that does not readily enter through leaves may still enter through roots and be carried upwards in the transpiration stream. This can lead to different views about how systemic a particular substance is. In practice, those substances which enter predominantly through roots often have aqueous solubilities at the lower end of the range for the group, and almost all substances that readily enter foliage can also enter through the roots (although not necessarily easily reach roots through soil).

Systemic poisons have several important advantages over other types of persistent poisons. They are not, for example, susceptible to wash-off by rain. Furthermore, by virtue of their movement within plants, they give protection in regions away from the locus of application, as well as protection at a time removed from the time of application. They often tend to be transported towards regions of growth, giving protection to the vital and vulnerable tissue near meristems – they may, indeed, give protection to tissue not even differentiated at the time of spraying.

Since many systemic compounds have only a weak contact action, beneficial organisms in the vicinity at the time of spraying may not succumb even if the spray hits them. This opens up a new dimension in pest control, for, as Carson (1963) stressed, one of the principal objections to chemical pest control is that indiscriminate destruction of pests and predators can result in explosive reproduction (flare-up) of surviving pests. Their limited contact toxicity arises from the fact that the *sprayed substance* is often a weak cholinesterase inhibitor, the *actual toxicant* being formed by metabolic action within the plant (or sometimes after oral uptake by the insect). The metabolic activation can take several possible forms but it involves oxidation by mono-oxygenases in a manner analogous to examples encountered in section 4.6.

Systemic insecticides are used against a wide range of phytophagous insects but they excel as aphicides and some are used against red spider and other mites. Vicarious control of some aphid-borne virus diseases such as cucumber mosaic virus and beet virus yellows is also possible. Other common targets are bulb flies, sawflies, codling moth, leaf miners, weevils and woolly aphids.

Many systemic compounds fall into chemical families characterised by the presence of thioether and carbamate groups.

The Thioether Family of Systemic Insecticides

The formulae of three members of the **thioether** family are shown in figure 4.9.

A major metabolic change undergone by compounds of this type is oxidation of the thioether sulphur atom to give sulphoxides and then sulphones:

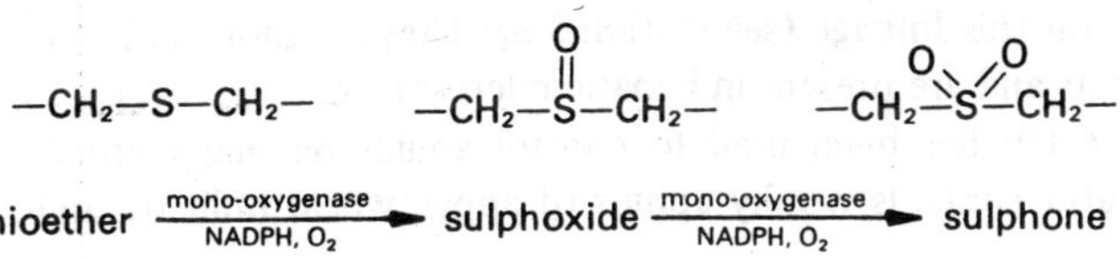

(4.10)

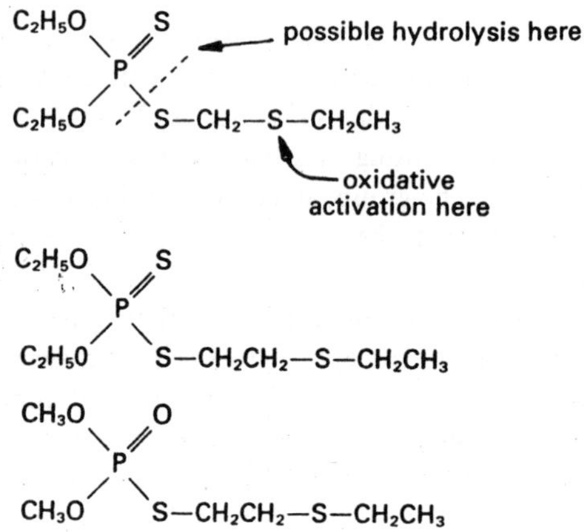

(a) Phorate

(b) Disulfoton

(c) Demeton–S–methyl

figure **4.9** The thioether family of systemic insecticides. ((a) diethyl S-(ethylthiomethyl) phosphorothiolothionate; (b) diethyl S-(ethylthioethyl) phosphorothiolothionate; (c) dimethyl S-(ethylthioethyl) phosphorothiolate)

This series of changes is usually accompanied by an increase in mammalian toxicity and with an increased inhibitory effect *in vitro* upon cholinesterase activity. For example, the dose of the sulphone of phorate which inhibits cholinesterase by 50 per cent is only one-twentieth of that of phorate itself. There is also some evidence that oxon formation, similar to that which converts parathion to paraoxon, can also activate thionphosphates of this family.

Detoxication of members of the thioether family appears to be by the hydrolytic removal of the thioether moiety, principally but not exclusively by cleavage of the P–S (rather than the S–C) bond.

The Carbamate Family of Systemic Insecticides

The second chemical family contains substances which are essentially amides, for they contain a substituted carbamic acid group, –CONHR, in the side chain (figure 4.10). Carbamic acid is aminoformic acid, $HOOC.NH_2$.

Activation by oxidation probably follows one of the two well-established patterns already discussed. Firstly, the P=S group, when present, can be converted to P=O and secondly the $-CH_2-S-CH<$ group, when present, can be converted to sulphoxide. Degradation, however, is rather different from that in earlier examples, for the substituted amide (carbamoyl) group adds a new feature which can be exploited to achieve selective toxicity. Special enzymes, **carboxylamidases,** probably act on this linkage (see section 3.3). They are more active in vertebrates than in insects and are present in hepatic microsomes.

Dimethoate (figure 4.10) has been used to control aphids on many crops, including tobacco. It also controls woolly aphids of apple, wheat bulb fly and

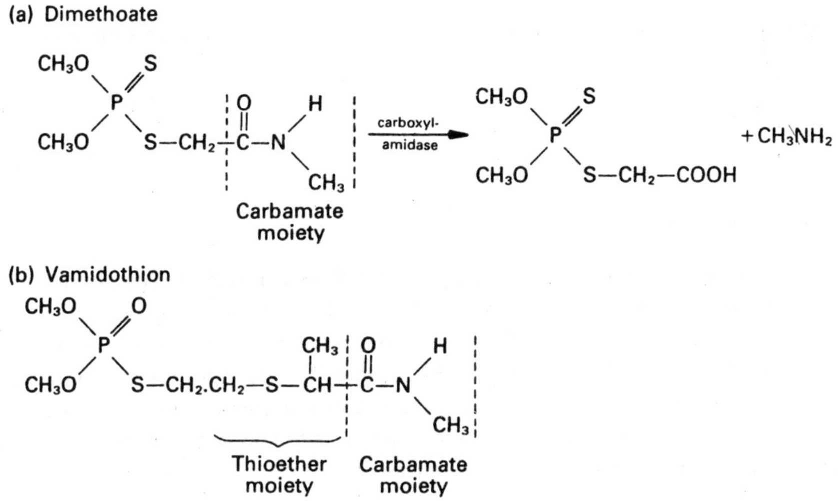

figure 4.10 Systemic organophosphorus insecticides containing the carbamate group. ((a) dimethyl S-(methylcarbamoylmethyl) phosphorothiolothionate; (b) dimethyl S-[2-(1-methylcarbamoylethylthio)ethyl] phosphorothiolate

red spider mite. It has found use in animal hygiene and, in accord with the observation about carboxylamidases in the previous paragraph, it is noteworthy that it has a relatively high LD_{50} for mammals compared to the LD_{50}s of most other organophosphorus insecticides that possess appreciable aqueous solubility (table 4.1).

Other Systemic Organophosphorus Insecticides

There are some systemic compounds which do not belong to either of the two families so far mentioned. Those containing a halogen and those containing a vinyl group are worthy of note. **Trichlorphon** (trichlorfon) is a phosphonate with the following structure:

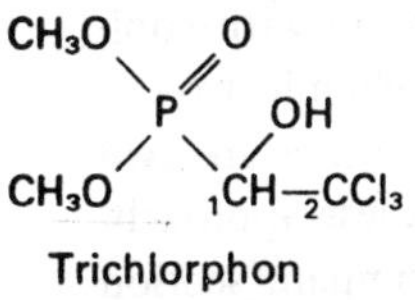

(4.11)

Trichlorphon

It is particularly useful against leaf miners and cabbage root fly larvae. Despite its phosphonate structure, it is rapidly destroyed, perhaps by *in vivo* conversion to dichlorvos (section 4.8).

Phosphamidon and **dicrotophos** are examples of substances containing a vinyl group. Phosphamidon is miscible with water and is rapidly decomposed

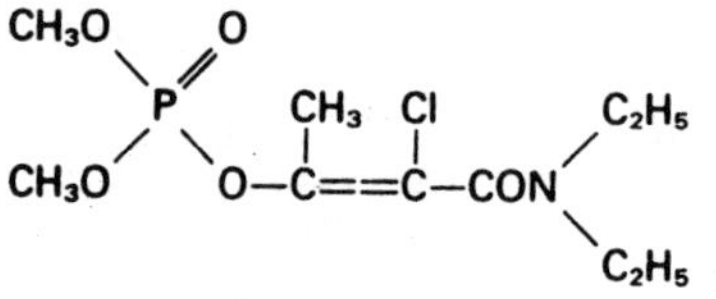 (4.12)

Phosphamidon

both in plants and in animals. It is mainly employed to control aphids on broad bean, field beans and peas. Animals appear to detoxify it by removing one of the methyl groups shown on the left, while plants probably activate it to a more potent cholinesterase inhibitor by removal of one of the ethyl groups shown on the right. Dicrotophos is similar to phosphamidon but has a hydrogen atom instead of chlorine and methyl groups instead of ethyl. It has been used to control the beetle that spreads Dutch elm disease.

4.8 ORGANOPHOSPHORUS COMPOUNDS WITH A FUMIGANT ACTION (Sub-group 4)

Dichlorvos is the main practical example of an organophosphorus compound with a vapour pressure sufficiently high to enable it to act in the vapour phase. Its vapour pressure of 1.2×10^{-2} mm Hg at 20°C is high compared to the lethal concentration needed to kill insects and is 10–1000 times higher than the vapour pressures of most other organophosphorus insecticides (table 4.1).

Dichlorvos was originally introduced to kill house flies but has found horticultural application in greenhouses and mushroom houses. It can be applied as an aerosol mist but, being a liquid, is more usually formulated as a solid solution in plastic strips from which it slowly evaporates. It is very toxic to human beings but its rapid decomposition usually enables a sprayed area to be entered safely within 24 hours. It is also used in animal hygiene to control flies and similar insects on livestock.

The relatively high volatility of dichlorvos has led to its use for the protection of food in *closed* storage areas. The suitable design of warehouses, so that pests cannot gain easy access, is the first line of defence, and ventilation with cool, dry air can further minimise losses (Howe, 1978). Nevertheless, chemical treatment, especially of stored grain, is often imperative, and then there are two methods of attack. Fumigants such as dichlorvos represent one of these approaches, for such materials diffuse readily around the warehouse (compare section 2.5). Dichlorvos kills free-living insects and can also reach most beetles living within the kernels of grain (Desmarchelier *et al.*, 1977). Other common fumigants include methyl chloride, ethylene dibromide and phosphine. Some fumigants are both highly toxic to man and absorbed into food products, so treated foods should always be cooked before use. The second method of attack includes the admixing of a relatively non-toxic and non-volatile insecticide with food material, usually grain, or depositing it on walls and sacking. The advantage of such materials (e.g. malathion, pirimiphos-methyl, section 4.6) is that they give protection against renewed invasion, but they have the disadvantage of leaving residues in the food which have proved to be psychologically embarrassing to governments if not always physically damaging to the consumer. Usually of greater scientific relevance, in the view of the writer, is the fact that insect resistance develops rather rapidly to some residual protectants, including malathion (Champ and Dyte, 1976).

The structure of dichlorvos and two probable routes of degradation are shown in figure 4.11. No activation system is known.

The hydrolytic removal of the dichlorovinyl group (as dichloroacetaldehyde) occurs in non-biological aqueous solutions and, more rapidly, in the presence of a water-soluble esterase. This route is characterised by the formation of dimethyl phosphate. The second route involves demethylation and the consequent production of demethyl dichlorvos. The latter is then further degraded to monomethylphosphoric acid. In the general case, the removal of methyl groups in this manner can be catalysed by glutathione-S-alkyl transferase (section 3.5) but whether this or simple hydrolysis occurs in the present instance is uncertain (Matsumura, 1975). Dichlorvos decomposes rapidly in animals, half of it being decomposed and excreted within 12 hours.

Another volatile compound occasionally used to fumigate enclosed areas is **sulfotep**. This is a close relative of the now withdrawn foliage insecticide, TEPP (section 4.5), but contains sulphur atoms in the two dative bond positions $[(C_2H_5O)_2.PS.O.PS(OC_2H_5)]$. It is used as a smoke-generating formulation for fumigation of greenhouses; special clothing, including a respirator, must be worn by anyone entering a greenhouse within 12 hours of treatment.

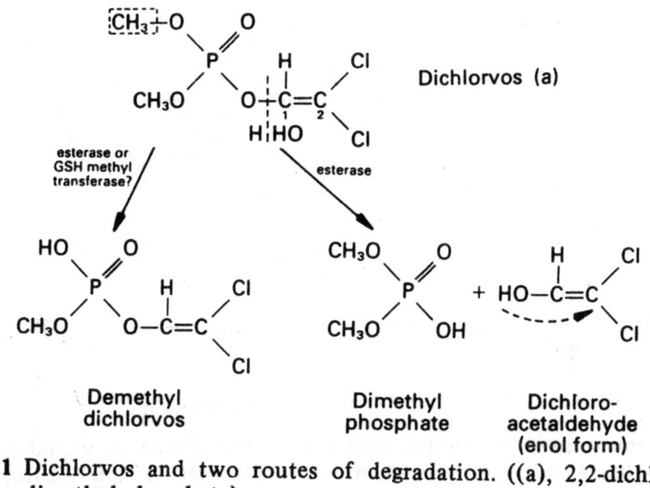

figure 4.11 Dichlorvos and two routes of degradation. ((a), 2,2-dichlorovinyl dimethyl phosphate)

4.9 ORGANOPHOSPHORUS COMPOUNDS USED IN GRANULAR FORM (Sub-group 5)

Several organophosphorus compounds have been formulated as granules often, but not invariably, to facilitate their application to soils. Details of granular formulations and their uses are given in section 2.6, but it will be recalled that a granule comprises a suitable absorbent mineral impregnated with the active ingredient. Advantages of granular products include the relative safety with which they can be handled and the ease with which they can be placed where they give most protection. Granules bounce off foliage so that (unlike sprays), most of the active ingredient penetrates to ground level. For many agricultural purposes they are placed by special applicators near to row crops. Soil-applied granules may be left to weather on the surface or they may be incorporated into the soil.

Several organophosphorus compounds are formulated as granules. One of these is **chlorfenvinphos** (figure 4.12), a vinyl compound distantly related to dichlorvos. It is used on brassicas, carrots and sweet corn for the control of cabbage root fly, carrot fly and fruit fly respectively, and is added to mushroom compost for the control of mushroom flies.

Diazinon (figure 4.7) and **disulfoton** (figure 4.9) are two organophosphorus compounds that are not only formulated for foliage spray application but also as granules. In granular form they are applied to soil for similar purposes to those for which chlorfenvinphos is employed.

Disulfoton and **phorate** are amongst several systemic compounds which can be applied near to the roots of seedling plants to give protection against aphids and the virus diseases they transmit. In addition, they are also used by potato growers to limit wireworm damage. **Bromophos** granules are widely used by the home gardener in Britain for application to the soil surface or to control root fly larvae, wireworms, cutworms and chafer grubs.

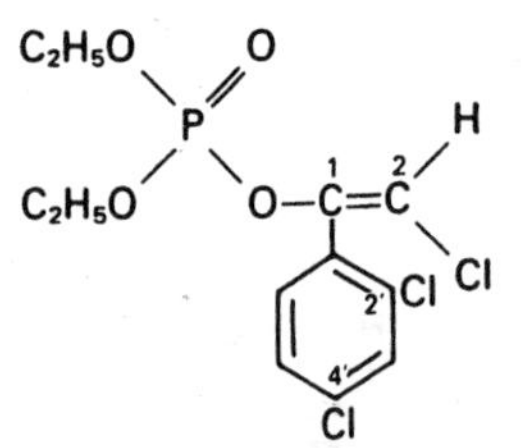

(a) Chlorfenvinphos

phorate—see fig. 4.9
disulfoton—see fig. 4.9
diazinon—see fig. 4.7

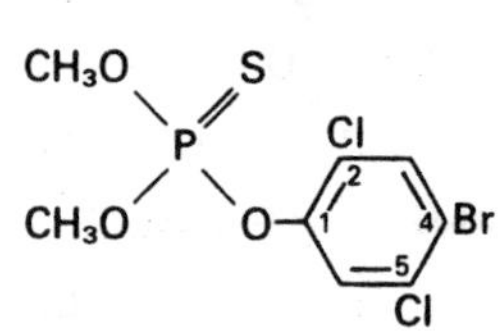

(b) Bromophos

figure 4.12 Some organophosphorus compounds formulated as granules. ((a) 2-chloro-1-(2′,4′-dichlorophenyl) vinyl diethyl phosphate; (b) 4-bromo-2,5-dichlorophenyl dimethyl phosphorothionate)

4.10 USE OF ORGANOPHOSPHORUS COMPOUNDS AGAINST PESTS OF ANIMALS

Agricultural pesticides are often thought of as an adjunct to crop production but several organophosphorus compounds have a place in animal production. Many of the compounds used to control insects and some other invertebrates that attack higher animals are amongst the less toxic members of the sub-groups already discussed. However, a few are marketed specifically for use in animal hygiene, being unsuitable for use on crops on account ot their high phytotoxicity. Some compounds used on animals are systemic (i.e. are carried around inside the animal) after addition to fodder or application to the skin. Others act by direct contact action on ectoparasites after application as drenches to the skin.

Two notable examples of compounds used against animal parasites are **fenchlorphos** and **coumaphos** (figure 4.13). Fenchlorphos (also known as ronnel in the USA) is closely related to bromophos (figure 4.12) and both of these compounds are of low mammalian toxicity (table 1.2). Fenchlorphos is an example of a substance that is too phytotoxic for use on crops but when it is fed to cattle it enters the bloodstream and acts systemically, killing a variety of parasites, including screw worms, lice and maggots of the warble fly. In some countries (e.g. Mauritius) the effect of the biting fly, *Stomoxys*, on milk production by cows is so great that the animals have to be incarcerated in sheds for most ot their lives. Fenchlorphos and **dimethoate** are two insecticides which show promise as control measures for this pest (Harris *et al.*, 1976). Studies by Sitkiewicz *et al.* (1977) suggest that fenchlorphos is converted to its oxon analogue in rats.

Coumaphos is somewhat different from fenchlorphos in that it is rather toxic when given by mouth. It is, however, relatively safe when applied to the skin (the dermal LD_{50} for rats is approximately 900 mg/kg). It is used on cattle and sheep to control ectoparasitic grubs and lice, and on poultry to protect them against mites. Other compounds used on animals for the control of various pests include trichlorphon and dichlorvos. Despite their oral toxicity, coumaphos and

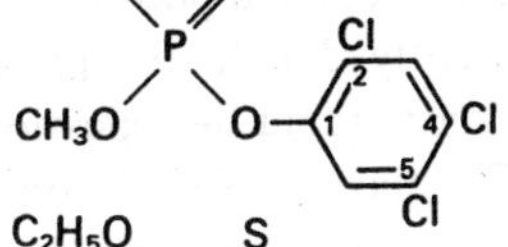

(a) Fenchlorphos
(cf. bromophos, fig. 4.12)

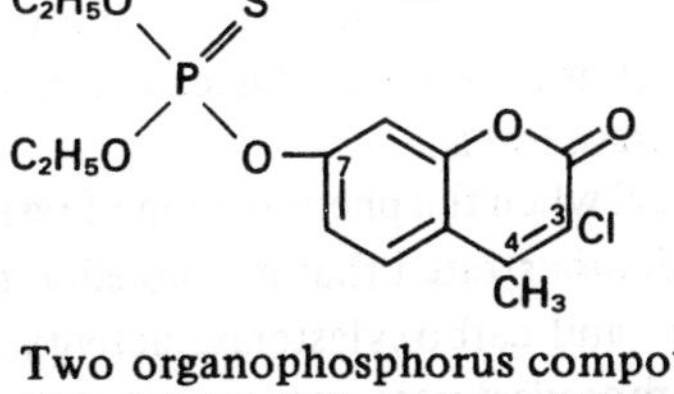

(b) Coumaphos

figure 4.13 Two organophosphorus compounds used to control pests of animals. ((a) 2,4,5-trichlorophenyl dimethyl phosphorothionate; (b) diethyl (3-chloro-4-methyl-7-coumarinyl) phosphorothionate) dichlorvos are amongst several compounds tested in animals for the control of helminth diseases. Although the chemotherapeutic index (the ratio of toxic dose to therapeutic dose) only ranges from 2 to 12, such is the difficulty of dealing with the filarial worms causing elephantiasis, river blindness (onchocerciasis) and Calabar swellings (loiasis) that their use is being actively studied (Lämmler, 1977). Together, these three forms of filarial infections affect about 300 million people.

4.11 INSECT RESISTANCE TO ORGANOPHOSPHORUS COMPOUNDS

The resistance of insects to organophosphorus compounds is a complex phenomenon and (except perhaps for the student) this complexity is a fortunate consequence of the molecular diversity mentioned in earlier sections. There is not a single cause of resistance such as, for example, some sort of adaptation to organophosphorus compounds at the site of action in the insect nervous system. The absence of a single cause greatly diminishes the likelihood of a simultaneous development of resistance to all organophosphorus compounds by one species or to one organophosphorus compound by numerous species.

Resistance can involve **avoidance** or **tolerance** (Hollingworth, 1976) and each of these may result from a number of causes. Avoidance is a term including all the behavioural or other changes which enable insects to keep clear of the poison. For example, flying insects may acquire the instinct not to land on a surface sprayed with a particular compound.

Genetic changes occurring over several generations can result in a population the individuals of which are statistically more tolerant than those of the original population. One explanation of such a change would be a decrease in the sensitivity to an organophosphorus insecticide of its target enzyme, the insect brain cholinesterase. One such case is described by Oppenoorth *et al.* (1977) but, as was mentioned above, this sort of change is seldom the major, or the single, cause of resistance. More usually, resistance is associated with a change in the level of some enzyme capable of attacking the insecticide or a change in the availability of the site of action to an externally-applied poison.

Resistance to malathion is often apparently due to the presence of raised levels of carboxylesterase and thus to the more rapid detoxication of the pesticide (section 4.6). Blackman *et al.* (1977) have provided some quantitative evidence for coinheritance of increased carboxylesterase activity and resistance in aphids. Their results suggest that resistance in this case is heterozygous for two alleles coding for increased esterase levels.

However, surprises are not unusual when the phenomenon of resistance is being examined and other studies have demonstrated that it is possible to have a **negative correlation** between resistance and carboxylesterase activity. One explanation is that a mutant form of carboxylesterase may occur in resistant insects such that, while *carboxyl* esterase action may be low, a modified form of the enzyme becomes capable of hydrolysing other kinds of ester linkages (e.g. R–O–P). Examples are to be found in the article by Plapp (1970).

Another seeming paradox may be resolved rather similarly. Beranek and Oppenoorth (1977) have shown that methyl parathion (which contains no carboxylic acid ester linkage) is hydrolysed by a resistant strain of peach potato aphid rich in carboxylesterase, but not by a susceptible strain poor in this enzyme. Of course, the possibility of simultaneous elevation of the levels of two enzymes – the known carboxylesterase and an unknown one which actually attacks methyl parathion – cannot be excluded. However, two enzyme preparations, each with a high specific activity for one of these two types of action, are both water soluble, very stable and have similar electrophoretic patterns.

Evidence relating to the contribution of glutathione-S-transferases to resistance is conflicting. Elevated levels of the enzyme(s) can undoubtedly occur in resistant strains. However, at least in the case of housefly resistance to diazinon, Motoyama *et al.* (1977) concluded that glutathione-dependent enzymes were not a major cause of the resistance. Oppenoorth *et al.* (1977), working with a strain of house flies resistant to tetrachlorvinphos, found that the resistant strain could dealkylate various compounds, including parathion and methyl parathion, up to 120 times faster than a susceptible strain and that the enzyme system responsible was glutathione-dependent. This seems to suggest that efficient glutathione-S-transferase systems could be an important cause of resistance in some resistant strains. However, later work (Oppenoorth *et al.*, 1979) on substrains showed that correlation of high resistance and high transferase activity was incomplete. Several different transferases were separated from resistant houseflies using carboxymethylcellulose columns. The transferase that degrades tetrachlorvinphos (figure 4.3) does this almost exclusively by O-demethylation. In summary, the role of glutathione transferase levels as a cause of insect resistance cannot as yet be regarded as resolved.

Correlation of resistance with increased levels of various oxidases has not been widely reported for organophosphorus compounds. This probably relates in part to the complicating effect of oxidative activation and to the fact that (the commonly-occurring) O-dealkylation reactions can be catalysed by oxidases, transferases or hydrolases. There is some evidence (Plapp, 1970) for the existence of a semidominant gene which leads to resistance to carbamate insecticides. Resistance associated with this gene can be reversed by methylene dioxyphenyl synergists. Georghiou (1965) has provided a useful review of the genetics of resistance.

REFERENCES

Baker, S. R., McGilliard, A. D. and Dahm, P. A. (1979). *Pestic. Biochem. Physiol.*, **11**, 20

Beranek, A. P. and Oppenoorth, F. J. (1977). *Pestic. Biochem. Physiol.*, **7**, 16

Blackman, R. L., Devonshire, A. L. and Sawicki, R. M. (1977). *Pestic. Sci.*, **8**, 163
British Crop Protection Council (1979). *The Pesticide Manual: a World Compendium* (ed. C. R. Worthing), 6th edn, 74 London Road, Croydon, Surrey, CR0 2TB, England
Carson, R. (1963). *Silent Spring*, Hamish Hamilton, London
Champ, B. R. and Dyte, C. E. (1976). *Report of the FAO Global Survey of Pesticide Susceptibility of Stored Grain Pests*, Rome, U.N. Fd Agric. Org., 297pp.
Dauterman, W. C. and Matsumura, F. (1962). *Science, N.Y.*, **138**, 694
Desmarchelier, J. M. (1978). *Pestic. Sci.*, **9**, 33
Desmarchelier, J. M., Banks, H. J., Williams, P. and Minett, W. (1977). *J. stored Prod. Res.*, **13**, 1
Dowson, R. J. (1977). *Pestic. Sci.*, **8**, 651
Dubois, K. P. (1971). *Bull. W.H.O.*, **44**, 233
Edson, E. F., Sanderson, D. M. and Noakes, D. N. (1966). *Wld Rev. Pest Control*, **5**, 143
Eto, M. (1974). *Organophosphorus Pesticides: Organic and Biological Chemistry*, CRC Press, Cleveland, pp. 131–133
Fishbein, L. (1976). *Pesticide Biochemistry and Physiology* (ed. C. F. Wilkinson), Heyden, New York, p. 555
Fukami, J. and Shishido, T. (1966). *J. econ. Entom.*, **59**, 1338
Georghiou, G. P. (1965). *Adv. Pest Control Res.*, **6**, 171
Harris, E. G., Grose, J. E. H., Saliba, L. J. and Zammit Lucia, M. (1976). *PANS*, **22**, 201
Hollingworth, R. (1976). *Pesticide Biochemistry and Physiology* (ed. C. F. Wilkinson), Plenum Press, New York, p. 580
Howe, R. W. (1978). *Outl. Agric.*, **9**, 198
Lämmler, G. (1977). *Pestic. Sci.*, **8**, 563
Lee, P. W., Allahyari, R. and Fukuto, T. R. (1978). *Pestic. Biochem. Physiol.*, **8**, 146, 158
Levin, H. S. and Rodnitzky, R. L. (1976). *Clin. Toxic.*, **9**, 391
Matsumura, F. (1975). *Toxicology of Insecticides*, Plenum Press, New York
McCullum-Deighton, J. (1978). *Outl. Agric.*, **9**, 240
Mitchell, J. W., Smale, B. C. and Metcalf, R. L. (1960). *Adv. Pest Control Res.*, **3**, 359
Morelli, M. A. and Nakatsugawa, T. (1979). *Pestic. Biochem. Physiol.*, **10**, 243
Motoyama, N., Dauterman, W. C. and Plapp, F. W., Jr., (1977). *Pestic. Biochem. Physiol.*, **7**, 443
Nomeir, A. A. and Dauterman, W. C. (1979). *Pestic. Biochem. Physiol.*, **10**, 121
Oppenoorth, F. J., Smissaert, H. R., Welling, W., van der Pas, L. J. T. and Hitman, K. T. (1977). *Pestic. Biochem. Physiol.*, **7**, 34
Oppenoorth, F. J., van der Pas, L. J. T. and Houx, N. W. H. (1979). *Pestic. Biochem. Physiol.*, **11**, 176
Plapp, F. W. (1970). *Biochemical Toxicology of Insecticides* (eds. R. D. O'Brien and I. Yamamoto), Academic Press, New York, p. 179
Price Jones, D. and Edgar, E. C. (1961). *Outl. Agric.*, **3**, 123
Rainsford, K. D. (1978). *Pestic. Biochem. Physiol.*, **8**, 302
Sitkiewicz, D., Rakowska, I., Skonieczna, M. and Chajdak, A. (1977). *Pestic. Biochem. Physiol.*, **7**, 501
Spiller, D. (1961). *Adv. Pest Control Res.*, **4**, 249
Townsend, M. G. and Busvine, J. R. (1969). *Entomologia exp. appl.*, **12**, 243
Usui, K., Fukami, J. and Shishido, T. (1977). *Pestic. Biochem. Physiol.*, **7**, 249

5 Carbamate Insecticides, Molluscicides and Nematicides

5.1 INTRODUCTION

Physostigmine is a naturally-occurring carbamate possessing powerful anti-cholinesterase activity. It is present in the calabar bean (*Physostigma benenosum*) which was used in West Africa in witchcraft trials by ordeal. The accused was made to drink a suspension in water of macerated bean seeds. The survival of the accused can be attributed more to his ability to vomit than to the extent of guilt – if he was not sick the physostigmine entered his body, inhibiting acetylcholin-esterase, so that he died of an excess of his own acetylcholine (section 4.4). The local name for the calabar bean is esere, which accounts for an alternative name, eserine, often used for physostigmine.

Physostigmine is one of many medically important carbamates. It is not very toxic to insects even though *in vitro* it is a powerful inhibitor of insect nerve cholinesterase. Its low *in vivo* toxicity is a consequence of its high degree of ionisation at pH 7 because it has difficulty in penetrating insect cuticle and reaching the nervous system. Insecticidal carbamates originated from attempts to retain the supposedly toxic N-methyl carbamate part of the molecule but to attach to it a more lipophilic group. This, it was argued, would enable the molecule to penetrate to its site of action and, in appropriate cases, would confer the right level of chemical stability yet allow the molecule to retain adequate affinity for the enzyme. Since the earliest studies (Fukuto *et al*., 1954) some thousands of carbamates have probably been screened for insecticidal potential, but probably less than twenty are currently marketed in commercially significant amounts. Some of these are listed in table 5.1, together with their aqueous solubilities and other properties.

5.2 STRUCTURAL DIVERSITY OF CARBAMATE INSECTICIDES

Carbamates with insecticidal (and related) properties possess the general structure:

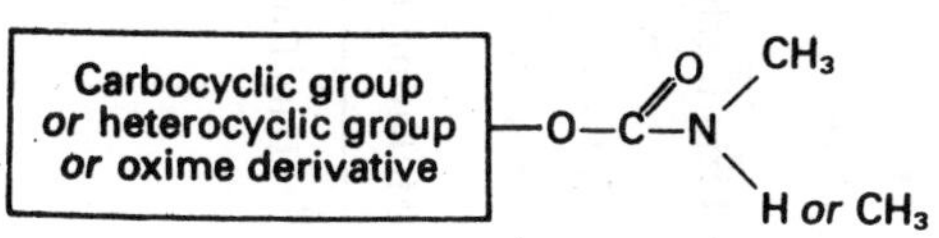

(5.1)

Table 5.1 Physicochemical and biological characteristics of some carbamate insecticides

Substance	Sub-group	Route of systemic action	[1]Oral toxicity to rats (LD_{50}, mg/kg)	Vapour pressure (mm Hg at 25–30°C)	Uses, other than as an insecticide	[1]Solubility in water (ppm at 25–30°C)
Carbaryl	1	none	700	3×10^{-3}	earthworm killer; fruit thinning	40
Propoxur	1	none	90	–	–	1000
Metalkamate	1	?	87	3×10^{-5}	–	<40
Methiocarb	1	none	130	–	molluscicide	–
Carbofuran	2	by root uptake	11	1×10^{-5}	(nematicidal at high dosage)	700
Pirimicarb	2	by leaves or roots but persistence very low	~100	3×10^{-5}	–	2700
Aldicarb	3	by root uptake	1	1×10^{-4}	nematicide	6000
Oxamyl	3	by root uptake	5	–	nematicide millipede killer	–
Methomyl	3	by leaf uptake	21	5×10^{-5}	–	58 000

[1]Data from commercial literature, from Kuhr and Dorough (1976) and from Matsumura (1975).

There are three major sub-groups. Sub-group 1 comprises N-methyl carbamate esters of phenols – that is, of compounds with a hydroxyl group attached directly to a phenyl or naphthyl ring*. Four members of the sub-group are listed in table 5.1 and the formula of **carbaryl**, as type example, is given in figure 5.1. Sub-group 2 similarly comprises N-methyl and N-dimethyl esters of heterocyclic phenols and **carbofuran** is an example. The third sub-group contains oxime derivatives (e.g. **aldicarb**) and will be considered separately.

Sub-group 1
Aryl N-methyl carbamates

e.g. 1-naphthyl N-methylcarbamate

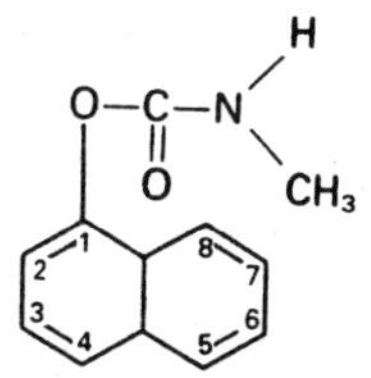

Carbaryl

Sub-group 2
Heterocyclic mono- or dimethyl carbamates

e.g. 2,3-dihydro-2,2-dimethyl benzofuran-7-yl N-methylcarbamate

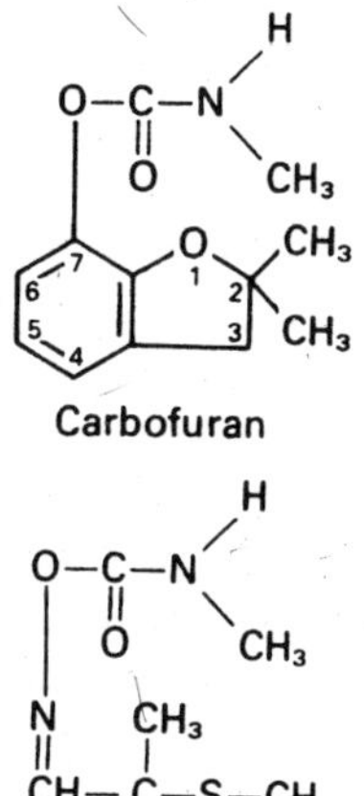

Carbofuran

Sub-group 3
Oximes, the OH group of which has been carbamylated

e.g. 2-methyl-2(methylthio)propionaldehyde O-(methylcarbamoyl) oxime

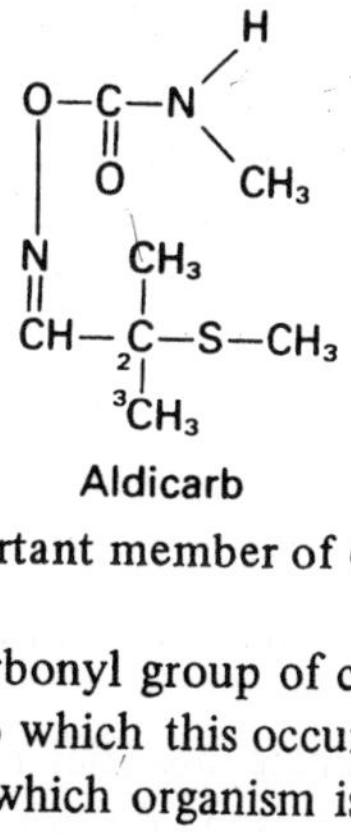

Aldicarb

figure 5.1 Carbamate insecticide sub-groups, with one important member of each

Esterases attack the bond to the oxygen side of the carbonyl group of compounds in the first two sub-groups, although the extent to which this occurs in practice varies greatly according to which carbamate and which organism is involved. Hydrolysis by esterases is probably less important in relation to longevity of carbamates than was originally supposed. **Amidases** may attack the bond on the nitrogen side of the carbonyl group, although, in contrast to such compounds as dimethoate (figure 4.10), this process seems usually to contribute little to

*Carbamic acid, HOOC.NH_2 is aminoformic acid; N-methyl carbamic acid is therefore HOOC.NH.CH_3. Carbamic acid should not be confused with the amide of formic acid, NH_2OC.H.

detoxication. Instead, the carbocyclic or heterocyclic rings are normally attacked by mono-oxygenases and the hydroxylated products are then either conjugated directly or hydrolysed to products which are then conjugated.

Compounds in sub-group 3 can be regarded as ester-like compounds formed between the carboxyl group of methylcarbamic acid and the hydroxyl group of an oxime. The latter is a compound formed when an aldehyde or ketone reacts with hydroxylamine, NH_2OH:

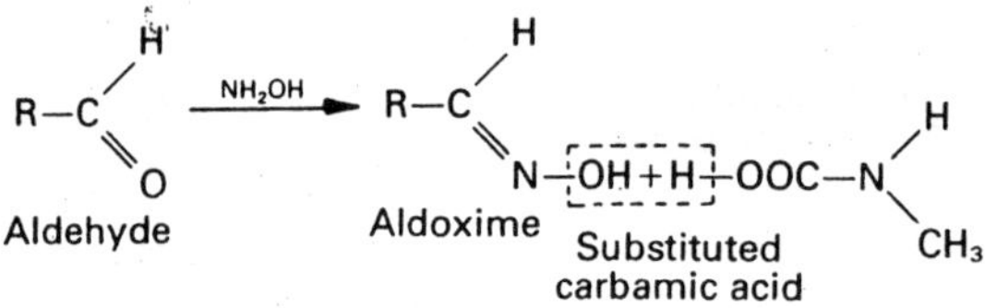

(5.2)

Compounds in sub-group 3 illustrate most clearly the way in which carbamates were designed specifically to resemble acetylcholine and yet to remain sufficiently lipophilic to penetrate to their site of action in insects. This similarity of structure to that of acetylcholine probably helps to explain the very high toxicity of aldicarb to both mammals and insects.

The structures of compounds in sub-groups 1 and 2 can hardly be regarded as resembling that of acetylcholine. On the other hand, such superficial dissimilarity may obscure similarities in certain inter-atomic distances or of electronic distribution which may be quite vital in regard to how well the molecule fits the active site of acetylcholinesterase. In practice, the ring structures confer on the molecules a balance of properties compatible with insecticidal action. These include sufficient penetration of lipid cuticle; transport in the haemolymph without too rapid decomposition; sufficiently good fitment to cholinesterase to enable an enzyme–carbamate complex to be formed at low carbamate concentration; and carbamylation of the enzyme at an adequate rate (section 4.4 and 5.4).

5.3 PRODUCTION FIGURES AND PRACTICAL USES OF CARBAMATES

Carbamate insecticides are frequently employed to control insects which, for some reason, do not readily respond to organophosphorus compounds. They will, of course, usually control more susceptible insects as well, but since they are twice as expensive as common organophosphorus compounds (and up to eight times dearer than DDT), their higher cost is usually a sufficient incentive to regard them as essentially 'heavy duty' insecticides, to be used in special situations or when other pesticides fail. Their production is steadily increasing and this could well lower the price differential. In a valuable monograph Kuhr and

Dorough (1976) quote the following figures for USA carbamate manufacture in 1971:

> carbaryl, 55m lb; carbofuran, 8m lb; metalkamate, 6m lb; aldicarb, 2m 1b. Figures for propoxur, pirimicarb, methomyl, oxamyl and methiocarb are not given.

Carbamate insecticides are particularly useful for dealing with aphids and other pests which have developed resistance to organophosphorus compounds or which are difficult to control for other reasons. Whiteflies, leaf miners, ants, mealy bugs, scale insects, cockroaches, earwigs and wasps are amongst the pests controlled by various carbamates (table 5.2). Some carbamates also possess special attributes which favour soil application, either for the control of soil organisms or as a means of getting systemic action following uptake by the root system of the crop. Appropriate substances can, for example, be incorporated into the soil to control woodlice, wireworms and millipedes, or placed on to the soil surface to control slugs and snails. Upward translocation following root absorption can be used not only to control leaf-sucking pests but vicariously to control virus diseases for which some such pests are vectors. Thus Drinkwater *et al.* (1979) found that carbofuran granules applied to planting furrows gave better control of maize streak virus than did similar applications of phorate or of endosulfan.

Soil application of certain carbamates has proved of particular value for the reduction of populations of phytophagous nematodes, for these organisms are both difficult and expensive to control. Thus Whitehead *et al.* (1979a) reported fair control of the stem nematode of onions, *Ditylenchus dipsaci*, by application of granules of aldicarb or oxamyl over the seed furrows during sowing. The same team of workers (Whitehead, 1979b) also found that invasion of crop plants by the beet cyst nematode, *Heterodera schachtii*, was minimised by incorporation of aldicarb or oxamyl into the top 15 cm of soil prior to sowing sugar beet seeds. Rather similarly, Muthukrishnan *et al.* (1977) observed that a carbofuran root dip decreased the population of the rice root nematode, *Hirschmanniella oryzae.*

Finally, it should be added that, in the medical field, several carbamates have been advocated for the control of the water snail which is the alternate host of the organism responsible for bilharzia, while Das and Rajagopalan (1979) have reported on the use of carbaryl and several organophosphorus compounds against the vectors of bancroftian filariasis, of malaria and of dengue.

5.4 MODE OF ACTION AND TOXIC EFFECTS OF CARBAMATE INSECTICIDES

The biological role of (acetyl)cholinesterase was outlined in section 4.4. In

Table 5.2 Some uses of carbamate pesticides for crop protection

Crop	Pest	Chemicals
Alfalfa	alfalfa weevil	methomyl
Apples and other top fruit	codling and tortrix moth mites on apples	carbaryl, methiocarb, oxamyl
Bananas	various pests	carbaryl
Beans	o/p resistant aphids	pirimicarb
Brassicas	cabbage root fly o/p resistant aphids	carbofuran pirimicarb
Citrus	mites fruit flies	oxamyl methiocarb
Coffee	American bollworm	carbaryl, carbofuran
Cotton	American bollworm	aldicarb, carbaryl, methiocarb
Hops	aphids	methomyl, aldicarb, methiocarb, propoxur
Maize and sweet corn	corn rootworm corn earworm various pests	carbofuran methomyl metalkamate
Onions	stem nematode	aldicarb
Peas	pea moth, pea aphid	carbaryl, carbofuran, pirimicarb
Peanuts	various soil pests	aldicarb
Pineapple	various pests	metalkamate
Potatoes	colorado beetle potato cystworms	oxamyl aldicarb, pirimicarb
Rice	army worm, stem borer, water weevil	metalkamate, carbofuran
Sugar beet	Docking disorder, leaf miners, millipedes o/p resistant aphids	aldicarb, carbofuran pirimicarb
Sugar cane	cane borer	carbaryl, carbofuran, aldicarb
Tobacco	tobacco hornworm root knot nematode	methomyl oxamyl
Vegetables, esp. carrots, onions	various pests slugs, snails	carbaryl, methomyl methiocarb

particular an account was given of the way its active site reacts with acetylcholine, organophosphorus insecticides and with carbamates. It will be recalled that whereas organophosphorus compounds appear to act by phosphorylating the enzyme, the carbamates seem experimentally to compete with acetylcholine for its active site. The reason for this difference will now be further examined.

The superficial difference between the action of organophosphorus and carbamate insecticides arises from differences in the relative magnitudes of the four rate constants in the expression:

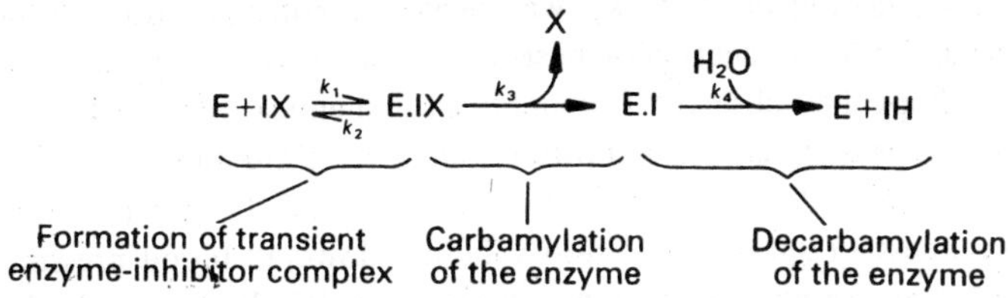

(5.3)

(the symbols are explained for eq. 4.4)

This equation is considered more fully on page 76, where the meanings of the symbols are explained.

In contrast to what happens when most organophosphorus compounds inhibit cholinesterase, the rate-limiting step when carbamates are inhibitors is the carbamylating reaction (which **blocks** the enzyme) and which is governed by the rate constant, k_3. For organophosphorus compounds the concentration of E.IX is usually kept low by the rapidity of its irreversible change to E.I. For most carbamates, however, the rate constant k_3 is rather low, so that E.IX tends to accumulate, the major factor determining its steady state concentration being the rate constant, k_2, of the back reaction. This last-named reaction leads to regeneration of free enzyme, E, as does the irreversible **unblocking** reaction governed by rate constant k_4. The relative sizes of these various rate constants are such that the reaction of most carbamates with cholinesterase resembles the competitive inhibition of succinate dehydrogenase by malonate. This competition is expressed in *in vitro* systems by the ability of added acetylcholine partly to reverse inhibition caused by carbamates, just as addition of succinate can partly reverse the inhibition caused by malonate.

In practice, hydrolysis of carbamylated cholinesterase (rate constant k_4) is 50 per cent complete in about 30 min whereas hydrolysis of phosphorylated enzyme is half complete in 80–480 min. Values of the carbamylation constant, k_3, and of the affinity constant, K_D, for several carbamates including carbaryl and propoxur, were determined by O'Brien *et al.* (1966). For the substances investigated using bovine erythrocyte acetylcholinesterase as the enzyme, they found K_D ranged from about 10^{-2}M to 10^{-6}M, whereas values for k_3 only varied between 0.2 and 2.0 min^{-1}. For comparison, the K_D value for acetylcholine and bovine erythrocyte acetylcholinesterase is about 2×10^{-5} and the acetylation constant, k_3, is about 3×10^5 min^{-1}. The magnitude of the decarbamylation constant, k_4, is rather similar for all common insecticidal carbamates, and ranges from 0.011 to 0.052 min^{-1} (Kuhr and Dorough, 1976). For organophosphorus compounds, k_4 is much lower than this, so that the phosphorylated enzyme takes a long time to recover (section 4.4), whereas with the acetylated enzyme, formed when acetylcholinesterase is functioning normally, deacetylation occurs with great rapidity. More recent investigations include those of Nishioka *et al.* (1977) and of Kamoshita *et al.* (1979). The former workers found that K_D varied about 100-fold for the group of phenyl N-methyl carbamates they were investigating but that k_3 was relatively constant. Kamoshita and colleagues, using fly-head acetylcholinesterase, attempted to explain the quantitative relationships they observed from a physical chemical viewpoint.

Severe carbamate poisoning causes symptoms similar to those mentioned for organophosphorus compounds. These include constriction of the pupils of the eyes, muscular weakness or spasms, respiratory failure, lowered blood pressure and cardiac arrest. Convulsions can be tonic, leading to rigid limbs, or clonic, leading to rapid, uncontrolled movements.

At sublethal dosage, humans suffer headaches, vomiting and diarrhoea. A study by Morse *et al.* (1979) of occupationally acquired disease by workers at a pesticide plant showed that almost half the packaging workers exposed to methomyl had suffered blurred vision or pupillary constriction. Indeed, at this plant where 11 per cent of the 102 workers had entered hospital for illnesses relating to some kind of chemical exposure, methomyl was the commonest cause of occupational medical problems. Next to methomyl a frequent cause of trouble was a dermatosis arising from a contaminant in dichloroaniline and propanil (chapter 13).

Mild carbamate poisoning, like mild poisoning by organophosphorus insecticides, can affect behavioural patterns, reducing mental concentration and slowing the ability to learn. Experiments with rats have suggested that this reduced learning ability is greatly accentuated if poisoned animals are simultaneously denied adequate amounts of protein. Intramuscular injection of carbaryl into monkeys at a level of 5 mg/kg apparently reduced their capacity to learn (Anger and Setzer, 1979). There are also reports of possible effects on the oestrus cycle and on the thyroid gland of various mammals.

Atropine can reverse the toxic effects of carbamates, but in view of the mode of action proposed above, it is of interest that PAM, an important antidote for poisoning by organophosphorus compounds, is often rather ineffective at reversing the toxic effects of carbamates. PAM is the methiodide of **pyridine-2-aldoxime**. The oxime hydroxyl group is a much more powerful nucleophilic reagent than the hydroxyl group of water (i.e. it attacks positively charged atoms more readily) and the stable phosphorylated enzyme therefore hydrolyses more rapidly to yield active enzyme when PAM is present. The reduced effectiveness of PAM in relation to carbamate poisoning is presumably a reflection of the fact that the carbamylated enzyme hydrolyses quite rapidly even when only water is present (i.e. k_4 is relatively large). Perhaps for this same reason, certain carbamates partly reverse the action of organophosphates, but only if they get to the enzyme surface before the organophosphorus compound.

5.5 ARYL METHYL CARBAMATES (Sub-group 1)

For general agricultural purposes, **carbaryl** (figure 5.2) is the most widely used carbamate insecticide. In Britain, it is recommended for use against caterpillars of the winter moth, as well as earwigs and capsid bugs. In the USA it is registered for use on about 90 crops, but some 40 per cent of its total American usage is on cotton. Other major uses are on maize and soybeans; of local importance is its use for pest control on pineapple, coffee, rice and sugar cane (table 5.2). Care has to be taken when it is being applied to foliage for as little as 1.5 μg per insect is lethal to bees. It has also been used to control earthworms, although this may be a mixed blessing. For example, Sharpley *et al.* (1979) observed that the volume of surface run-off water doubled when a permanent pasture was treated with carbaryl. They attributed this to a three-fold reduction in infiltration caused by an accumulation of litter at the soil surface in the absence of surface casting.

The American legal tolerance figures for carbaryl are often quite high, enabling it to be used on some crops up to near the time of harvesting. This high tolerance is in part a reflection of the large oral LD_{50} to vertebrates compared to the LD_{50}s of most other carbamates and, indeed, of many frequently-used organophosphorus and organochlorine compounds. On the other hand, it has only a moderate residual action, the half life in soil being about 9 days. At normal levels for spray application it only persists for a few days in plant leaves.

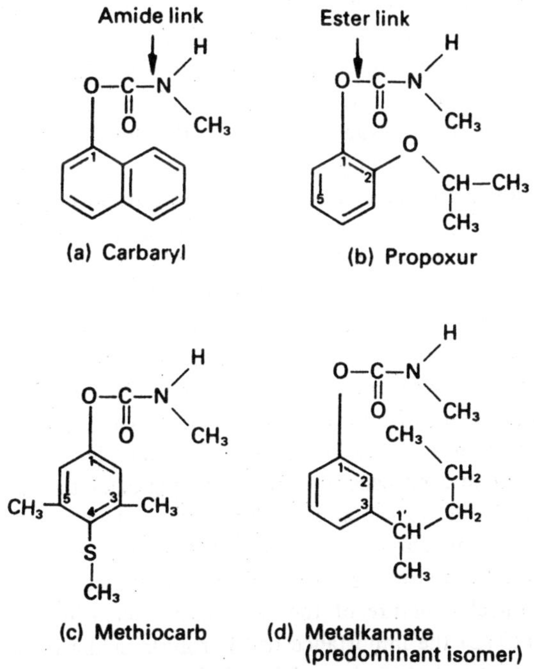

figure 5.2 Sub-group 1: aryl methyl carbamates. ((a) 1-naphthyl N-methylcarbamate; (b) 2-isopropoxyphenyl N-methylcarbamate; (c) 3,5-dimethyl-4-methylthiophenyl N-methylcarbamate; (d) 3-(1′-methylbutyl) phenyl N-methylcarbamate)

The chief danger to mammals associated with the use of carbaryl (and probably some other carbamates) arises from the ease with which it is absorbed through the skin. It is absorbed into the human forearm some ten times faster than compounds such as parathion and lindane. Although no danger appears to exist for man at the residue level or if sensibly handled at the time of application, experiments on animals suggest that some danger exists that both carcinogenic and teratogenic effects could occur at high concentration. Thus Beraud *et al.* (1979) have shown that N-nitrosocarbaryl, a known carcinogen, is readily formed in rat gastric juice in the presence of traces of nitrite and carbaryl. Teratogenic effects have been investigated by Murray *et al.* (1979). Although no effects on offspring of pregnant mice were observed even at doses that were maternally toxic, female rabbits given 200 mg/kg carbaryl from day 6 to day 18 of gestation gave birth to young with an increased incidence of omphalocele (umbilical hernia).

For carbamates, as for other important groups of insecticides, there is evidence that metabolism plays an important role in determining effective toxic dose as well as the time the poison persists in a living organism. In relation to the first of these, it has been shown that the dose, I_{50}, which reduces the activity of an *in vitro* preparation of house fly head acetylcholinesterase to half the control level, is not closely correlated with the LD_{50} of that same carbamate when it is applied topically to house flies. If, however, the LD_{50} is measured when a dose of carbamate, *synergised with piperonyl butoxide*, is applied to the flies, the correlation of LD_{50} and I_{50} for a range of carbamates is quite high (Fukuto *et al.*, 1962). This strongly suggests not only that metabolism materially reduces the effective toxicity *in vivo* (section 5.8) but that, for at least some carbamates, the main degradative route involves one or more important oxidative steps. An important difference thus exists in the role of oxidative metabolism in the carbamate and in the organophosphorus groups of insecticides.

Hydrolysis appears to play a less important initial role in metabolism of carbamates than it does for organophosphorus compounds, although it should be recalled that many degradative reactions of organophosphorus compounds once regarded as hydrolytic are now believed to be either oxidative or glutathione-dependent. The extent of hydrolysis varies from one species to another; a considerable proportion of a dose of carbaryl, for example, is hydrolysed by the rat and by the German cockroach but not by the pig or by the American cockroach (Kuhr, 1971). Plants do not appear to hydrolyse it significantly. Some species of bacteria tend to metabolise it predominantly by oxidation while others appear to do so largely by hydrolysis. The relative importance of hydrolysis also varies with the chemical structure of the carbamate. According to Schlagbauer and Schlagbauer (1972) the importance of hydrolysis in the metabolism by the rat of four prominent carbamate insecticides follows the order

$$\text{carbofuran} < \text{propoxur} < \text{carbaryl} < \text{aldicarb}$$

The carbamate linkage can theoretically be hydrolysed either on the ester side or on the amide side of the carbonyl carbon atom. In practice, the ester side seems usually to be the more vulnerable. Carbaryl is hydrolysed at the ester linkage by an enzyme present in the albumen fraction of plasma (Reiner and Skrinjaric-Spoljar, 1968). In the rabbit the proportion of a dose of carbaryl hydrolysed in the plasma is greater than that in the liver, where metabolism tends to be by the oxidative route. Practically no work has been done for nearly two decades on the hydrolases which attack carbamate insecticides and the present state of knowledge cannot be regarded as being very satisfactory.

The principal routes of initial metabolism of carbaryl are summarised in figure 5.3. Oxidative routes lead to 4-hydroxy and 5-hydroxy derivatives of carbaryl. In addition, epoxidation and N-hydroxymethyl carbaryl formation are characteristic of reactions normally catalysed by NADPH-dependent mono-oxygenases. Hydrolysis of carbaryl epoxide by an epoxide hydrolase (section 3.3) leads to the formation of a diol. Since hydrolysis of the carbamyl ester linkage (to

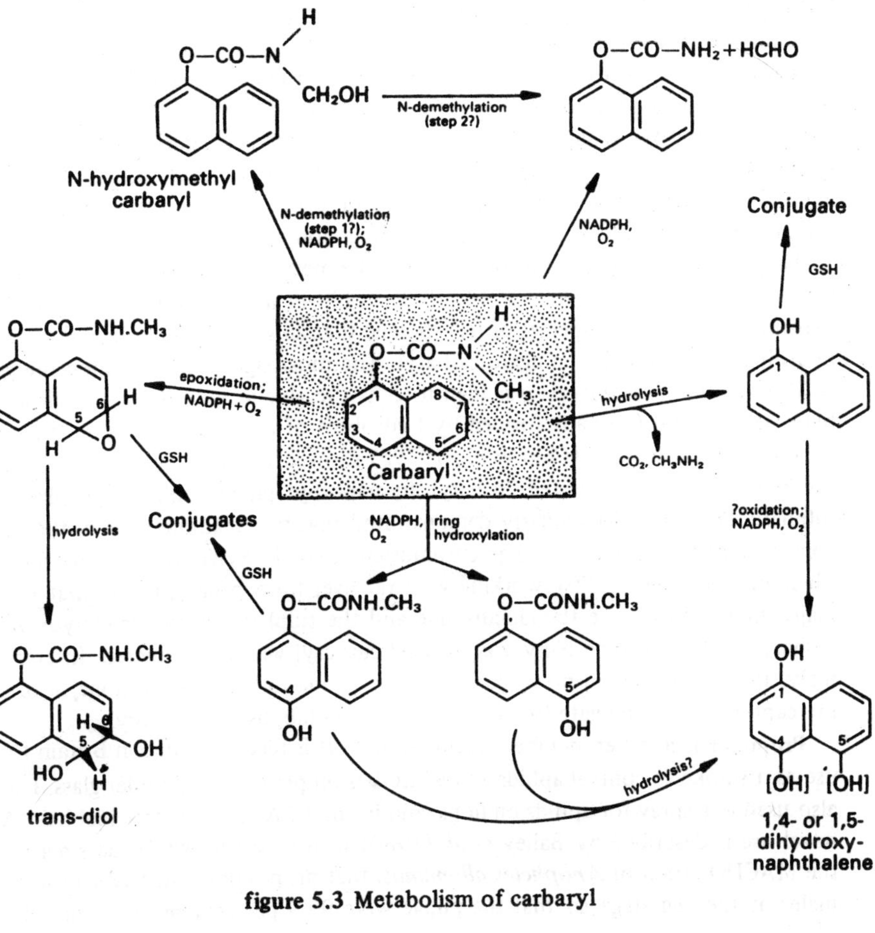

figure 5.3 Metabolism of carbaryl

expose the 1-naphthyl hydroxyl group) may occur more readily after initial oxidation in ring positions 4 and 5, the isolation of 1,4- or 1,5-dihydroxy-naphthols (and of their conjugation products) does not necessarily imply that the initial attack on the molecule is hydrolytic.

The relative order in which oxidation and hydrolysis occurs in the metabolic sequence of carbamate insecticides is of more than academic interest. If, for example, hydrolysis is slow unless oxidation precedes it, substances such as piperonyl butoxide can indirectly reduce loss by hydrolysis by retarding the initial oxidative step and in consequence act as synergists in carbamate formulations.

Microsomal mono-oxygenases activate many organophosphorus insecticides but in the case of carbamates the effect of oxidative metabolism depends on which carbamate is involved. 5-hydroxy-carbaryl is about twice as active against cholinesterase as is carbaryl itself (Oonnithan and Casida, 1968) so mild activation can indeed occur *in vitro.* However, since hydroxylation may initiate rapid hydrolysis and so decrease persistence, the net effect of these opposing tendencies may not be predictable *in vivo.*

Many of the hydroxy compounds, whether formed by oxidation or by hydrolysis, can be conjugated in a secondary stage which always results in detoxication. As expected (chapter 3), interspecies variation exists in the types of conjugates formed; in plants, glucosides usually predominate while in many higher animals, glucuronides and sulphates are most evident. In insects, glucosides, sulphates or phosphates are often major conjugation products.

The best compendium of information on the metabolism of carbaryl has been produced by Menzie (1969, 1974, 1978). Chin *et al.* (1979a), working with explanted livers of five species of fish showed that all species could demethylate, hydrolyse, oxidise and conjugate carbaryl. The most significant common metabolite in all of these species was identified as the glucuronide of 5,6-dihydrodihydroxy carbaryl (perhaps formed by hydrolysis of a transient epoxide, figure 5.3). The same workers (Chin *et al.*, 1979b) showed that kidney explants from the human foetus converted carbaryl to naphthyl glucuronide and naphthyl sulphate. Pekas (1979) similarly demonstrated that the small intestine of the rat converted carbaryl to naphthyl glucuronide and to hydroxynaphthyl glucuronide. Chen and Dorough (1979), working with rat liver homogenate, found that two major metabolites were the glucuronide and the sulphate of dihydrodihydroxy carbaryl, although some N-hydroxymethyl carbaryl was also detected. Interestingly, they could obtain no evidence for the formation of a glutathione or of a mercapturic acid conjugate (section 3.5) in the metabolism of carbaryl.

Propoxur is another member of sub-group 1. It is recommended in Britain for use as a smoke to control aphids and whiteflies on plants grown under glass. It is also used as a spray for aphids on hops, and in the USA, as an insecticidal bait. A novel use is described by Bailey *et al.* (1980) who observed, while mass rearing the MACHO strain of *Anopheles albimanus*, that propoxur eliminated all the females at the egg stage so that the pupae were 99.9 per cent males. It has also been used in Kentucky by aerial application to give good control of *Aedes sollicitans* (Williams *et al.*, 1979). Propoxur has a relatively long residual action – it can persist for as long as six weeks – and is an example of one of the 'heavier armoury' of insecticides. It is used against refractory pests such as cockroaches, ants, earwigs, crickets and millipedes. It is much more toxic than carbaryl to higher animals (table 5.1).

Compared with carbaryl, the isopropoxy group of propoxur is the main feature of the molecule (figure 5.2). A certain similarity with the heterocyclic ring structure of carbofuran can be detected (figure 5.4). The molecule appears to fit the surface of acetylcholinesterase more snugly than does the molecule of carbaryl, for the dissociation constant, K_D, of the enzyme–propoxur complex, E.IX (equation (5.3)) is far smaller than that for the enzyme–carbaryl complex, a difference which implies a high affinity of inhibitor and enzyme. On the other hand, k_3 is not particularly large, so carbamylation of the enzyme occurs rather slowly (sections 4.4 and 5.4).

The isopropoxy group introduces a new metabolic feature into the molecule for its removal exposes the hydroxyl group of 2-hydroxphenyl N-methyl carbamate. This hydrolytic dealkylation probably occurs in bean plants. The major routes of metabolism in most organisms are, however, hydroxylation in position 5 of the ring (figure 5.2) and hydroxylation of the N-methyl group. Oonnithan and Casida (1968) found that the product of the latter process is five times as active against cholinesterase as is propoxur itself. 5-hydroxy propoxur is a major product both in house flies and in rat liver preparations, although Hama *et al.*, (1979) detected the N-hydroxymethyl metabolite in both a susceptible and a resistant strain of the green rice leaf hopper, *Nephotettix cincticeps*. Little work has been done on metabolism by humans, but Dawson (1964) obtained evidence that hydrolysis of the carbamate ester linkage probably occurred.

Metalkamate (figure 5.2) is a mixture of two compounds with isomeric side-chains in position 3. In 1971 it was the third most important carbamate in the USA in terms of the amount produced (Kuhr and Dorough, 1976). This reflects its use against pests of various types of corn, but elsewhere it is used against pests of rice and pineapple (table 5.2). Metabolically, an interesting new feature is that its structure allows the possibility for mono-oxygenases to attack the hydrogen atom on the aliphatic carbon atom attached to position C-3 of the benzene ring:

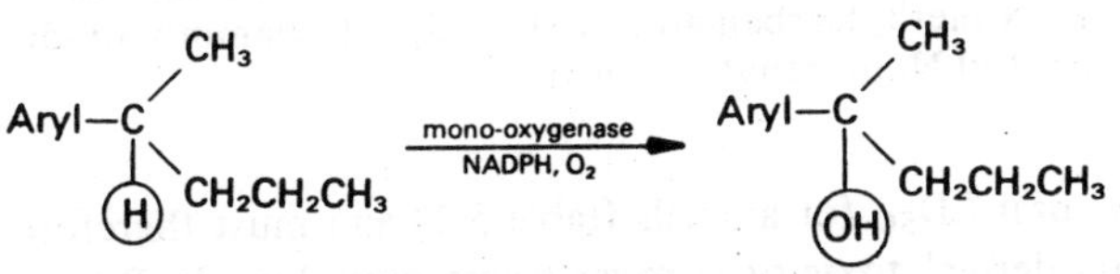

(5.4)

For many types of molecules this type of reaction takes place more readily than hydroxylation of the ring.

Methiocarb was introduced as an insecticide in 1965 but has since been found additionally to be a valuable molluscicide. For this purpose it is broadcast on the surface of the soil. It is used on cotton, vegetables and hops for the control of fruit flies, codling moth, leaf hoppers and mites. It has also been used in the form of a seed dressing to act as a bird repellant.

As can be seen from figure 5.2, the feature of the methiocarb molecule which distinguishes it most clearly from other members of sub-group 1 is the methylthio group in position 4. In this respect it should be compared with aldicarb and methomyl (figure 5.6), with organophosphorus insecticides possessing an alkylthio group (figure 4.7), and with the acaricide, chlorbenside (figure 7.8) which contains a substituted methylthio group. As with these other compounds, metabolism can lead to the oxidation of the sulphide sulphur to give sulphoxide or sulphone. This is brought about by microsomal mono-oxygenases in the presence of NADPH. Whether there is a net activating effect is uncertain, but it is undoubtedly a major factor contributing to the low half-life of methiocarb in animals (it is often less than two days). In soil, it is more persistent, the half-life normally varying from one to five weeks according to soil type.

5.6 HETEROCYCLIC MONOMETHYL AND DIMETHYL CARBAMATES (Sub-group 2)

The two best-known members of this sub-group are **carbofuran** and **pirimicarb** (figure 5.4). The similarity between the structures of carbofuran and propoxur has already been noted, and a second feature of interest is that pirimicarb is one of the few examples of an insecticidal N-dimethyl carbamate.

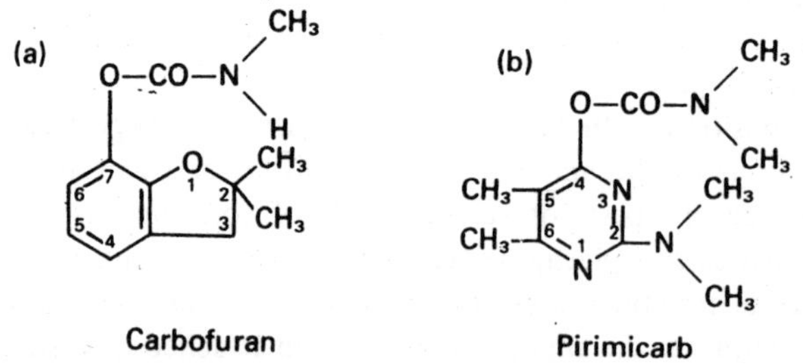

figure 5.4 Sub-group 2: heterocyclic carbamates. ((a) 2,3-dihydro-2,2-dimethyl-benzofuran-7-yl N-methylcarbamate; (b) 2-dimethylamino-5,6-di-methyl-pyrimidin-4-yl N-dimethylcarbamate)

Carbofuran has a low oral LD_{50} for animals (table 5.1) and must therefore be handled with care. Its dermal toxicity is nevertheless quite low. In Britain it is recommended that its use should be restricted to uptake from soil by plant roots and it is marketed as a granular formulation for this purpose. Its principal use is for the control of cabbage root fly larvae. In the USA, however, it has quite widespread application and in 1971 it was second only to carbaryl in terms of tonnage manufactured there. It is used in various parts of the world for pests of sugar cane, sugar beet, maize, rice and coffee; some of the insects controlled are listed in table 5.2. It has been used to protect seed (for planting, not eating) and Chundurwar and Karanjkar (1979) reported that sorghum seed treated with carbofuran and stored 180 days germinated normally when sown. Carbofuran also has some nematicidal action, although usually only at a threefold higher dosage than is customarily employed for the control of insects in soil. Its half-life in soil is of the order of one or two months but in plants it persists for a much shorter time.

A curious interaction with the herbicide atrazine (chapter 13) has been reported by Lichtenstein *et al.* (1979). On applying a non-toxic dose of 15 μg atrazine per animal to flies, the LD_{50} of carbofuran fell to half its former value. Moreover, twice as much carbofuran was left outside the fly and less 3-hydroxy carbofuran and its phenol derivative was excreted when atrazine was present. The herbicide thus seems to inhibit the penetration of carbofuran into the insect and to delay its metabolism once it has gained access. The precise mechanism of this **synergism** has not been resolved.

Metabolism of carbofuran is predominantly oxidative at least when done by microsomal preparations from the livers of higher animals, but once it is modified, conjugation or hydrolysis (or both) can occur *in vivo*. Four common metabolites are shown in figure 5.5. In rat liver preparations, 3-hydroxy carbofuran is formed about three times as abundantly as is the N-hydroxymethyl oxidation product.

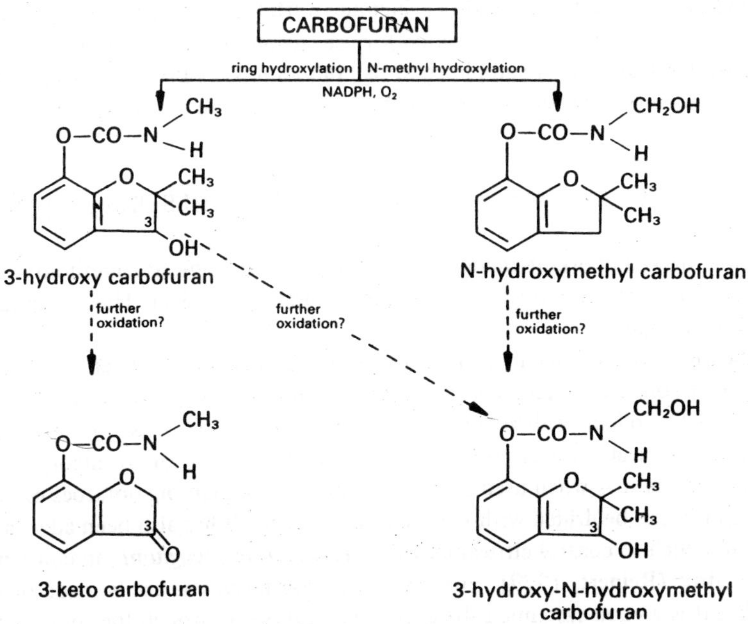

figure 5.5 Oxidative metabolites of carbofuran

A small amount of the doubly hydroxylated 3-hydroxy-N-hydroxymethyl carbofuran is also formed. Using an injected dose of ^{14}C-labelled carbofuran, Marshall and Dorough (1979) showed that up to 45 per cent passed into rat bile. Since up to 70 per cent of the ^{14}C eventually appeared in the urine and only 3 per cent in the faeces, this implies that much of the metabolite excreted in the bile must re-enter the body (probably after hydrolysis in the gut) and so recirculate. In the present instance there is some evidence that 3-hydroxy carbofuran glucuronide is present in the bile and that this is hydrolysed to the aglycone before absorption. Since the aglycone is quite toxic, **biliary excretion** followed by **enterohepatic circulation** is not necessarily in the animal's best interest.

The major products produced by living house flies are 3-hydroxy carbofuran and its glucoside. Other products include N-hydroxymethyl carbofuran and 3-keto-carbofuran.

The National Research Council of Canada has issued a Report which considers the criteria for interpreting the effects of the use of carbofuran on environmental quality (NRC Canada, 1979). Carbofuran is discussed in this Report with respect to its toxicity, its possible mutogenicity, teratogenicity and carcinogenicity. The Report also describes its effects on plants, invertebrates, birds, mammals and micro-organisms.

Pirimicarb is the least persistent of the common insecticidal carbamates. It was marketed in 1969 as a fast-acting aphicide and it kills by fumigant as well as by contact action. When sprayed on foliage there is a rapid translaminar movement, but systemic action is only achieved by uptake through crop plant roots following soil application. Spray formulations include wettable powders and dispersible granules, and these are mainly intended for use against strains of aphids that have developed resistance to organophosphorus compounds. In Britain, it is also marketed in smoke generators to control aphids on greenhouse crops.

5.7 N-METHYL CARBAMATE DERIVATIVES OF OXIMES (Sub-group 3)

The three most important members of this group are aldicarb, methomyl and oxamyl. All three are very toxic to higher animals (table 5.1). Their structures are shown in figure 5.6.

Aldicarb is one of the most potentially toxic substances currently used in crop protection. In Britain it is exclusively formulated as granules, since these are much safer to handle than are liquids or dusts. It has found use in the control of leaf miners, scale insects, mealy bugs and millipedes. In can be incorporated into the soil, as it is when used as a treatment prior to planting potatoes or bulb onions, or it can be drilled with seed (e.g. sugar beet). It has also been advocated as a soil drench to control citrus blackfly, *Aleurocanthus woglumi*, on container-grown citrus (Reinert, 1979). For many of these purposes, use is made of the fact that it is a good systemic substance when uptake is through the roots of the crop. It is usually used at a rate of between two and ten kg/ha, and at this level

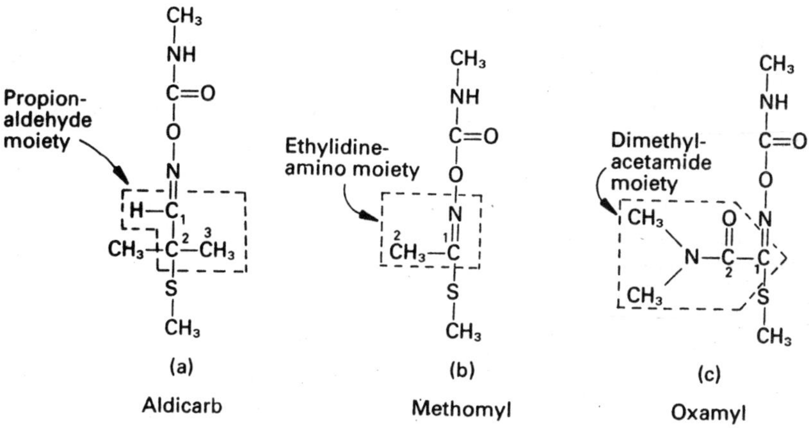

figure 5.6 Sub-group 3: N-methylcarbamates of oximes. ((a) 2-methyl-2-(methylthio)propionaldehyde O-methylcarbamoyl oxime; (b) 1-(methylthio)-ethylidineamino N-methylcarbamate; (c) 1-methylcarbamoyloxyimino-1-methylthio-N,N-dimethylacetamide)

it usually persists for between one and three months. For example, when used as a side dressing in cotton furrows Malik and Yadar (1979) found that its disappearance followed the course of a first order reaction. This indicated a half-life of about 20 days, although estimates as low as seven days have been reported by other workers. At a very low concentration (2 ppm) it has been reported to enhance the growth and respiration of *Rhizobium* spp in the cowpea, *Vigna sinensis* (Seka and Balasubramanian, 1979).

The most important property of aldicarb is that it is in some circumstances a moderately good **nematicide**, sometimes being nearly as effective as the extremely expensive volatile nematicides such as **dichloropropene**. For example, Boag (1979), in a study of soil nematodes in farms where carrots were grown, found that in a site heavily infested with the nematode *Rotylenchus robustus* aldicarb increased carrot yield by 17 per cent. Sometimes, however, a marked reduction of numbers of nematodes does not lead to a marked increase in crop yield. Thus, in only two out of five trials of harvested spring barley was the yield of barley increased, although the numbers of *Pratylenchus* spp in the roots (and, after harvest, in the soil), was in all cases considerably diminished (Saynor, 1979). In the case of field corn, *Zea mays*, yield was increased in the first year, and greatly increased in the second year, following the use, *inter alia*, of aldicarb and oxamyl to control *Belonolaimus longicaudatus* (Rhoades, 1979).

In the USA, the legal residue tolerance of aldicarb varies from zero to 1 ppm on different crops, its main use being on cotton, sugar beet and sugar cane. It is, however, also used on onions, peanuts and tobacco, nearly always by systemic action following soil application.

In common with other carbamates, it is a potent inhibitor of cholinesterase.

It can apparently cross the placental barrier, for Cambon *et al.* (1979) found that as little as 0.01 mg/kg given by gastric intubation to pregnant rats decreased the activity of the enzyme in both maternal and foetal blood and brain.

In animal liver preparations, the principal primary metabolite is the sulphoxide. A small proportion of the latter is normally further oxidised to the sulphone, although reduction of the amide moiety of the sulphoxide to a sulphoxide nitrile (figure 5.7) has also been reported. In consequence, the bulk of the identified metabolites often comprise the sulphoxide, its hydrolysis products, and the hydrolysis products of the sulphoxide nitrile (Kuhr and Dorough, 1976). Cotton plants convert substantial amounts of absorbed aldicarb into water-soluble substances devoid of the carbamate linkage (Bartley *et al.*, 1970).

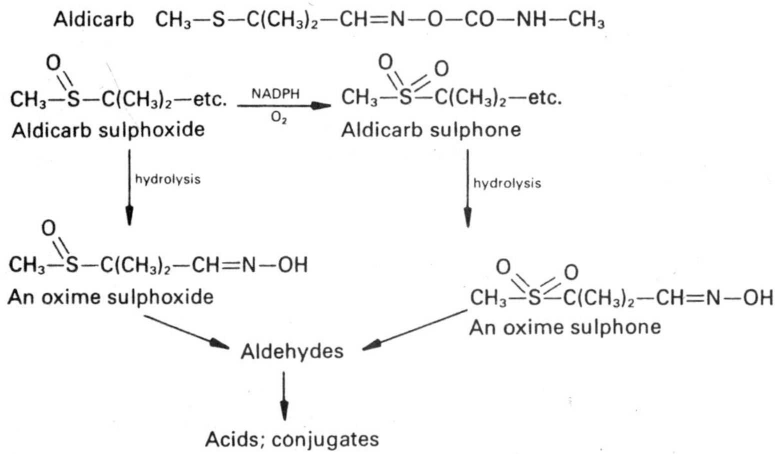

figure 5.7 Principal metabolites of aldicarb

Methomyl is a contact and systemic insecticide that is often applied to foliage. It is less toxic by mouth than is aldicarb (table 5.1) and its dermal LD_{50} is unexpectedly high (500 mg/kg for rabbits). However, for substances with oral toxicities as high as that of methomyl, constant care is essential, as is evidenced by a report by Liddle *et al.* (1979). These workers report three fatalities from accidental ingestion of methomyl stored in an unlabelled tin* when it was used to make an Indian dish called roti. Post-mortem studies suggested that the lethal dose of methomyl to these human beings was of the order of 12–15 mg/kg human body weight.

In Britain, the main use of methomyl is as a foliage application for the control of aphids on hops. For this purpose up to ten applications are made per season.

* As Marlene Dietrich would no doubt have observed had she been singing of poisons rather than flowers, 'Wann wird man je verstehen?' – 'won't they ever learn?'.

Elsewhere, it is used to control the alfalfa weevil, the tobacco hornworm as well as the codling moth and the diamond back moth on fruit. On foliage it has a half-life of about seven days, and this is also the safety period that must elapse between the final application and harvesting under British conditions.

The metabolism of methomyl seems to be quantitatively different from that of aldicarb in that neither the sulphoxide nor the sulphone accumulate in the animals and plants so far investigated. Instead, if they are formed at all, they rapidly break down to give volatile and water soluble components; the former include carbon dioxide (exhaled by animals) and methyl cyanide whilst most of the water-soluble compounds have not been identified.

Oxamyl is a broad-spectrum insecticide with marked nematicidal properties. It is used in Britain as a systemic foliage spray to control the potato cyst nematode and the onion stem nematode as well as Docking disorder of beet. The latter is caused by root damage arising from the predations of soil-living nematodes of the genera *Trichodorus* and *Longidorus*. It is also used against millipedes and the sugar beet leaf miner, while in the USA it is applied to foliage to control tobacco root knot nematodes. In the latter instance use is made of its somewhat unusual property of being able to move downwards after leaf application (Peterson *et al.*, 1978). It is also used for mite control on apples and on citrus fruit.

Some effects of oxamyl on the potato cyst nematode *Globodera rostochiensis*, have been described by Hague (1979). Oxamyl delayed emergence from cysts, disorientated second stage larvae in the soil and delayed their invasion of roots. There was also a reduction in the number of eggs and a marked reduction in the final population of cysts. The half-life of oxamyl in soil varies both with the soil type and with its moisture content (Smelt *et al.*, 1979). At a moisture content above the wilting point the half-life was 13 days in clay loam but 39 days in humic loamy sand.

5.8 INSECT RESISTANCE TO CARBAMATE INSECTICIDES

According to Brown (1971), few of the 225 known instances of field resistance to pesticides have arisen as a result of exposure to carbamates, although some cross-resistance to them may follow the development of resistance to organochlorine or to organophosphorus insecticides.

As was mentioned earlier, 'resistance' may result from a behavioural change, a structural change or a biochemical change. As an example of the second of these, there may be an alteration in cuticular structure such that penetration is reduced or internal movement is restricted. Biochemical changes may relate to a loss of susceptibility at the site of action or to an increase in the rate of metabolism or rate of elimination from the body.

In the case of carbamates change in the rate of metabolic degradation is usually, although apparently not invariably, the major cause of resistance. Considering first the exceptional cases, where decrease in sensitivity of acetylcholinesterase appears to have occurred, most of the evidence relates specifically to resistance to propoxur. Working on the green leafhopper of rice, *Nephotettix cincticeps*, Iwata and Hama (1972) observed in one strain that loss in cholinesterase-inhibiting ability paralleled a reduction of *in vivo* toxicity to the organism. This strain was resistant to both carbamates and to organophosphates. In later work the same team of workers (Hama *et al.*, 1979) attributed resistance both to insensitivity of cholinesterase and to a rate of degradation greater than that typical of susceptible strains. Ayad and Georghiou (1975), working on resistance of larvae of the mosquito *Anopheles albimanus* to propoxur, reported a 10^4-fold decrease in sensitivity of cholinesterase in the propoxur-resistant strain.

On the other hand, a study by Green and Dorough (1970) using a strain of house flies showing a 20-fold increase in resistance to propoxur, led to the conclusion that resistant and susceptible strains had acetylcholinesterase in head homogenates that was almost equally sensitive in the two strains ($I_{50} = 4 \times 10^{-9}$M). Moreover, the enzyme from resistant flies had the same decarbamylation constant, k_4, as did that from susceptible house flies. The situation is therefore far from clear; resistance probably arises in different ways in different species and perhaps with different carbamates, and in any one species may be caused by more than one contributory factor. Georghiou (1972), in a review of the evolution of resistance to pesticides, suggested that sometimes several factors, each one of which by itself would not have had a decisive effect, have actions which multiply together to provide the total observed resistance to the insecticide.

The importance of oxidative enzymes in relation to the mechanism of resistance to carbamates was first recognised by Moorefield. He demonstrated that pyrethroid synergists (sections 3.4 and 7.4) such as piperonyl butoxide, enhanced the activity of several carbamates (Moorefield, 1958) and that strains of house flies which were resistant to carbaryl nevertheless responded to a carbaryl–piperonyl butoxide mixture. He also showed (Moorefield, 1960) that the build-up of resistance which occurred when house flies were exposed to carbaryl was largely prevented when selection was attempted in the presence of carbaryl–piperonyl butoxide mixtures.

Shortly afterwards, Fukuto *et al.* (1962) showed that there was a large discrepancy between the 50 per cent inhibitory doses of several carbamates to house fly cholinesterase *in vitro* and the corresponding LD_{50}s of those same substances to house flies *in vivo*. In fact, the 50 per cent inhibitory doses only varied four-fold, whereas the LD_{50}s varied about a hundred-fold. When these same carbamates were then applied together with piperonyl butoxide, the synergised LD_{50}s also only varied about four-fold. Such results strongly suggest that observed LD_{50} values are a composite between intrinsic activity at the site of action and metabolic or other factors which tend to decrease the amount of substance that reaches the site of action (or to decrease the time of contact with the site of action). Upon suppression of the latter factors, the LD_{50}s reflect the intrinsic toxicity as measured by *in vitro* anticholinesterase activity.

Piperonyl butoxide is one of the **methylene dioxyphenyl** group of synergists. These compounds are, so far as is known, specific inhibitors of the microsomal mono-oxygenases (chapter 3). It is known that the major degradative routes of most carbamates in insects are essentially oxidative and it thus seems likely that the development of insect resistance to carbamates is partly, and often predominantly, a result of gene selection leading to more effective microsomal oxidation. In the house fly, microsomal enzyme enhancement is controlled by partially dominant genes. The precise nature of the enhancement is not certain, for although there is a higher level of cytochrome P_{450} in resistant insects than in susceptible ones, the difference is insufficient to account for the increase in metabolism unless the enzymes present are also more efficient than those in susceptible house flies. Yu and Terriere (1979) provided evidence that the cytochrome P_{450} in resistant strains of houseflies was in some respects different from that in susceptible strains and, in particular, contained a higher complement of high-spin haemoprotein. It is possible that the newly-formed enzyme binds the carbamate more readily than does the enzyme in susceptible strains (Philpot and Hodgson, 1971).

It will be recalled that resistance to organophosphorus compounds is often associated with the presence of a different or of a more efficient esterase in resistant strains (section 4.11). In the case of DDT and of some other organochlorine compounds resistance is associated with an elevated level of **dehydrochlorinase** (section 6.9). Although in both of these two groups, as well as in the carbamate group, mono-oxygenases can participate in degradative processes, nevertheless the existence of more than one degradative mechanism suggests that cross-resistance between pesticides in different groups is not inevitable. This is a matter of crucial importance, for the armoury of different groups of pesticides suitable for use on crops is very limited. All four major groups of insecticides (organophosphorus and carbamate compounds, organochlorine insecticides and pyrethroids) are neurotoxic. Were most or all of these groups to fail simultaneously because of insensitivity of the site of action or because of elevated levels of degradative enzymes, a most serious situation would arise.

Within the carbamate group itself, cross resistance often occurs, since most are degraded by pathways containing major oxidative steps. There is some evidence that cross-resistance diminishes as the molecular structures become more disparate. For example, carbaryl-resistant leaf hoppers were in one test also very resistant to propoxur but only moderately resistant to a carbamate of the oxime group. Evidence about cross-resistance between carbamates and organochlorine compounds is conflicting. Usually, DDT-resistant strains are not particularly resistant to carbamates, but exceptions do exist, especially for some strains of house flies and of German cockroaches (Kuhr and Dorough, 1976).

A further complication is that resistance to one group of insecticides can 'condition' an insect population to rapid adaptation to survive in the presence of another pesticide to which it is initially susceptible. Thus a strain resistant to dieldrin may initially be as susceptible to carbaryl as a dieldrin-sensitive strain of that insect. Nevertheless, under carbaryl pressure, a strain that is resistant to dieldrin often develops carbaryl resistance very readily.

REFERENCES

Anger, W. K. and Setzer, J. V. (1979). *J. Toxic. environ. Health*, **5**, 793

Ayad, H. and Georghiou, G. P. (1975). *J. econ. Ent.*, **68**, 295

Bailey, D. L., Lowe, R. E., Dame, D. A. and Seawright, J. A. (1980). *Am. J. trop. Med. Hyg.*, **29**, 141

Bartley, W. J., Andrawes, N. R., Chancey, E. L., Bagley, W. P. and Spurr, H. W. (1970). *J. agric. Fd Chem.*, **18**, 446

Beraud, M., Pipy, B., Derache, R. and Gaillard, D. (1979). *Fd Cosmet. Toxic.*, **17**, 579

Boag, B. (1979). *Ann. appl. Biol.*, **93**, 199

Brown, A. W. A. (1971). *Pesticides in the Environment*, Vol. 1, Part 2 (ed. R. White-Stevens), Marcel Dekker, New York, chapter 4

Cambon, C., Declume, C. and Derache, R. (1979). *Toxic. appl. Pharmac.*, **49**, 203

Chen, K.-C. and Dorough, H. W. (1979). *Drug Chem. Toxic.*, **2**, 331

Chin, B. H., Sullivan, L. J., Eldridge, J. E. and Tallant, M. J. (1979a). *Clin. Toxic.*, **14**, 489

Chin, B. H., Sullivan, L. J. and Eldridge, J. E. (1979b). *J. agric. Fd Chem.*, **27**, 1395

Chundurwar, R. D. and Karanjkar, R. R. (1979). *Seed Res., New Delhi*, **7**, 34

Das, P. K. and Rajagopalan, P. K. (1979). *Indian J. med. Res.*, **70**, 412

Dawson, J. A., Heath, D. F., Rose, J. A., Thain, E. M. and Ward, J. B. (1964). *Bull. Wld Health Org.*, **30**, 127

Drinkwater, T. W., Walters, M. C. and Van Rensburg, J. B. J. (1979). *Phytophylactica*, **11**, 5

Fukuto, T. R., Metcalf, R. L., March, R. B. and Maxon, M. G. (1954). *J. Am. chem. Soc.*, **76**, 5103

Fukuto, T. R., Metcalf, R. L., Winton, M. Y. and Roberts, P. A. (1962). *J. econ. Entom.*, **55**, 341

Georghiou, G. P. (1972). *A. Rev. ecol. Syst.*, **3**, 133

Green, L. R. and Dorough, H. W. (1970). *Texas Agric. exptl Station Progress Rept. No. 2835*

Hague, N. (1979). *Ann. appl. Biol.*, **93**, 205

Hama, H., Iwata, T. and Tomizawa, C. (1979). *Appl. ent. Zool.*, **14**, 333

Iwata, T. and Hama, H. (1972). *J. econ. Ent.*, **65**, 643

Kamoshita, K., Ohno, I., Fujita, T., Nishioka, T. and Nakajima, M. (1979). *Pestic. Biochem. Physiol.*, **11**, 83

Kuhr, R. J. (1971). *Pure appl. Chem. Suppl.*, p. 199

Kurh, R. J. and Dorough, H. W. (1976). *Carbamate Insecticides: Chemistry, Biochemistry and Toxicology*, CRC Press, Cleveland, Ohio

Lichtenstein, E. P., Kunstman, J. L., Fuhremann, T. W. and Liang, T. T. (1979). *J. econ. Ent.*, **72**, 785

Liddle, J. A., Kimbrough, R. D., Needham, L. L., Cline, R. E., Smrek, A. L., Yert, L. W., Bayse, D. D., Ellington, A. C. and Dennis, P. A. (1979). *Clin. Toxic.*, **15**, 159

Malik, S. K. and Yadar, P. R. (1979). *Indian J. agric. Sci.*, **49**, 745

Marshall, T. C. and Dorough, H. W. (1979). *Pestic. Biochem. Physiol.*, **11**, 56
Matsumura, F. (1975). *Toxicology of Insecticides*, Plenum Press, New York
Menzie, C. M. (1969). *Metabolism of Pesticides*, United States Department of the Interior: Fish and Wildlife Service; Bureau of Sport Fisheries and Wildlife, Division of Pesticides Registration. Special Scientific Report, Wildlife No. 127, Washington, D.C.
Menzie, C. M. (1974). *Metabolism of Pesticides: an Update*, United States Department of the Interior: Fish and Wildlife Service; Office of Environmental Assistance. Special Scientific Report, Wildlife No. 184, Washington, D.C.
Menzie, C. M. (1978). *Metabolism of Pesticides: Update II*. United States Department of the Interior: Fish and Wildlife Service. Special Scientific Report, Wildlife No. 212, Washington, D.C.
Moorefield, H. H. (1958). *Contr. Boyce Thompson Inst. Pl. Res.*, **19**, 501
Moorefield, H. H. (1960). *Misc. Publs ent. Soc. Am.*, 2, 145
Morse, D. L., Baker, E. L., Kimbrough, R. D. and Wisseman, C. L. (1979). *Clin. Toxic.*, **15**, 13
Murray, F. J., Staples, R. E. and Schwetz, B. A. (1979). *Toxic. appl. Pharmac.*, **51**, 81
Muthukrishnan, T. S., Rajendran, G., Ramamurthy, V. V. and Chandrasekaran, J. (1977). *Indian J. Nemat.*, 7, 8
Nishioka, T., Fujita, T., Kamoshita, K. and Nakajima, M. (1977). *Pestic. Biochem. Physiol.*, 7, 107
NRC Canada (1979). National Research Council of Canada Associate Committee on Scientific Criteria for Environmental Quality. *Carbofuran: Criteria for interpreting the effects of its use on environmental quality*, Publication 0 (16740), pp. 1–191
O'Brien, R. D., Hilton, B. D. and Gilmour, L. (1966). *Molec. Pharmac.*, **2**, 593
Oonnithan, E. S. and Casida, J. E. (1968). *J. agric. Fd Chem.*, **16**, 28
Pekas, J. C. (1979). *Pestic. Biochem. Physiol.*, **11**, 166
Peterson, C. A., de Wildt, P. P. Q. and Edgington, L. V. (1978). *Pestic. Biochem. Physiol.*, 8, 1
Philpot, R. M. and Hodgson, E. (1971). *Chem. Biol. Interactions*, **4**, 399
Reiner, E. and Skrinjaric-Spoljar, M. (1968). *Croat. chem. Acta*, **40**, 87
Reinert, J. A. (1979). *Folia Entom. Mex.*, 0 (41): 41 (quoted from *Biological Abstracts*, **69**, 43277)
Rhoades, H. L. (1979). *Nematropica*, **9**, 43
Saynor, M. (1979). *Pl. Path.*, **28**, 27
Schlagbauer, A. W. J. and Schlagbauer, B. G. L. (1972). *Residue Rev.*, **42**, 1
Sekar, T. and Balasubramanian, A. (1979). *Pl. Soil*, **51**, 355
Sharpley, A. N., Syers, J. K. and Springett, J. A. (1979). *Soil Biol. Biochem.*, **11**, 459
Smelt, J. H., Dekker, A. and Leistra, M. (1979). *Neth. J. agric. Sci.*, **27**, 191
Whitehead, A. G., Tite, D. J., Fraser, J. E. and French, E. M. (1979a). *Ann. appl. Biol.*, **93**, 213
Whitehead, A. G., Tite, D. J., Finch, P. H., Fraser, J. E. and French, E. M. (1979b). *Ann. appl. Biol.*, **92**, 73
Williams, R. E., Knapp, F. W. and Clarke, J. L., Jr. (1979). *Mosquito News*, **39**, 622
Yu, S. J. and Terriere, L. C. (1979). *Pestic. Biochem. Physiol.*, **12**, 239

6 Organochlorine Insecticides

6.1 INTRODUCTION

The insecticidal potential of certain organochlorine (chlorinated hydrocarbon) insecticides was discovered during the 1939–1945 war. DDT and γ-BHC (benzene hexachloride; hexachlorocyclohexane) were at first praised as almost perfect insecticides, being relatively cheap to manufacture, strikingly effective against numerous insect pests and yet they were apparently safe to man and other warm-blooded animals. With the benefit of hindsight, it is easy to make accusations of complacency, but death came to many millions in more self-evident forms at that time. Furthermore, there was ample scientific evidence that acute toxicity of many members of the group was acceptably low (table 1.2). For example, occupying armies are neither renowned for their over-considerate treatment of civilian populations nor for their scientific training in pest control, yet over a million Neapolitans were deloused by the United States army with no reported untoward effects. In consequence, a typhus epidemic was brought under control, and many lives probably saved.

Since that time, several billions of people have been exposed to DDT, yet outside the range of industrial accidents or attempted suicides, one has to search assiduously for even disputed evidence of serious medical side effects from this compound. Conversely, probably no chemical made by man – not even penicillin, streptomycin and the sulphonamides – has saved so many lives as has DDT. The reason for this is that it was, for a time, outstandingly successful in controlling the vectors of organisms responsible for such mortal and debilitating diseases as malaria, river blindness and yellow fever.

It is therefore ironic that organochlorine compounds have possibly been the target of more emotional criticism as well as learned controversy than almost any other biological discovery. Some are notoriously persistent in the environment and so can be a real hazard to wildlife and, many believe, a potential hazard to man. In addition, their use has led to the indiscriminate killing of beneficial as well as harmful insects and, increasingly, to the development of **insect resistance** (Carson, 1963). The latter reason, in itself, is sufficient justification for the current policy of using organochlorine compounds for specific purposes rather than as general all-purpose insecticides. The organophosphorus and carbamate insecticides (chapters 4 and 5) have largely replaced them for many aspects of crop protection in numerous countries.

6.2 STRUCTURAL DIVERSITY AND PROPERTIES OF ORGANOCHLORINE INSECTICIDES

The organochlorine insecticides are usually divided into three main families comprising respectively compounds related to DDT, γ-BHC and compounds related to aldrin (figure 6.1). Some members of the first and third of these families contain elements other than carbon, hydrogen and chlorine (e.g. oxygen or sulphur, or both). Structurally, one of the few features members of the three families have in common is that the molecules contain one or more chlorinated carbocyclic rings.

Members of the different families do, however, have important physicochemical characteristics in common. The chemical stability of many members of the group (or of their immediate and often toxic metabolites) is high because their molecules are constructed, entirely or largely, from C–C, C–H and C–Cl bonds, all of which tend to be chemically rather inactive under normal environmental conditions. In consequence, traces of organochlorine compounds are to be found in air and water throughout the world. Kenaga (1972) quoted data assembled by Edwards (1970) indicating that rainwater contained (at that time) approximately 2×10^{-4} ppm, air contained about 4×10^{-6} ppm and that sea water contained 1×10^{-6} ppm.

A second physicochemical feature of great importance is the low solubility in water of most members of the group. This, coupled with their strongly lipophilic character, means that they possess partition coefficients which strongly favour accumulation in biolipids. In effect, the bodies of animals and plants behave like molecular lobster traps – lipophilic molecules often enter more or less readily but, once inside, have difficulty in escaping. Thus, unless the organism's defence mechanisms can degrade them to excretable products at a rate that is sufficiently rapid to outmatch the rate of entry, accumulation in body lipids will take place. Kenaga (1972) lists typical concentrations of DDT and its derivatives in various organisms – for example, 3×10^{-4} ppm in plankton, 1×10^{-3} ppm in aqueous invertebrates and 5×10^{-1} ppm in marine fish. These figures, incidentally, illustrate the so-called food chain effect (section 6.8). Similarly, vegetables contained 2×10^{-2} ppm, meat contained 2×10^{-1} ppm and human adipose tissue contained 6 ppm. (It is interesting to observe that, at the time of maximal use of

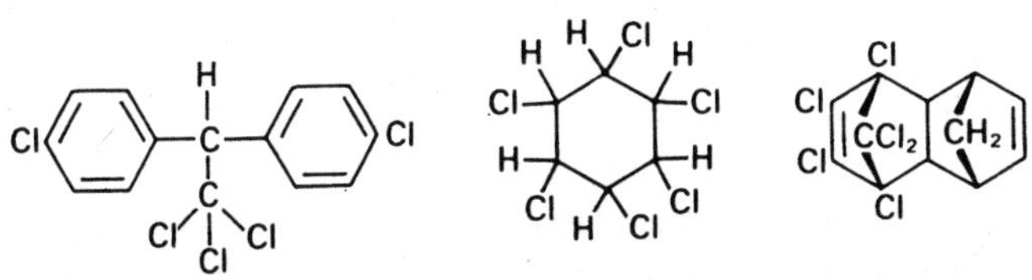

(1) DDT family, e.g., DDT.

(2) BHC family (only γ-BHC)

(3) Chlorinated cyclodiene family, e.g. aldrin

figure 6.1 The three families of organochlorine insecticides (projection formulae)

DDT, Americans were ingesting about 0.184 mg DDT per day and had 12 ppm (plus about 12 ppm DDE) in their body fat. Some humorist pointed out that, according to the American laws governing residue tolerances, flesh from Americans was unfit for human consumption!)

Most organochlorine insecticides are waxy solids at room temperature. Their vapour pressures are low but not negligible, for, as was implied above, their volatilisation leads to atmospheric contamination, while the volatility of such compounds as aldrin and γ-BHC enables them to be used as soil fumigants (section 6.3).

γ-BHC and most members of the DDT family undergo dehydrohalogenation in the presence of ethanolic KOH but they are otherwise rather inactive. Some organochlorine compounds are slowly destroyed by ultraviolet radiation but only when they are in a finely divided state.

Acute oral toxicities of organochlorine insecticides are variable but mostly of intermediate magnitudes (table 1.2). The approximate LD_{50} of DDT to rats is 200–300 mg/kg body weight. If the lower figure is roughly applicable to man, then a 70 kg man would need to eat some 14 g to have a 50 per cent chance of dying. Remembering DDT is a waxy solid, to swallow such an amount would be a somewhat difficult enterprise. However, convulsions have been known to occur in man after ingesting as little as one gram, i.e. 16 mg/kg body weight (Matsumura, 1975).

6.3 PRODUCTION FIGURES AND PRACTICAL USES OF ORGANOCHLORINE INSECTICIDES

Uncertainty about the long-term environmental effects of highly persistent insecticides, as well as political pressures arising as a response to public concern about contaminated food, have led to voluntary or compulsory control of the use of organochlorine insecticides in most countries. In some of these there has been a total ban on the use of specified members of the group, in others their use has been limited to special crop protection situations where no suitable alternatives exist. In Britain uses of DDT have been restricted under the Pesticide Safety Precautions Scheme; for example, it should not be used on grass for grazing, silage or hay. In the USA large quantities of organochlorine pesticides were once used on cotton but restrictions were placed upon its use in 1968. In 1970 the USA was, however, still manufacturing large amounts for export, especially for the control of malaria.

The increasing reluctance of manufacturers to disclose production statistics makes it impossible to assess the present status of organochlorine insecticide production but table 1.1 shows that in 1970–1971 the quantity of organophosphorus insecticides manufactured in the USA was greater than that of organochlorine compounds. At that time DDT and toxaphene accounted for the bulk of chlorinated hydrocarbon insecticides still manufactured there. In all probability, their worldwide use, especially for agricultural purposes, has continued to decline.

Organochlorine compounds have so many potential uses (despite restriction or proscription in many countries) that a simple list of crops, pests and control measures is impracticable. Instead, table 6.1 indicates some types of uses for which different compounds are most suited.

DDT is effective against a wide variety of pests, but it is relatively ineffective against aphids and spider mites. In the case of aphids, part of the problem is that the stationary colonies may escape spray droplets of DDT whereas their mobile predators can move into the contaminated area. In consequence, DDT treatment can sometimes result in a final situation worse than that which originally existed. Mites are inherently less susceptible to DDT than are insects, although the DDT analogue, dicofol, has been used to control them. DDT and other organochlorine insecticides have no systemic action.

The vapour pressure of γ-BHC is about fifty times as great as that of DDT. This substance is therefore preferable to DDT where a fumigant action is desirable and it is usually less persistent on a crop. The relatively high vapour pressure contributes to its usefulness for the control of soil organisms, especially in warmer climates. In addition, γ-BHC is also often used as a seed dressing.

Certain members of the cyclodiene family are amongst the most powerful of insecticides, possessing both a strong action and a long persistence. Aldrin has a relatively high vapour pressure and is therefore useful where a fumigant action in the soil is required, e.g. for wireworm control in potatoes. Dieldrin is a very effective insecticide for the control of certain insect parasites on animals, including lice, blowfly larvae and ticks. Unfortunately, the effluent from sheep dips has contributed significantly to total environmental pollution by chlorinated hydrocarbons. In Britain, its principal use in crop protection is as a dip for roots of cabbages at the time of transplanting and, under clearly specified conditions, for the treatment of seed of sugar beet, onions, beans and spinach.

All the organochlorine insecticides are phytotoxic to members of the cucumber family. In addition, DDT is toxic to some, but not to all, varieties of barley. Aldrin has sometimes been used on DDT-sensitive varieties of spring barley. All organochlorine compounds are very toxic to bees and to fish.

6.4 MODE OF ACTION OF ORGANOCHLORINE INSECTICIDES

The toxic effects of members of the three families of organochlorine compounds differ in detail but all are **neurotoxic** substances. One major difference is that the initial effect of DDT is upon the peripheral nervous system, whereas γ-BHC and aldrin appear to attack the central nervous system. At the physiological level, the effect of DDT on nerve–muscle systems has been investigated intensively, the action of the other two groups rather less so. Despite some differences, the general effect of all of them is to destabilise neural activity and this is mani-

Table 6.1 Biological spectrum of some organochlorine insecticides (The uses of these compounds are so numerous that it is easier to say what each is *not* used for and a short list covering individual crops is not possible)

DDT	*Poor* – when there is heavy sclerotinisation; when the life cycle is rapid; for soil organisms; for organisms living in crevices. Examples: larger *Hemiptera*; aphids; boll weevils; most soil pests and root-attacking insects; white flies; mites. *Medium* – cutworms; froghoppers, leaf miners, fruit flies, weevils. *Good* – Most *Hymenoptera*; most *Lepidoptera*; blossom weevils; codling moths; European corn borers; gipsy moths; hoppers; plant bugs
Methoxychlor	*Good* – Colorado beetles; elm bark beetles; flea beetles
BHC	Useful where fumigant action, stomach poisoning or high speed of action required. *Better than DDT* for – cotton boll weevils (penetrates bolls); house mosquitoes (does not excite insects to leave deposit rapidly); some mites on man and livestock; tsetse fly larvae. *Special uses* – seed dressings; apple sawflies; army worms; blowflies on sheep; cockroaches; crickets; grain weevils; grasshoppers; leatherjackets; wireworms
Aldrin	Useful for short non-persistent foliage action; for use in soils. *Good* – boll and other weevils; cutworms; dipterous larvae that eat roots; Japanese beetle larvae; mealy bugs; Mediterranean fruit fly; millipedes; scale insects; stalk borers, white grubs; wireworms
Dieldrin	Useful for surface treatment of soil; for seed dressings; for root dipping; for bran baits; for sheep dips. *Poor* – caterpillars; sucking insects *Good* – cabbage and similar root fly larvae; crickets; cutworms; grasshoppers; leatherjackets. (In public health, cockroaches; fleas; some locusts; mosquito larvae; scorpions; tsetse flies)
Endrin	*Good* – sucking and boring insects of many tropical crops
Endosulfan	*Good* – ants; beetles; grasshoppers; many lepidopteran larvae; pod midges; tsetse flies; wasps
Toxaphene	*Good* – boll weevils and other spp. on cotton
Chlordane	*Better than DDT* – aphids, Colorado beetle larvae, grasshoppers *Special uses* – ants, beetle larvae, chafer larvae; earthworms; earwigs; household insects; moth larvae; soil insects; termites; wireworms

Sources of information: Brooks (1974, Vol. 1); MAFF (1980); Martin and Worthing (1976).

fested by a hyper-excitability of nerves and muscles. Before describing these abnormal effects it is necessary to consider briefly the conduction along normal non-myelinated nerve axons.

A nerve axon can be regarded as a cable of aqueous fluid insulated from the extracellular fluid by a lipoprotein membrane, the plasmalemma of the nerve cell. The fluid inside is richer in potassium ions than is the fluid outside, indicating that the membrane is relatively impermeable to ions when the nerve is resting. The inside of the membrane is normally some 70 mV negative relative to the outside, a figure which can be predicted from the **Nernst equation** (equation (6.1)), if it is assumed that this resting potential is a function of the difference between the potassium ion concentrations on the inside and the outside of the membrane.

$$E_{\mathrm{mV}} = \frac{1000\,RT}{F} \ln \frac{[K^+]_{\mathrm{external}}}{[K^+]_{\mathrm{internal}}} \qquad (6.1)$$

In this equation E is the potential across the membrane expressed as mV, R is the gas constant (8.3 joules per degree per mole), T is the absolute temperature and F is the Faraday in coulombs per g equivalent (see p. 359 of Ref. 2 in Bibliography).

The passage of an impulse along a non-myelinated axon is a rapid wave of depolarisation moving at some 1–20 m/s. As the impulse passes any particular position along the axon, the inside of the plasmalemma at that point rapidly changes from being negatively charged compared to the outside of the membrane and becomes positively charged instead. This is caused by a rapid influx of sodium ions. Thereafter, the potential inside the membrane rapidly returns to normal for two reasons. These are that the membrane's normally low permeability to sodium ions is restored and because there is an efflux of potassium ions (figure 6.2).

The reversal of electrical charge across the nerve membrane can be studied by amplifying the signal so that it drives the pen of a chart recorder, whereupon the alteration is recorded as a pen movement or 'spike'. The **depolarisation** which occurs when a stimulus arrives, and the subsequent restoration of the resting

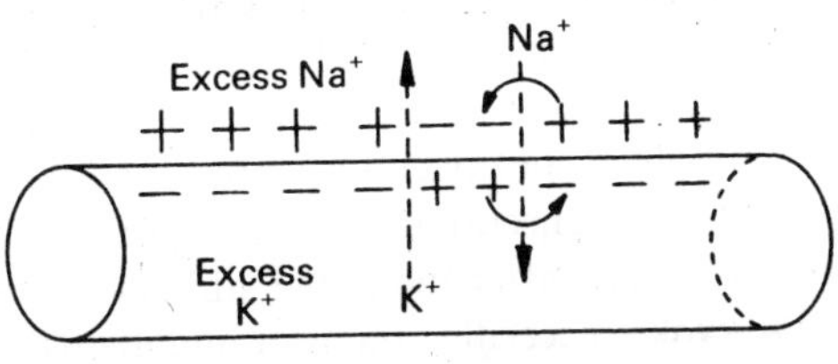

figure 6.2 Resting and action potentials. (The outside of the plasmalemma of the axon is, when resting, positive with respect to the inside. When a wave of action potential moves along the axon, an entry of Na^+ precedes an efflux of K^+)

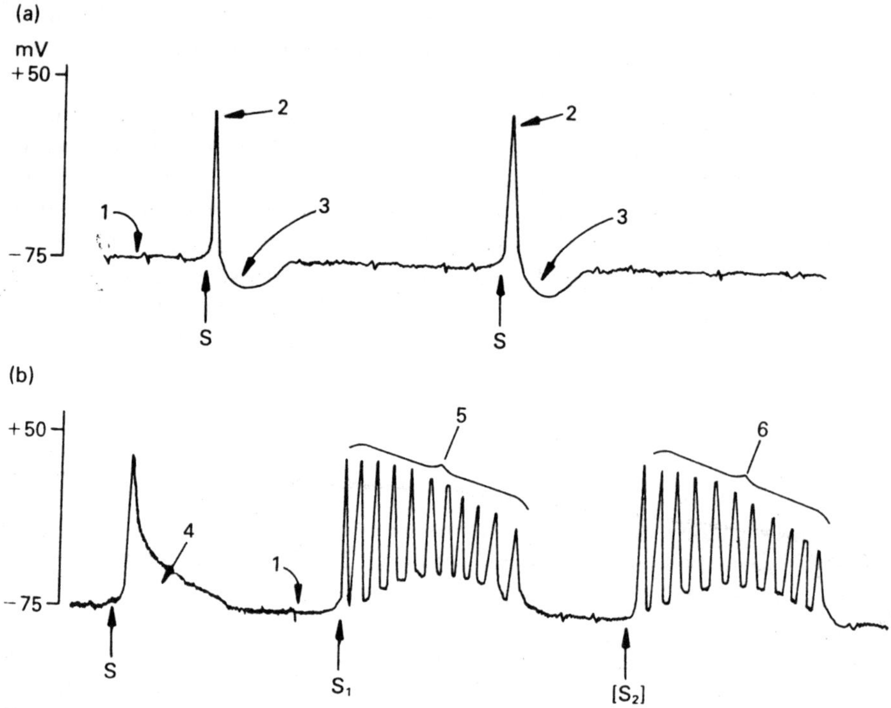

figure 6.3 Discharges in (a) a normal nerve and (b) in a DDT-poisoned nerve. ((a) Electrical stimulation, S, of an unpoisoned nerve converts the resting potential, 1, to a single action potential 'spike', 2, which is followed by a short afterpotential of *hyper*polarisation. (b) After DDT poisoning there may be a single spike followed by a prolonged afterpotential, 4, of *hypo*polarisation. In addition, a train of repetitive discharges, 5, may occur after a single impulse, S_1 or S_2. At high DDT concentration, a train of trains 5,6, may follow a single shock, S_1 (i.e. S_2 is unnecessary))

potential, are illustrated in figure 6.3a. Over a longer time scale the slow depletion of internal potassium ions is prevented by the operation of the 'sodium–potassium pump'. Calcium ions are also important for the correct functioning of axons, apparently because some sort of surface recalcification is a pre-requisite to the restoration of a stable resting potential.

DDT has a **destabilising effect** on the nerve membrane, a single electrical stimulus leading to multiple uncontrolled responses of poisoned nerves, and thus to muscular spasms. At high concentrations, trains of repetitive discharges arise spontaneously – that is to say, after the initial stimulus, electrical impulses are no longer necessary to evoke succeeding trains of impulses (figure 6.3b).

The membrane-destabilising effects of γ-BHC and of dieldrin are similar to those of DDT, although not entirely the same. Nevertheless, how these effects are brought about by organochlorine insecticides remains obscure. There are two apparently diverse types of explanations, the first of which is that the insecticides may inhibit one or several enzymes of importance to the mechanism of membrane stability or of nerve impulse transmission. An alternative possibility is that they may initiate some physical change in the structure of the nerve membrane which rather literally causes it to misfire (e.g. by alteration of its permeability to ions or to other solutes). Paradoxically, at the near-molecular level of a lipoprotein membrane, these two possibilities are not necessarily distinct, for insoluble enzymes are *integral components* of such a membrane. The remainder of this section will illustrate the application of these alternative approaches to the unsolved problem of why organochlorine compounds are neurotoxic.

It has been known for many years that membrane-bound enzymes exist which, *in vitro*, appear to split ATP to give ADP and inorganic phosphate. *In vivo*, such splitting is probably an overall effect, the reactants and products really representing fuel of, and waste from, a reaction which drives a vital mechanical process. Thus ATPases are probably associated with muscular activity, mitochondrial processes and solute movement through tonoplast and plasmalemma as well as with the activities of nerve membranes. It now appears that one sort of ATPase is dependent on sodium and potassium ions and is actually a part of the sodium pump; another type is dependent on the presence of calcium ions and a third type requires the presence of both calcium ions and magnesium ions. It was shown in 1969 that DDT can affect the activity of ATPase but more recently it has become apparent that the calcium-dependent ATPase that is located on the outer surface of axons (ecto-ATPase) is much more sensitive to DDT than is the ATPase associated with the working of the sodium pump (Mullins and Brinley, 1975). Ecto-ATPase appears to regulate the binding of calcium ions to the surface of the membranes. If there is a shortage of calcium ions, axonic excitation occurs which closely resembles DDT poisoning and it is perhaps therefore significant that a calcium-dependent ATPase in the peripheral nerves of the lobster *Homarus americanus* is affected by 10^{-9}M DDT (Ghiasuddin and Matsumura, 1979). Moreover, the neurotoxic symptoms of DDT poisoning are enhanced if the calcium ion concentration is reduced below 20 mM and they are greatly diminished if the calcium ion concentration is raised above 100 mM (Matsumura and Narahashi, 1971).

In contrast to this highly specific 'chemical' theory of toxic action, it has been proposed that the mere presence of lipophilic molecules at (any) lipoprotein membrane may so distort its structure that its permeability is altered or the integrity of the reactions occurring on or in it is destroyed. Candidate examples are axonic conduction, photosynthesis, oxidative phosphorylation as well as ion pumps in membranes of many sorts. A simple mechanical analogy might be the effect of stretching a thick band of rubber to which inter-geared cog-wheels of a construction set are attached.

The physical toxicity hypothesis is further discussed in section 7.8. Its disadvantage in the present context is that it fails to explain why numerous compounds with rather similar structures and partition coefficients to those of powerfull insecticides are of low toxicity (e.g. m,m′-DDT, or α-BHC). Compared with such similarities, insecticidal members of the three organochlorine families are dramatically different from one another in structure, a difference which has so far not been satisfactorily explained by proponents of either physical or chemical theories of action.

In its simplest form the physical toxicity hypothesis is evidently inadequate. Partition coefficients must be such as to get a molecule to its site of action and general reactivity must be so low that highly specific chemical interaction does not occur at a concentration below that at which 'physical' toxicity is manifest. For the substance to be neurotoxic, however, the sorbed molecule must have a structure capable of eliciting a particular sort of distortion of the membrane. This distortion can then be imagined to trigger toxic action (e.g. to alter permeability to ions or to modify the fixation of calcium ions). This problem was discussed by Mullins (1954) in relation to the mode of action of DDT and later the same author explained the toxicity of the γ-isomer of BHC in terms of the fitment of its molecules into inter-membrane spaces (Mullins, 1955). Soloway (1965) introduced a new requirement, postulating the need for two opposing centres of electronegativity positioned across the plane of symmetry of the molecule.

6.5 DDT AND ITS ANALOGUES

The remarkable insecticidal properties of **p,p′-DDT** (figure 6.4) were recognised in 1939 and the Swiss firm of Geigy patented it for this purpose in 1942. It proved its worth in public hygiene and was soon to become of vital importance in the control of malaria and other diseases carried by insect vectors. Mass screening of organochlorine compounds for potential insecticidal action followed, leading to the discovery of benzene hexachloride (section 6.6) and of various chlorinated cyclodienes (section 6.7).

Technical DDT is a white waxy solid which melts at about 90°C and has a characteristic sickly-sweet smell. The pure p,p′-isomer melts at 110°C and has a vapour pressure of 3×10^{-7} mm Hg at 25°C. It is so insoluble in water that accurate values have been difficult to determine. Biggar, Dutt and Riggs (1967) used a centrifugation technique to avoid spuriously high results and concluded that the most likely value was about 1.7 ppb. It is moderately soluble in many organic solvents, including natural oils and body lipids, although solubility is generally higher in olefinic and aromatic hydrocarbons than it is in saturated aliphatic hydrocarbons.

The primary insecticidal effect of DDT, except at high doses, is not upon the motor nerve which innervates a contracting muscle but upon a distant sensory nerve. Thus the whole reflex arc (figure 6.5) appears to be involved in the transmission of the multiple discharges (section 6.4) characteristic of its action.

Removal of one chlorine atom from the trichloromethyl group of p,p′-DDT

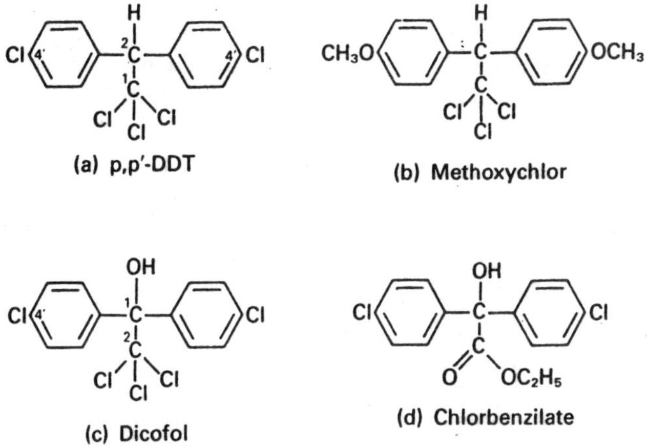

figure 6.4 DDT and its analogues. ((a) 1,1,1-trichloro-2,2-bis(4′-chlorophenyl) ethane; (b) 1,1,1-trichloro-2,2-bis(4′-methoxyphenyl)ethane; (c) 2,2,2-trichloro-1, 1-bis(4′-chlorophenyl)ethanol; (d) ethyl p,p′-dichloro-benzilate)

and its replacement with a hydrogen atom gives the structure of p,p′-DDD (figure 6.6). This substance is a minor insecticide called rhothane and is frequently referred to in US literature as TDE (tetrachlorodiphenylethane). In certain circumstances DDD is also a metabolic product of DDT and is, next to DDE, the commonest breakdown product of DDT to be found in the adipose tissue of man (Robinson, 1969). DDD is, in general, inferior to DDT as an insecticide, but it has proved particularly useful for the control of some hornworms and as a mosquito larvicide. It is only about one-fifth as toxic as DDT to higher animals.

Methoxychlor (figure 6.4) has a lower tendency than has DDT to accumulate in animal fat. It also possesses a much lower acute oral toxicity than DDT to higher animals and is therefore sometimes used for veterinary treatment on or

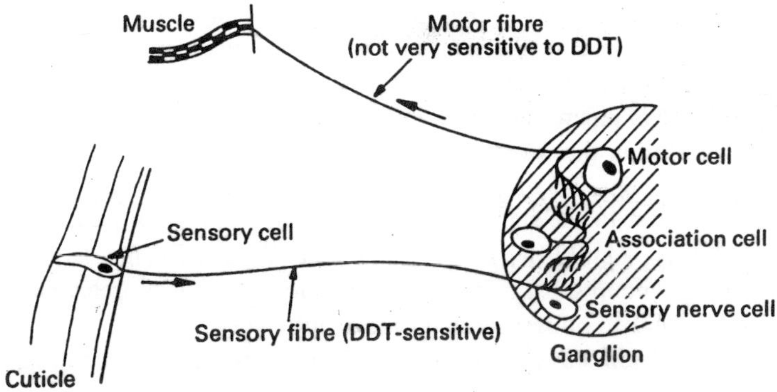

figure 6.5 Diagram of a reflex arc

near livestock. Its biological spectrum is rather similar to that of DDT, but rather more of it is usually needed to achieve a given level of control. This, together with its greater manufacturing cost, has limited its use in crop protection. This is unfortunate, for the methoxy groups, unlike the p,p'-Cl groups of DDT, provide a point of attack which dramatically changes the pattern of its persistence and accumulation compared to DDT. In addition, the methoxy groups render the molecule somewhat more water soluble than DDT and such differences can have a crucial bearing on biological magnification. This is well illustrated by the work of Kapoor *et al.* (1970) using a model ecosystem. For DDT, a concentration of 0.22 ppb in water led to a concentration of 54 ppm in *Gambusia*, a magnification of 10^5. The corresponding figures for methoxychlor were 0.11 ppb, 0.17 ppm and 10^3 respectively.

Dicofol (figure 6.4) is another interesting member of the DDT family, for it is a metabolic product of DDT in some species of insects and yet it is itself marketed to control certain ticks and mites. The latter include the agriculturally important red spider mite and the medically important vector of scrub typhus (Trombiculid species, or chiggers) which can infest clothing under poor hygienic conditions. Dicofol is a brown oil of moderate toxicity to higher animals. In its molecule the solitary ethane hydrogen atom of DDT is replaced by an hydroxyl group, a modification that greatly decreases toxicity to insects.

Chlorbenzilate (figure 6.4) is another substance which can perhaps be regarded as a relative of DDT. It was marketed in 1956 as an acaricide. It is a yellowish solid of low melting point, not very toxic to insects and of much lower mammalian toxicity than DDT. The rapid life cycle of phytophagous mites and the moderate persistence of chlorbenzilate (and of many other acaricides) provide conditions that favour the development of resistance. Thus the red spider mite, *Tetranichus urticae*, rapidly developed resistance to chlorbenzilate when bred under chlorbenzilate 'pressure' (Brown, 1958).

The DDT family of organohalogen compounds is an excellent group for the study of structure–activity relations. Numerous investigations have been made over three decades and the reader is referred to reviews by Metcalf (1955), Metcalf and Fukuto (1968) and by Brooks (1973). The p,p'-difluoro analogue of DDT possesses nearly as high a toxicity as DDT to some insects. Several compounds with small molecular weight alkyl, alkoxy or alkylthio groups in the p,p' positions of the benzene rings are moderately insecticidal, especially in the presence of piperonyl butoxide. The frequent effectiveness of this synergist is a sure indication that overall insecticidal power is often limited by oxidative degradation (section 3.3) and reflects the biodegradability of these p,p' substituents in contrast to the stability of the C–Cl bonds in these positions in DDT itself. Recently, Abu-El-Haj *et al.* (1979) have reported that a DDT analogue with a methyl group substituted for the solitary ethane hydrogen atom of DDT is as toxic as DDT to house flies and to mosquito larvae.

The three major pathways by which DDT is degraded are shown in figure 6.6.

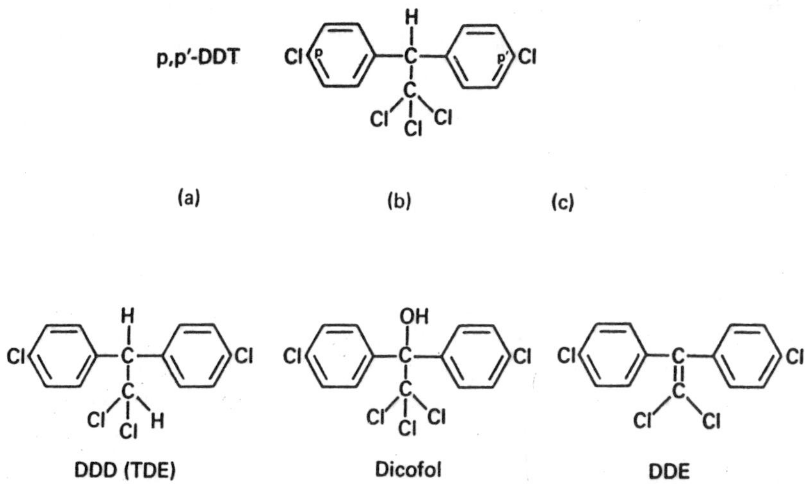

figure 6.6 Metabolism of DDT. (a) Reductive dechlorination; by anaerobic liver + NADPH, by dead tissues and by some micro-organisms; (b) oxidation, by some insects, e.g. *Drosophila*; (c) dehydrochlorination, by most insects, birds and mammals (the three pathways are not mutually exclusive)

One involves dehydrochlorination to DDE, a second proceeds via reductive dechlorination to DDD and a third is oxidative and leads to dicofol.

DDE is the major DDT-derived residue normally found in autopsy or biopsy samples of animal tissues (Robinson, 1969). The enzyme, DDT dehydrochlorinase (DDTase), which catalyses its formation, is a glutathione-dependent enzyme which was first separated by Lipke and Kearns (1960). Isolated microsomes from animal liver do not convert DDT to DDE. The development of resistance to DDT in some species of house flies and mosquitoes is closely correlated with the greater capacity of resistant organisms to convert DDT to the less insecticidal DDE. Some insect species have an inherent resistance to DDT when it is administered orally but not when it is injected, a difference that can be attributed to the presence of a dehydrochlorinase in the cells of the alimentary canal which destroys DDT before it can get to the (otherwise susceptible) nervous system. DDE is particularly effective at inducing microsomal enzymes in mammalian liver. This is not necessarily a harmful effect, but does alter the body's capacity to metabolise many types of foreign compounds, including pollutants and medical drugs.

A few insects metabolise DDT by a predominantly oxidative route which results in the formation of dicofol. The route seems to be a typical NADPH-dependent microsomal oxidation (section 3.3). This aliphatic hydroxylation occurs in *Drosophila* and pehaps in resistant strains of the cockroach (Agosin *et al.*, 1961).

Reductive dechlorination of DDT occurs widely in nature and DDD is an almost ubiquitous residue in animal fat (Robinson, 1969). However, the bulk of this residue may not be formed under normal conditions by enzymes located in vertebrate liver for DDD can be readily produced by bacteria living anaerobically in the gut. Several workers, including Walker (1969) have shown that NADPH-supplemented liver preparations convert DDT to DDD under anaerobic conditions and Hassall (1975) observed that if riboflavin was also present, even microsomes preheated to 70°C were able, in the presence of NADPH, to convert DDT to DDD. A similar heat-stable and flavin-dependent reduction system has been reported by Esaac and Matsumura (1979) to be able to bring about the demethylation of a carbamate insecticide.

Further or secondary metabolism probably takes place in many species, for water-soluble metabolites occur both in insects and in higher animals. In mammals, the most important of these is DDA, an acid substance formed by the replacement of the $-CCl_3$ part of the DDT molecule by $-COOH$. This acid may be excreted in free or conjugated form, in faeces or in bile, according to the species. Rather surprisingly, it is still not clear by what route or routes DDA is formed; in particular it is not clear whether DDD is an obligatory intermediate or whether it can be formed via DDE.

The primary metabolism of methoxychlor is summarised in figure 6.7. Kapoor *et al.* (1970) showed that an oxidative change resulted in one or both of the methyl groups being removed. The change is consistent with it being brought about by microsomal O-demethylation reactions (section 3.4). The same workers have shown that the methoxychlor analogue of DDE is also produced although it does not appear to be a prominent metabolite.

Some of the situations in which DDT is more or less useful are shown in table 6.1. In Britain, the Wilson Committee recommended in 1965 that the use of DDT should be confined to a number of specified purposes, but in 1971 it

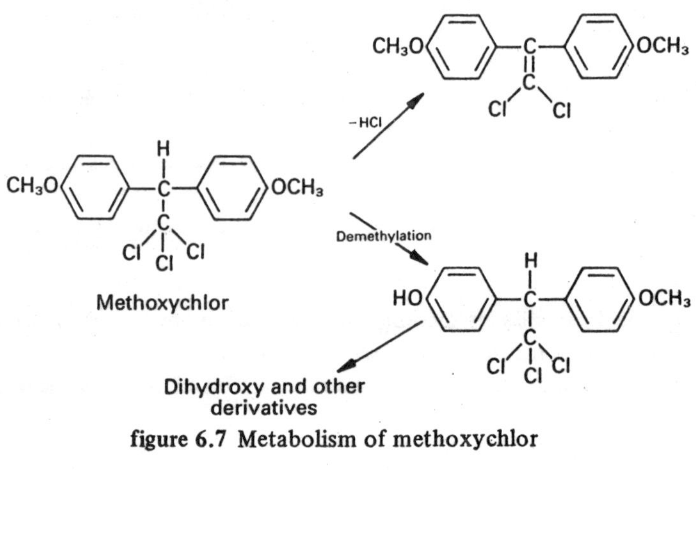

figure 6.7 Metabolism of methoxychlor

was still used to control some cereal pests, pea and bean weevils and mangold flies for which various objections to alternative control measures exist (Brooks, 1972). By 1969 various American states were placing restrictions on the use of DDT and other organochlorine compounds and the DDT Advisory Committee of the Environmental Protection Agency recommended that the use of DDT in the USA should be drastically curtailed. As in Britain, however, some insects (e.g. the gipsy moth) are difficult to control at an economic price using DDT substitutes. The use of DDT for crop protection ceased in Norway in 1970 and its domestic use was banned in Sweden in the same year. Its outdoor use is also restricted in Canada, Denmark and Italy (Brooks, 1974).

6.6 BENZENE HEXACHLORIDE (HEXACHLOROCYCLOHEXANE)

There is only one important member of this sub-group, namely, the gamma (γ) isomer of hexachlorocyclohexane, HCH. Worldwide, the latter is probably better known by its older name, benzene hexachloride, BHC. The ring which forms the framework of BHC is not planar (as is an unsaturated benzene ring) but chair-shaped, and it is possible by constructing models to show that eight forms of BHC can exist. Of these isomers, only the gamma isomer is highly toxic to insects, being for various insects from fifty to several thousand times as toxic as the α or δ isomers. The β and ϵ isomers are usually almost inert. The purified gamma isomer is often called **lindane**, a tribute to Van der Linden who discovered some of the BHC isomers.

The cyclohexane ring is shown in figure 6.8; in the left-hand diagram it is represented as it appears after making a molecular model, and on the right it is portrayed in the conventional form. Normally, hydrogen atoms and their valency bonds are not shown in cyclic formulae, but in the right-hand portrayal small

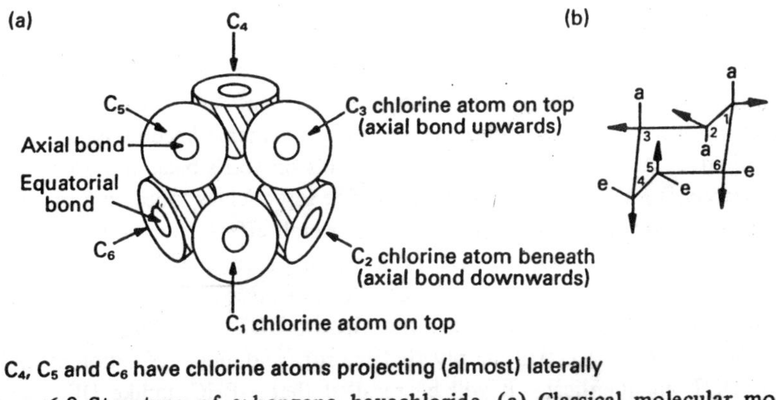

figure 6.8 Structure of γ-benzene hexachloride. (a) Classical molecular model representation of γ-BHC; (b) conventional 'chair' representation of γ-BHC. (γ-BHC or γ-HCH: 1,2,3,4,5,6-hexachlorocyclohexane)

arrows are inserted to show the direction in which they point, the bonds a, e, being occupied by chlorine atoms.

It will be observed that three carbon atoms (marked C_1, C_3, C_5) each have a bond directed upwards out of the plane of the ring. The three carbon atoms alternating with them (C_2, C_4, C_6) each have a corresponding bond directed downwards. Such bonds, whether directed up or down, are termed *axial* (a) bonds. The remaining bonds (one per carbon atom) project almost, but not quite, sideways out of the ring, and are termed *equatorial* (e) bonds. To construct a model of the gamma isomer proceed around the ring putting chlorine atoms in the following positions, a, a, a, e, e, e. Note that, if the first axial chlorine is 'up', the second must be 'down'.

Before purification, the technical material contains about 13 per cent of the gamma isomer, some 68 per cent of two alpha stereo isomers and small quantities only of the beta and delta isomers. From this mixture, lindane can be concentrated by exploiting differences in the solubility of the isomers in various solvents. Such purification is desirable, especially for treatment of crops with tap roots or storage organs, because impure BHC tends to have an unpleasant musty smell and to taint foodstuffs. Lindane (which by definition, contains >99 per cent of the gamma isomer) is an almost odourless white solid, slightly soluble in water (10 ppm), moderately soluble in kerosene (2 per cent) and very soluble in acetone (43 per cent). It has a vapour pressure of about 10^{-5} mm Hg at 20°C and is stable to heat and light. The only significant chemical reaction it undergoes is the loss of three molecules of HCl under mildly alkaline conditions, the product being 1,3,5-trichlorobenzene. It is about twice as toxic as DDT to higher animals (table 1.2).

It is not known why only one of the eight isomers of BHC is outstandingly insecticidal; there is no obvious correlation of toxicity with either physical properties or chemical reactivity, for the vapour pressure, the solubility in oil and the chemical reactivity of the gamma isomer all lie between those of various inactive isomers. Reference to the membrane interspace theory was made on page 128.

The insecticidal action of lindane is somewhat similar to that of DDT in that the insect starts to tremble, with telescopic movements of the abdomen. With passage of time, hyperexcitation increases and movements become increasingly uncoordinated. At higher dosage, the insect falls on to its back and eventually suffers paralysis. As with other organochlorine insecticides, poisoning causes a pronounced increase in respiratory rate. The overall effect on action potentials is somewhat similar to that of DDT (section 6.4) a complete reflex arc being essential for the action to be fully evident. However, insect resistance to lindane is linked to resistance to cyclodienes rather than to DDT, so a similarity of toxic symptoms does not necessarily imply that the sites and modes of action of BHC and of DDT are identical – it will be recalled that γ-BHC, unlike DDT, acts primarily on the central nervous system (page 123).

Precise details of the routes of degradation of lindane are not known, although both microsomal oxidases and glutathione-dependent enzymes participate in different reactions. One route involves dehydrochlorination leading eventually to 1,2,4,-trichlorobenzene. The latter may then form a labile epoxide in the presence of a NADPH-dependent mono-oxygenase, the epoxide re-arranging to form a chlorinated phenol. On the other hand Tanaka *et al.* (1979) proposed a mechanism whereby direct oxygenation of the lindane ring by a cytochrome P_{450}-dependent reaction leads to the formation of 2,4,6-trichlorophenol. Various chlorinated phenols have in fact been identified in the urine and faeces of various organisms. Glutathione conjugates have also been identified, as have mercapturic acid derivatives in rats (Kurihara *et al.*, 1979). Thiochlorophenols have been observed, especially in insects. These could be formed by hydrolysis of glutathione or mercapturic acid conjugates in such a way that the sulphur atom remains attached to the phenol ring.

6.7 THE CHLORINATED CYCLODIENE FAMILY

Numerous organochlorine compounds can be made by condensing hexachlorocyclopentadiene with suitable cyclic hydrocarbons containing five or seven atoms of carbon. The reaction, involving a diene and a dienophile ('diene-liker') is termed the Diels–Alder reaction. The reaction is named after the two chemists who won a Nobel Prize in 1950 for discovering the diene synthesis. So also are dieldrin and aldrin, two of the best-known insecticidal members of the group. Their structures are shown in figure 6.9.

As the figure shows, the structural chemistry of cyclodiene insecticides is complex, because the molecule is not planar. The molecular geometry recalls the 'chair' configuration of BHC, but here two such rings (which can exist in alternative 'boat' forms) are fused together, and the fusion can occur in more than one way with respect to (a) the 'bridges' in the rings and (b) the curvature of one ring with respect to the curvature of the other. In addition, if an epoxide is present, this too can occupy two possible positions in space. For a more detailed account the review of Riemschneider (1963) or the monograph of Brooks (1974) should be consulted. Figure 6.9 also shows the rather remarkable structure of **mirex**, a stomach insecticide used to kill ants, wireworms and snails. It is not formed by a Diels–Alder reaction but is a dimer of hexachlorocyclopentadiene.

$$2C_5Cl_6 \xrightarrow{\text{dimerisation}} C_{10}Cl_{12} \qquad (6.2)$$

Aldrin and **dieldrin** are stereochemically related and aldrin is rapidly metabolised to dieldrin by an NADPH-dependent enzyme in hepatic microsomes (section 3.4). Similarly, **isodrin** is rapidly epoxidised to **endrin**. In each case, the olefin is less polar and has a higher vapour pressure than the epoxide. This difference accounts for the use of aldrin as a soil fumigant to control wireworms and the larvae of root flies whereas dieldrin has been more often employed as a root dip

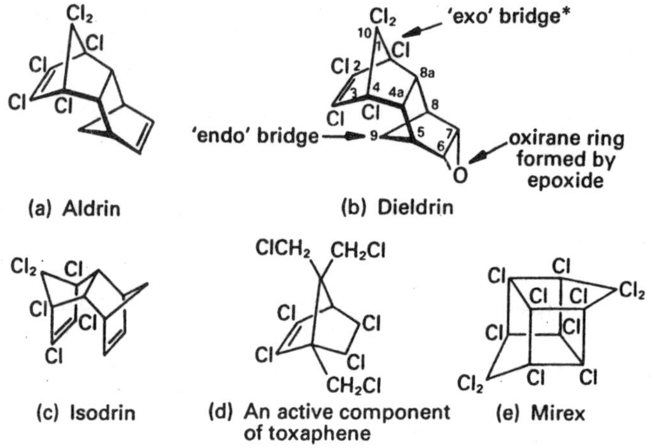

figure 6.9 Common insecticides of the cyclodiene family. (a) 1,2,3,4,10,10-hexachloro-1,4,4a,5,8,8a-hexahydro-*exo*-1,4-*endo*-5,8-dimethanonaphthalene. (b) 1,2,3,4,10,10-hexachloro-6,7-epoxy-1,4,4a,5,6,7,8,8a-octahydro-*exo*-1,4-*endo*-5,8-dimethanonaphthalene (* more than one convention exists for exo-endo naming; a common American system is used here). (c) 1,2,3,4,10,10-hexachloro-1,4,4a,5,8,8a-hexahydro-*endo*-1,4-*endo*-5,8-dimethanonaphthalene. (d) 2,2,5-endo-6-*exo*-8,9,10-heptachlorobornene (one of several active components of toxaphene). (e) Dodecachloro-octahydro-1,3,4-metheno-2H-cyclobuta[c,d]-pentalene

and as a seed dressing. Dieldrin has also been of great importance in locust control and is very effective against animal ectoparasites such as blowflies and lice; the problem is to know how to dispose of the spent liquid from sheep and cattle dips without contaminating water courses.

Chlordane, in technical form, is a mixture of variously chlorinated dicyclopentadienes, pure chlordane comprising two geometric isomers with eight chlorine atoms in their molecules (six in one ring and two adjacent to each other in the smaller ring). The *cis*-isomer is by far the more toxic of the two geometric isomers (see β-chlordane, p. 234, Metcalf, 1955). Some of its uses are shown in table 6.1. Both isomers can apparently be dehydrogenated to an olefine (analogous to aldrin) and then epoxidised to give oxychlordane (which is analogous to dieldrin).

Of the remaining compounds, **toxaphene** is worthy of mention because, in 1971, it accounted for 80 per cent of all chlorinated cyclodiene insecticides still used at that time by farmers in the United States (Green *et al.*, 1977, p. 16). It is not, strictly speaking, a cyclodiene and is not made by the Diels–Alder synthesis. Instead, it is prepared by chlorinating the terpene, camphene, and it contains heptachlorobornane as the principal active ingredient. It was manufactured in large quantities because it was used extensively to control pests of cotton.

Chlorinated cyclodiene insecticides are amongst the most toxic and persistent of all pesticides. They can be used safely and there are circumstances where their use is probably essential, but unfortunately some of the justified opprobrium their overuse has engendered has rubbed off on other organochlorine compounds, such as methoxychlor and lindane. The use of chlorinated cyclodienes is now subject to control in most countries; in Britain, the Cook Report in 1964 recommended that, wherever possible, DDT, BHC or organophosphorus insecticides should be employed instead of them. Their use in seed dressings was regarded as undesirable because of their high toxicity to birds although the Committee accepted that, in Britain, treatments with aldrin and dieldrin saved the equivalent of about 4000 acres of winter sown wheat at that time. In practice, the use of aldrin and dieldrin has been virtually restricted to root dips to protect crops from the larvae of the cabbage root fly and, in the case of aldrin, as a dust for the control of wireworms and leatherjackets in designated crops (MAFF, 1980). Resistant strains of cabbage root fly have, however, been reported in several regions of Britain.

Chlorinated cyclodienes evoke the same general neurotoxic symptoms as DDT, although there are several important differences. Both families of compounds cause uncoordinated body movements and, in isolated nerve preparations, repetitive discharges upon stimulation, but whereas the toxicity of a given dose of DDT is enhanced by lowering the temperature, most chlorinated cyclodienes show a positive correlation between temperature and level of response. Again, some chlorinated cyclodienes elicit a fanning movement of the wings of house flies. This abnormality is characteristic of the action of γ-BHC but not of DDT. Another peculiarity of chlorinated cyclodienes is that a latent period exceeding two hours usually elapses between application to an insect and the appearance of a toxic response. For dieldrin, it has been suggested that metabolism to the more toxic aldrin 6,7-*trans*diol (figure 6.10) occurs during this period (Wang *et al.*, 1971).

Little is known about the biochemistry of the neurotoxic action of chlorinated cyclodienes although, as for DDT, the accumulation on axonic membranes in some way alters the binding of ions to membranes or the permeability of membranes to ions. It has long been known that appropriate concentratons of chlorinated cyclodienes can inhibit an ATPase associated with the mechanism of the sodium pump (Koch, 1969). More recently, Yamaguchi *et al.* (1979) have shown that an ATPase dependent not on sodium and potassium ions but on magnesium and calcium ions is inhibited by less than 10^{-5}M heptachlor epoxide. These workers concluded that this inhibition led to a reduction in the binding of calcium ions to the axonic membrane (a conclusion which can be compared with the postulated effect of DDT described in section 6.4). Butler and Crowder (1977) demonstrated that toxaphene caused the level of cyclic AMP in poisoned nerves to be abnormally high, and suggested that such a change could result in an alteration of ion fluxes across the axonic membrane.

In addition to their neurotoxic action, several chlorinated cyclodienes affect the activity of the Malpighian tubules in insects, excretory activity almost ceasing in insects poisoned with endrin. In vertebrates, acute poisoning leads to anorexia, convulsions and coma while prolonged absorption has been known to cause degenerative lesions of the brain, liver and kidneys (Treon and Cleveland, 1955). Birds are several times more susceptible to these compounds than are mammals. Moderately high doses of several members of this family induce drug metabolising enzymes in vertebrate liver microsomes.

The first step in the metabolism of the olefinic cyclodienes aldrin, isodrin and heptachlor involves oxidation by microsomal oxidases to the corresponding epoxides. These epoxides are relatively stable, especially in insects, although slow metabolism to a variety of products occurs. The latter are variously excreted in free or combined form in the urine or, via the biliary route, in the faeces. In the case of dieldrin, for example, a secondary alcohol group can be formed on the unchlorinated bridge by the action of mono-oxygenases present in the livers of higher animals, whereas hydrolysis of the oxirane ring by epoxide hydrolase (section 3.3) usually leads to the formation of a *trans* diol (figure 6.10). Another characteristic product, 2-ketodieldrin, is formed by oxidation, dechlorination and molecular rearrangement. In many species this substance is the metabolite present in largest amount in the urine.

Despite these changes, the general picture is one of rather slow metabolism of epoxide cyclodienes in the majority of organisms and, in consequence, residues of chlorinated cyclodienes are almost ubiquitous in animal fats. This notorious persistence led to a search for biodegradable cyclodiene compounds. If the unchlorinated bridge is omitted from the ring, an epoxide, HEOM (figure 6.11) is obtained which is rapidly metabolised, especially by the action of epoxide hydrolase (section 3.3), which converts it to a diol. In contrast, Hassall and Addala (1979) discovered that a dimethylated epoxide analogue of HEOM was not measurably attacked by epoxide hydrolase, the methyl groups in the

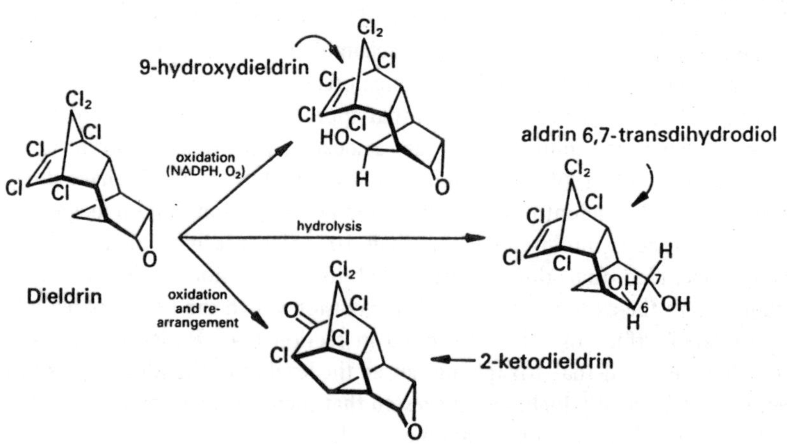

figure 6.10 Three metabolites of dieldrin

region of the oxirane ring causing steric hindrance of enzyme action. Instead, it is hydroxylated by NADPH-dependent mono-oxygenases. This dimethylated epoxide (DME, figure 6.11) turns out to be a poor insecticide but has proved useful as a substrate for demonstrating the existence of more than one sort of microsomal mono-oxygenase system in the rat. In particular, it has been used to explore the role of rat testicular and pituitary hormones in the initiation or control of some kinds of mono-oxygenase activity (Finnen and Hassall, 1980).

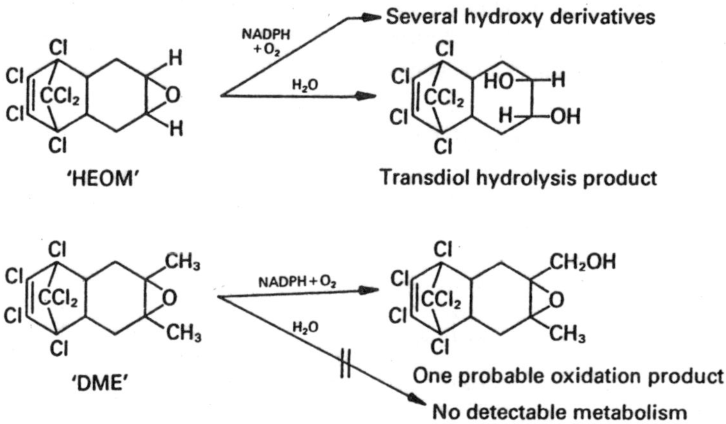

figure 6.11 Biodegradable dieldrin analogues (projection formulae)

6.8 PERSISTENCE AND SIDE EFFECTS OF ORGANOCHLORINE INSECTICIDES

In **soil**, most organochlorine compounds persist for months or even years. This is especially so when high doses are applied to clay soils or to soils rich in organic matter. Volatility of such compounds as γ-BHC may limit persistence in light soils, but for aldrin the situation may be more complex. Kushwaha *et al.* (1978) found that this substance persisted a shorter time in sandy loam at 25°C than it did at 35°C (half-life, 49 and 177 days respectively, at 6 ppm). The probable explanation for this observation is that soil organisms probably convert aldrin to the less volatile dieldrin more rapidly at the higher temperature.

There is little evidence that a long-term effect on soil fungi or bacteria takes place when organochlorine compounds are applied at normal concentration. However, there may be temporary disturbance in the equilibrium existing between the various species of soil micro-arthropods when DDT is applied at 1 lb per acre. The number of Collembola (a group of insects which feed on organic matter) suffers a sharp initial decline but then increases again to a higher steady state level, apparently because of the destruction of soil mites and other predators. Earthworms, which account for a biomass of some 500 g/m^2 of soil surface in pastures, are not greatly affected except perhaps by chlordane.

Indirect problems may also arise from the presence of persistent substances in or on soil. Predators may be threatened indirectly if they eat earthworms that have been living in soil contaminated by pesticides, because worms, like many other organisms, often have more pesticide in them than exists in the neighbouring environment. The ratio of the concentration in a named organism to the concentration in its surroundings is known as the **biomagnification factor.** For earthworms this factor is, in fact, often rather low compared to that observed for many organisms living in an aqueous environment (table 6.2).

Table 6.2 Biomagnification of DDT and its metabolites (From Kenaga (1972))

Environment	Organism	Maximum observed biomagnification(1)
Soil	roots of crops	0.1
	slugs	4
	earthworms	73
Water	algae	33
	crabs	144
	shrimps	2800
	clams	70 000
	fish	829 300

(1) Biomagnification is a concentration factor, representing the concentration within the organism compared to that in the immediate soil or water environment in which the analysed organisms lived. Figures extracted, with permission, from Kenaga (1972), Table 2, p. 201, which in turn contained data collated by Edwards (1970).

The general level of organochlorine pesticide contamination of soils has been discussed by Matsumura (1972, 1975). In the first of these references, tables are included which give a general indication of duration of persistence in soils, mean concentrations in water systems and levels stored in human body fat of citizens of several countries. Another useful source of information is the book by Rudd (1964).

In **plants** that are rapidly transpiring, substantial amounts of even very sparingly soluble substances may enter the roots over a period of time, and may then move up to accumulate in the edible parts of food crops. γ-BHC is probably taken up more readily than most other members of the group, especially at high rates of application and tainting of root crops has often followed the use of technical BHC. It can also cause abnormal nuclear division in meristematic tissue. On the other hand, Lichtenstein and Schulz (1960) could not detect any translocation of DDT from roots to the aerial parts of pea plants.

Residues in **water** can be locally high, due to factory effluent, animal dips and run-off from agricultural land (Holden, 1966). Drainage water from treated fields is almost free from organochlorine contamination because of the high adsorption on to soil particles. Of possibly greater long-term concern (because no national government can control world-wide contamination) is the flowing of contaminated rivers into the oceans of the world and general wash down from the air in rain water. Intermediate between these is the urgent shorter term problem of inflow of contaminated rivers into lakes and into inland or nearly-landlocked water (e.g. the Great Lakes, the Rift Valley and the Baltic, Black, Caspian and Mediterranean Seas). Some figures for the concentrations of organochlorine compounds in various stretches of water during the period 1964–1970 are given by Matsumura (1972). In the case of DDT and its persistent metabolites 112 parts were found in 10^{12} parts of water in the Mississippi delta, 10 parts per 10^{12} in US river basins and 1.6 parts per 10^{12} in British rivers. Figures for contamination by other organochlorine compounds were in general lower than those for DDT, an exception being a much higher figure (18.7 parts per 10^{12}) for γ-BHC in British rivers.

Work on the possible chronic effects of DDT on **animals** has been reviewed by Matsumura (1975). In almost all cases where side effects were reported the dosage applied was unrealistically high compared to the exposure of the normal human population. President Kennedy's Scientific Advisory Committee in 1963 stated that the level of DDT in American body fat was about 12 ppm in 1951 and the daily ingestion at that time was about 0.18 mg per person per day. The latter figure corresponds to some 0.003 mg/kg human body weight (the amount ingested has greatly diminished since then). Experiments conducted by Tarjan and Kemeny on several generations of mice showed an increased incidence of leukaemia and pulmonary carcinoma but the animals concerned had received at least 0.4 mg/kg per day. More convincingly, Shabad *et al.* (1972) obtained evidence of hyperplasia in mice with a daily dose of 0.020 mg/kg, which is only seven times the intake of humans in 1951. In short, although constant vigilance is essential in such matters, the present writer believes it is equally necessary to be on one's guard against a common scientific *non-sequitur* – namely, to extrapolate from the observed effects of high doses on experimental animals to the danger inherent in man's exposure to much lower doses.

There is, however, an effect of medium or medium-low doses of certain organochlorine compounds (e.g. DDE) to which reference has been made in earlier sections. Exposure of higher animals to appropriate doses of organochlorine insecticides, polychlorinated biphenyls, phenobarbital and many other lipid soluble substances can cause the **induction** of liver enzymes which metabolise potential poisons. Exactly how this *de novo* synthesis is brought about is not clear. Whether this effect, and the enlargement of the liver that often accompanies it, is beneficial or harmful probably depends on the circumstances – it can result in quicker metabolism of pesticide residues, for example, but also may interfere with medication. Induction may, however, be associated with cytological abnormalities, for De La Pena de Torres (1978) observed that DDT not only caused a proliferation of the smooth endoplasmic reticulum (the site of enzyme induction) but the nuclei of liver cells of mice also contained large intranuclear cytoplasmic invaginations and an increase in the number of nucleoli.

The level of contamination of **meat, milk and eggs** with organochlorine compounds is of importance to man. In particular, the reliance of the young upon cow's milk places this food in a very special category. Zweig *et al.* (1961) observed that when DDT was fed to dairy cows, the concentration of DDT in the milk came within a fortnight into steady state with the amount in the diet. When the cows were taken off the diet, the concentration in the milk fell rapidly. When methoxychlor is used instead of DDT, little or none of it or its metabolites appears in milk. Several surveys have been made of organochlorine levels in human milk, and most of them indicate that the levels are higher than in cow's milk. Moreover, placental transfer seems to favour higher amounts (0.07 ppm) in the milk than in the maternal blood. Matsumura (1975) has pointed out that if, as is probable, infants are more susceptible to organochlorine toxicity than are adults, they are probably exposed to an insult some 50 times greater than that suffered by a normal adult (on the assumption that the normal adult now only takes in about 0.005 mg/kg per day of DDT).

Another aspect of the consequence of high stability and lipid solubility is often called the **food chain effect**. When organisms are dependent on one another in a biological sequence, many members of a species of one trophic level may be eaten by one organism of the next higher trophic level. If, then, a small quantity of a stable lipid-soluble pesticide should be present in each of the lowlier organisms, the possibility exists that lipid concentration may rise with the trophic level of the organisms (figure 6.12).

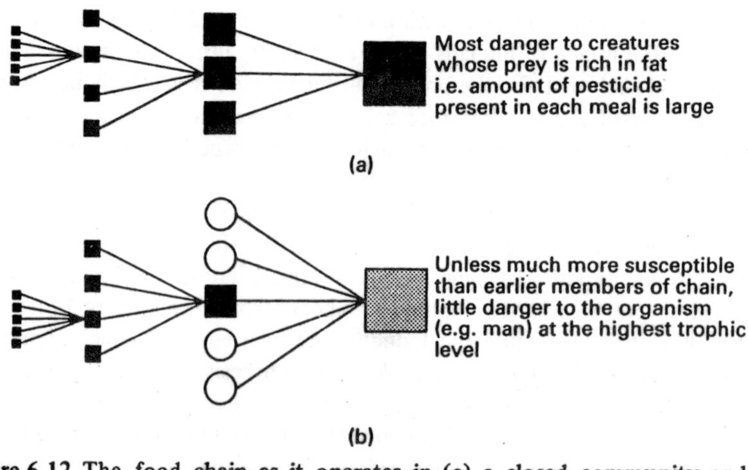

figure 6.12 The food chain as it operates in (a) a closed community and (b) when a less specific feeder participates

The most notorious example of a chain operating in this way is one terminating in *fish-eating birds*. An illustration is provided by Robinson *et al.* (1967) who sampled marine organisms off the Northumberland coast for their content of dieldrin and of DDT. Samples of the seaweeds *Fucus serratus* and *Laminaria digitata* (trophic level 1) contained only 0.001 ppm of dieldrin, whereas *Echinus esculentis* and *Mytilus edulis* (trophic level 2) contained 0.027 and 0.023 ppm respectively. Herrings (*Clupea harengus*) in trophic level 3 contained 0.057 ppm and some fish-eating birds (trophic level 4) contained 0.5 ppm. Other examples from data collated by Edwards (1970) are provided in table 6.2. The harm to prize avian species arising from this effect has, however, sometimes been overstated, for the operation of a food chain effect requires either that a closed system exists such that the fish-eating birds have no alternative source of food or that the whole local environment is equally contaminated. The first is, fortunately, an uncommon special case for avian species (but not for the fish) and the second is now being controlled by a combination of legislation and of public awareness. In addition, man is an omnivorous species which eats foods originating from different parts of the world. He is therefore less at risk than a species which has an obligatory requirement for one or a few kinds of foods obtained from a highly localised source.

The analysis of tissues of **birds and of birds' eggs** carried out by Walker *et al.* (1967) showed that residues of organochlorine insecticides were particularly high in some species of birds that feed on fish or on other birds. In some such species there was circumstantial evidence that organochlorine compounds were a cause of a serious decline in numbers. The plight of many birds of prey is serious in certain localities of Britain, probably because they preferentially attack birds which, having acquired far higher than average levels of the poison, are so weakened that they then lag behind the main flock (Mellanby, 1967). On the other hand, there is little evidence for a significant decline in numbers of herons and shags, despite occasionally quite high levels of residues. The situation as it affects British birds has been described by Walker (1975).

Birds are frequently more susceptible to organochlorine compounds than are mammals. In addition, some species select one type of seed for their main diet, and should that seed be the farmer's newly sown seed-dressed cereal, the birds may well pay dearly for their predations. This problem has declined since the restriction of use of organochlorine (and especially of chlorinated cyclodiene) compounds as seed dressings.

Eggs laid by birds which have eaten sub-lethal amounts of organochlorine compounds frequently have thin egg shells (Ratcliffe, 1967). Experiments have shown that at realistic intake levels of organochlorine compounds the thickness may decrease by ten per cent, but whether this is, in nature, a cause of declining populations is debatable. The thin egg shells may be a consequence of the inhibition of carbonic anhydrase, an enzyme involved in egg shell formation. More recently, however, Kolaja and Hinton (1978), working with the mallard duck (*Anas platyrhyncos*) observed that chronic ingestion of DDT caused oedema of type II epithelial cells of the eggshell gland, with vacuolation of the endoplasmic reticulum. Since these cells are responsible for calcium transport, it is possible that decreased availability of calcium may be responsible for egg shell thinning.

Organochlorine residues in birds or their eggs also seem sometimes to cause decreased egg production, lower fertility and lower hatchability of eggs. A delay in ovulation caused by DDT may be of practical importance, for Jefferies (1967) pointed out that large sub-lethal doses in adult birds may so delay egg-laying that double-brooding birds produce a late brood just as the food supply is failing. Less well grounded is the fear of Carson (1963) that the fertility of animals and man might have been affected. A trite yet fundamental observation is that the large-scale use of pesticides was precipitated by the explosion in the human population and unbridled fertility is currently a major threat to the social, political and financial stability of the human race, especially in poorer countries.

6.9 DEVELOPMENT OF INSECT RESISTANCE TO ORGANOCHLORINE INSECTICIDES

Resistant strains of insects tend to develop whenever a species survives in a locality despite the apparently successful use of pesticides there. The more rapid the life cycle of the organism and the more persistent the poison, the greater is the risk that the development of resistance will be rapid. By 1960, instances of insect resistance were numerous enough to cause concern, but the misuse of organochlorine compounds during the next decade exacerbated the problem.

In the laboratory, selective breeding, under 'pressure' of an organochlorine compound, is possible if a captive population is exposed to a dose which kills most but not all of the population. By mating the survivors it is possible in some cases to achieve a 100–1000 fold increase in tolerance. Some of the experimental techniques for the detection and measurement of resistance have been discussed by Busvine (1968).

Under field conditions, the effect of a persistent pesticide is to lead to 'unnatural selection'. It kills off the more susceptible organisms, leaving a few naturally resistant individuals to recolonise the area. Breeding rapidly in the absence of competitors, these resistant organisms produce offspring which have a greater mean tolerance to the poison than had the earlier population.

In theory, resistance could be avoided by destroying every individual of the species in the sprayed area. In practice, this is normally quite impracticable. To kill off as many organisms as possible is usually *not* the next best thing; rather, it is to apply that concentration of poison which gives the *minimal control of the pest consistent with economic considerations*. This approach does not prevent the development of resistance but it reduces its rate of increase. When the treatment leading to selection ends, a partial reversion towards the susceptible state occurs over a number of generations. For example, Betzios (1977) bred strains of the mosquito *Culex pipiens* that were highly resistant to dieldrin but resistance regressed towards normal levels when selection pressure was removed.

Resistance is a phenomenon which can occur for several reasons, the chief factors being behavioural, morphological, biochemical and physiological. Examples of some of these have been given elsewhere (sections 4.11 and 6.8). In the case of the DDT family, there is evidence that resistant strains of some, but not all, insect species contain a much more effective complement of **dehydrochlorinase** than the corresponding susceptible strains (Lipke and Kearns, 1960). However, dehydrochlorination is not a major detoxication mechanism for cyclodienes, yet a thousand-fold increase in resistance of insect strains to dieldrin has been recorded. Attempts using house flies and mosquitoes to detect a difference in the pattern of uptake, metabolism and excretion of dieldrin between susceptible and resistant strains were unsuccessful (Winteringham, 1962). So were similar attempts by Bridges and Cox (1959) to detect a difference in the rate of metabolism of γ-BHC in resistant and susceptible strains of house flies. It has been postulated that, in such cases as these, there is some physiological or structural modification of the site of action in the nervous system.

When insects are exposed to any one organochlorine compound, there is a tendency for resistance to develop not only to that one compound but also to certain others to which they have never been exposed. It is possible to divide organochlorine compounds into two groups, according to the nature of **cross-resistance** that occurs. The first is the DDT family; if resistance develops to DDT, then it is very likely that there will be simultaneous development of resistance to all the other members of this family. BHC and the cyclodienes form the second group, for exposure to BHC often results in a greater or lesser development of resistance to cyclodienes, and vice versa. Conversely, selection pressure from BHC or from dieldrin does not usually result in immediate resistance to DDT. An example is the work of Betzios (1977) mentioned above, for the dieldrin-resistant strain of *Culex pipiens* obtained by dieldrin selection pressure was also resistant to γ-BHC but not to DDD. However, insects already resistant to one of these groups (e.g. to BHC/dieldrin) may develop resistance to DDT much more rapidly *on exposure* to DDT than do insects that are not dieldrin-resistant. The phenomenon of 'conditioning' can even cross group frontiers; thus dieldrin-resistant strains can rapidly develop resistance to carbaryl (section 5.8).

REFERENCES

Abu-el-Haj, S., Fahmy, M. A. H. and Fukuto, T. R. (1979). *J. agric. Fd Chem.*, 27, 258

Agosin, M., Michaeli, D., Miskus, R., Nagasawa, S. and Hoskins, W. M. (1961). *J. econ. Ent.*, 54, 340

Betzios, B. C. (1977). *Annls Inst. phytopath. Benaki*, 11, 284

Biggar, J. W., Dutt, G. R. and Riggs, R. L. (1967). *Bull. envir. Contam. Toxic.*, 2, 90

Bridges, R. G. and Cox, J. T. (1959). *Nature, Lond.*, **184**, 1740
Brooks, G. T. (1972). *Environmental Toxicology of Pesticides* (eds. F. Matsumura, G. M. Bousch and T. Misato), Academic Press, New York, p. 61
Brooks, G. T. (1973). *Drug Design, Vol. 4* (ed. E. J. Ariëns), Academic Press, London
Brooks, G. T. (1974). *Chlorinated Insecticides, Vol. 1: Technology and Applications; Vol. 2: Biological and Environmental Aspects*, CRC Press Inc., Ohio
Brown, A. W. A. (1958). *Adv. Pest Control Res.*, **2**, 351
Busvine, J. R. (1968). *Wld Rev. Pest Control*, **7**, 27
Butler, K. D. and Crowder, L. A. (1977). *Pestic. Biochem. Physiol.*, **7**, 474
Carson, R. (1963). *Silent Spring*, Hamish Hamilton, London
De la Pena de Torres (1978). *Archos. Farmac. Toxic.*, **4**, 339
Edwards, C. A. (1970). *Critical Reviews in Environmental Control*, Vol. 1, CRC Press Inc., Ohio
Esaac, E. G. and Matsumura, F. (1979). *Pestic. Biochem. Physiol.*, **10**, 67
Finnen, M. and Hassall, K. A. (1980). *Biochem. Pharmac.*, **29**, 3133; 3139
Ghiasuddin, S. M. and Matsumura, F. (1979). *Pestic. Biochem. Physiol.*, **10**, 151
Green, M. B., Hartley, G. S. and West, T. F. (1977). *Chemicals for Crop Protection and Pest Control*, Pergamon Press, Oxford
Hassall, K. A. (1975). *Pestic. Biochem. Physiol.*, **5**, 126
Hassall, K. A. and Addala, S. A. (1979). *Biochem. Pharmac.*, **28**, 3199
Holden, A. V. (1966). *J. appl. Ecol.*, **3** (Suppl.), 45
Jefferies, D. J. (1967), *Ibis*, **109**, 266
Kapoor, I. P., Metcalf, R. L., Nystrom, R. F. and Sangha, G. K. (1970). *J. agric. Fd Chem.*, **18**, 1145
Kenaga, E. E. (1972). *Environmental Toxicology of Pesticides* (eds. F. Matsumura, G. M. Bousch and T. Misato), Academic Press, New York
Koch, R. B. (1969). *J. Neurochem.*, **16**, 269
Kolaja, G. J. and Hinton, D. E. (1978). *Envir. Pollution*, **17**, 237
Kurihara, N., Tanaka, K. and Nakajima, M. (1979). *Pestic. Biochem. Physiol.*, **10**, 137
Kushwaha, K. S., Gupta, H. C. L. and Kavadia, V. S. (1978). *Ann. Arid Zone*, **17**, 200
Lichtenstein, E. P. and Schulz, K. R. (1960). *J. agric. Fd Chem.*, **8**, 452
Lipke, H. and Kearns, C. W. (1960). *Adv. Pest Control Res.*, **3**, 253
MAFF (1980). *Ministry of Agriculture, Fisheries and Food, UK, List of Approved Products for Farmers and Growers*, H.M.S.O., London
Martin, H. and Worthing, C. R. (1976). *Insecticide and Fungicide Handbook*, 5th edn, Blackwell, Oxford
Matsumura, F. (1972). *Environmental Toxicology of Pesticides* (eds. F. Matsumura, G. M. Bousch and T. Misato), Academic Press, New York, p. 33
Matsumura, F. (1975). *Toxicology of Insecticides*, Plenum Press, New York
Matsumura, F. and Narahashi, T. (1971). *Biochem. Pharmac.*, **20**, 825
Mellanby, K. (1967). *Pesticides and Pollution*, Collins, London
Metcalf, R. L. (1955). *Organic Insecticides*, Interscience, New York
Metcalf, R. L. and Fukuto, T. R. (1968), *Bull. Wld Hlth Org.*, **38**, 633
Mullins, L. J. (1954). *Chem. Rev.*, **54**, 289
Mullins, L. J. (1955). *Science, N.Y.*, **122**, 118
Mullins, L. J. and Brinley, F. J., Jr. (1975). *J. gen. Physiol.*, **65**, 135
Ratcliffe, D. A. (1967). *Nature, Lond.*, **215**, 208
Riemschneider, R. (1963). *Wld Rev. Pest Control*, **2**, 29
Robinson, J. (1969). *Can. med. Assoc. J.*, **100**, 180

Robinson, J., Richardson, A., Crabtree, A. N., Coulson, J. C. and Potts, G. R. (1967). *Nature, Lond.*, **214**, 1307

Rudd, R. L. (1964). *Pesticides and the Living Landscape*, Faber and Faber, London

Shabad, L. M., Kolesnichenko, T. S. and Nikonova, T. V. (1972). *Int. J. Cancer*, **9**, 365

Soloway, S. B. (1965). *Adv. Pest Control Res.*, **6**, 85

Tanaka, K., Kurihara, N. and Nakajima, M. (1979). *Pestic. Biochem. Physiol.*, **10**, 96

Treon, J. F. and Cleveland, F. P. (1955). *J. agric. Fd. Chem.*, **3**, 402

Walker, C. H. (1969). *Life Sciences*, **8**, 1111

Walker, C. H. (1975). *Environmental Pollution by Chemicals*, 2nd edn, Hutchinson, London

Walker, C. H., Hamilton, G. A. and Harrison, R. B. (1967). *J. Sci. Fd Agric.*, **18**, 123

Wang, C. M., Narahashi, T. and Yamada, M. (1971). *Pestic. Biochem. Physiol.*, **1**, 84

Winteringham, F. P. W. (1962). *Jl R. Soc. Arts*, **110**, 719

Yamaguchi, I., Matsumura, F. and Kadous, A. A. (1979). *Pestic. Biochem. Physiol.*, **11**, 285

Zweig, G., Smith, L. M., Peoples, S. A. and Cox, R. (1961). *J. agric. Fd Chem.*, **9**, 481

7 Other Insecticides and Related Compounds

Organophosphorus, organochlorine and carbamate insecticides currently account for most of the tonnage of insecticides used in agriculture, in locust control, and in food storage. However, other types of insecticides are used on a smaller scale and for specific purposes and some of these are considered in the present chapter.

7.1 NATURAL AND SYNTHETIC PYRETHROIDS

At least six closely related compounds occur (1–2 per cent w/w) in the dried inflorescences of *Chrysanthemum cinerariaefolium*, a plant which is grown commercially as a source of these insecticides in Kenya, Tanzania and, to a lesser extent, in several other countries (figure 7.1). Characterised by a remarkable 'knock-down' effect on flying insects, the natural pyrethroids have long been used as active ingredients in domestic sprays employed against flies and mosquitoes. This spectacular knock-down effect, coupled with their extremely low toxicity to warm-blooded animals, renders them ideal for this purpose. Knockdown, however, does not necessarily imply eventual death and many a housewife's jubilation would have been premature but for the customary presence of a second poison, usually an organophosphorus or organochlorine compound with a slower but more reliably lethal action. Natural pyrethroids have been little used for crop protection since they are rapidly destroyed by light. More recently, however, some synthetic pyrethroids have proved to be sufficiently light stable for use in field crop protection, and it is probable that some of these compounds will find increasing use in the near future.

Historical aspects of **natural pyrethrum** production have been described in an International Trade Centre publication (ITC, 1976). *C. cinerariaefolium* probably originated in Persia (now Iran) but by the mid-nineteenth century was grown and processed in Yugoslavia. By 1914, some 1500 tons of dried flowers were sold to the United States. Between the two World Wars, Japan became the world's main producer, some 1400 tons of dried flowers being produced each year before trade was halted by the 1939 war. Japanese production is now being phased out, for the yield of flowers per acre in the Kenyan highlands is such as to make Japanese production uneconomic. Tanzania, Zaire and Rwanda followed

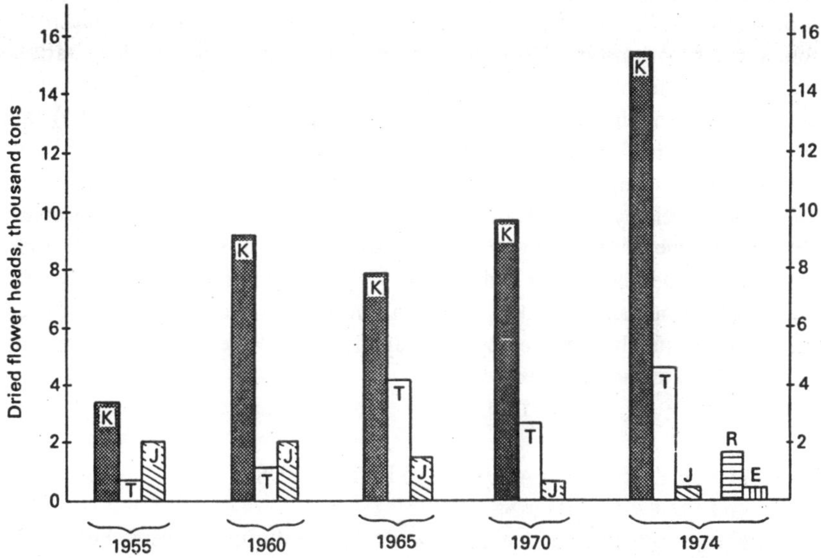

figure 7.1 Production of dried pyrethrum flowers, 1955–1974 (K, Kenya; T, Tanzania; J, Japan; R, Rwanda; E, Ecuador). (Drawn using figures from Table 1, International Trade Centre, UNCTAD/GATT, Geneva, 1976; 'Pyrethrum, a natural insecticide with growth potential'; figures for Rwanda and Ecuador only plotted for 1974)

Kenya's example and started to grow pyrethrum, although production in Zaire has been disappointingly low since Independence. Tanzanian production, however, is now about 25 per cent of that of Kenya (figure 7.1). Total world production is now probably in the region of 25 000 tons of dried flower heads.

Pyrethrum plants when grown on volcanic ash at altitudes of 1500–3500 m in tropical zones give larger overall yields of pyrethrin than do plants at low altitudes. The use of fertilisers, irrigation, herbicides – and insecticides – is often essential. Harvesting is usually done by hand, the flowers being frequently, but not always, dried in the sun. Nowadays the dried flowers (about 350 kg/ha) are processed in the producing countries. In terms of tonnage, extracts represent about 12 per cent of total pyrethrin exports, but a ton of extract is worth the same as about 35 tons of dried flowers. The solvent used for extraction is often petroleum ether or acetone, the extract being concentrated by vacuum distillation. Formulation is often done outside the pyrethrum-growing countries and involves forming wettable powders and dusts from the dried flowers or aerosols from the extract. In the latter case, spray supplements and a synergist are also added (section 7.4).

The first synthetic pyrethroid, **allethrin**, was introduced in 1949, and the brilliant work of Elliott and his colleagues at Rothamsted Experimental Station culminated in the development of light-stable pyrethroids such as **permethrin** (Elliott *et al.*, 1973). Modifications to the molecule of the natural pyrethroids included the substitution of 3-furylmethyl derivatives or of m-benzyl derivatives for the cyclopentenolone moiety. Examples of synthetic pyrethroids containing these two types of substitutions are respectively **bioresmethrin** (figure 7.6b) and **permethrin** (figure 7.6c). In 1974, Japanese and other workers replaced the chrysanthemic acid moiety without loss of activity, and thus introduced a new family of synthetic pyrethroids of which **fenvalerate** (figure 7.6d) is an example.

Excellent reviews of the progress in the development of synthetic pyrethroids have been provided by Elliott and Janes (1978) and by Elliott *et al.* (1978). The former deals principally with the structural diversity in the group and the latter considers their stability, toxicology and applications. References to about 150 papers are given in each of these reviews.

7.2 CHEMISTRY OF THE NATURAL PYRETHROIDS

There are four principal active ingredients in pyrethrum flowers, known as **pyrethrins** I and II and **cinerins** I and II (figure 7.2). All four are esters comprising an acid containing a three-carbon ring joined to an alcohol containing a five-carbon ring. In addition, small quantities of jasmolins I and II are present; these differ from the pyrethrins only in that one double bond in the side chain of the alcohol moiety of pyrethrins is saturated.

The acid present in compounds designated by I is called chrysanthemic acid

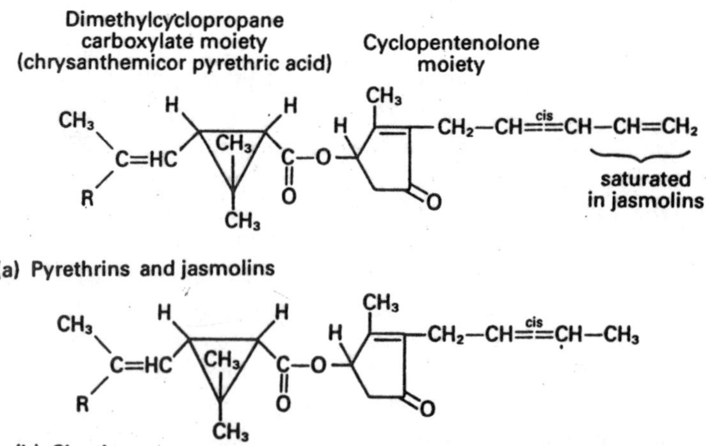

(b) Cinerins

figure 7.2 Natural pyrethroids ((a) Pyrethrin I, R = CH_3; pyrethrin II, R = $COOCH_3$. Jasmolins I and II resemble the corresponding pyrethrins but the side chain is saturated as indicated. (b) Cinerin I, R = CH_3; cinerin II, R = $COOCH_3$)

while that in those designated by II is called pyrethric acid. The difference between them is explained in the legend to figure 7.2. It will be observed that if chrysanthemic acid is called chrysanthemum monocarboxylic acid, then pyrethric acid would be chrysanthemum dicarboxylic acid monomethyl ester (a relationship of some interest in relation to metabolism, q.v.). These two acids can be joined to alcohols possessing an unsaturated side chain containing five carbon atoms (pyrethrolone and jasmololone) or four carbon atoms (cinerolone). These differences, and the relative proportions of each normally present, are summarised in table 7.1.

Table 7.1 Proportions of the six active ingredients in a typical pyrethrum product (names of the esters appear in the body of the table)

Alcohol present	Acid present	
	Chrysanthemic	Pyrethric
Pyrethrolone	pyrethrin I (35%)	pyrethrin II (32%)
Cinerolone	cinerin I (10%)	cinerin II (14%)
Jasmololone	jasmolin I (5%)	jasmolin II (4%)

The acid component of natural pyrethroids is capable of existing in *cis* and *trans* forms, and each of these could exist as optical enantiomorphs (+ or −). The naturally-occurring acid is, in fact, the trans (+) form and its esters are more toxic than those formed from the three other stereoisomers produced synthetically.

Similarly, each of the alcohols can be made as a *cis* and as a *trans* form, and each of these as two optical isomers. Of these, only the cis (+) isomer occurs naturally. The geometrical isomerism occurs at the double bond marked in the formulae (figure 7.2) while the asymmetric carbon atom is the one carrying the hydroxyl group when the alcohol is uncombined.

The pyrethroid esters are oily liquids soluble in alcohol, acetone and petroleum but insoluble in water. The esters are unstable, even in the dust from the dried flower heads (there is about 20 per cent loss of activity per year on storage) but can be protected by antioxidants so long as light and water are excluded. In water they are hydrolysed, the reaction being both acid and base catalysed. In consequence they are unable to kill insects as stomach poisons and are also very ephemeral contact poisons.

7.3 USES OF NATURAL AND SYNTHETIC PYRETHROIDS

As was mentioned earlier, natural pyrethroids are of limited use for crop protection, although they have been used to give protection near harvest time and during transit and storage (their low toxicity to higher animals and their limited persistence make them ideal for such use). However, 20 per cent of all insecticides sold in the USA have non-agricultural uses, including use in the home and garden, as well as for animal health, for public and commercial buildings, for medical purposes and for forestry. Pyrethroids are ideal for some of these uses, the principal objection being that they are several times more costly than many other equally effective compounds even after the effect of synergists (see below) has been taken into consideration. However, a dozen or more light-stable synthetic pyrethroids are currently being used, or tested for use, in agriculture and many have proved to be particularly effective against larvae of the Lepidoptera. Several of the latter are important pests of such major food or cash crops as maize, rice, sugar cane, sugar beet and cotton. Although the toxicity of photostable pyrethroids to different types of insects varies greatly almost all of them are more insecticidal than are natural pyrethroids (Elliott *et al.*, 1978).

Although safe to higher animals, it should be recalled that pyrethroids of all kinds are toxic to fish. Moreover, they seldom discriminate between useful insects and pests (permethrin and fenvalerate are much less toxic to bees than is bioresmethrin). Spraying should therefore be done circumspectly, and at a time of day that minimises danger to pollinators, and special care should be taken not to pollute water courses.

7.4 METABOLISM AND MODE OF ACTION OF PYRETHROIDS

The ester linkage of natural pyrethroids is unstable when exposed to enzymes in insect gut. It is therefore of interest that there is considerable evidence that once they have gained access to the tissues, pyrethroids are often destroyed by oxidation rather than by hydrolysis.

It has been known since 1940 that certain methylene dioxyphenyl compounds powerfully synergise pyrethroids, i.e. increase the killing potential, of a given amount of pyrethroids. For example, one part of natural pyrethroid with two parts of piperonyl butoxide produces a mixture which is as toxic as seven parts of pyrethroid used alone (Chadwick, 1963). For many years, in fact, most formulations of pyrethroids have included a synergist such as piperonyl butoxide or

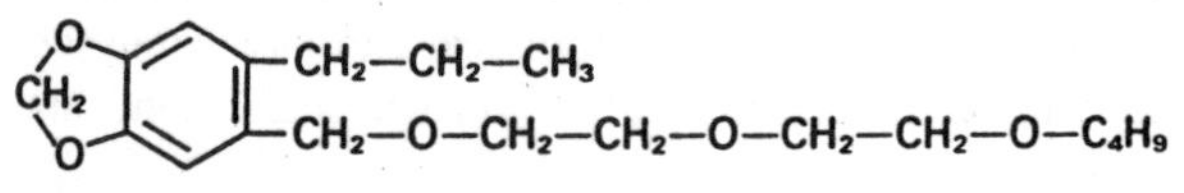

(a) Piperonyl butoxide

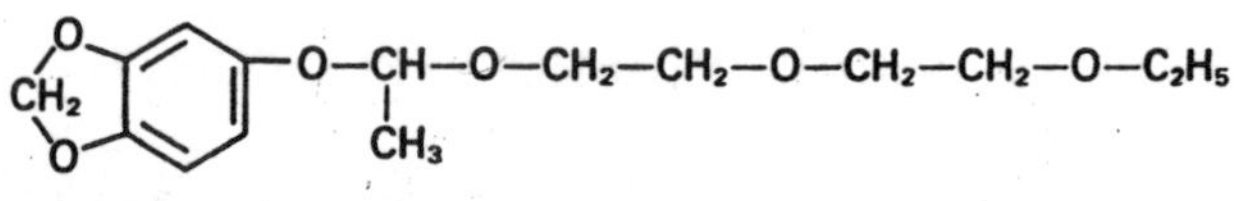

(b) Sesamex

figure 7.3 Structures of two methylenedioxyphenyl synergists

sesamex (figure 7.3). This synergism provides very strong circumstantial evidence that microsomal mono-oxygenases participate in the degradation of pyrethroids, for methylene dioxyphenyl compounds are probably specific inhibitors of these enzymes. Thus, if the enzymes which normally destroy pyrethroids are themselves incapacitated, the pyrethroids persist longer and, in consequence, less of each is needed to have some standard insecticidal effect. A similar conclusion about the importance of mono-oxygenases is reached from the NADPH-dependence of pyrethroid metabolism by vertebrate liver microsomes and by preparations from the abdomen of house flies (e.g. Casida and Yamamoto, 1967).

According to the Berkeley workers mentioned above, pyrethrin oxidation occurs without initial hydrolysis, i.e. the ester linkage remains intact. Instead, in the major oxidation product of pyrethrin I the asterisked methyl group (figure 7.4) in the *iso*butenyl side chain of the chrysanthemic acid is oxidised to a hydroxymethyl group in a NADPH-dependent reaction. The primary alcohol group formed in this way from pyrethrin I is readily converted to a carboxylic acid group. Pyrethrin II is metabolised by a similar NADPH-dependent system, the major metabolites being identical to those formed from pyrethrin I. In this case, O-demethylation, another typical mono-oxygenase reaction appears to occur (figure 7.4). Oxidative changes can also take place in the five carbon atom side-chain of the alcohol moiety, and altogether some ten oxidative metabolites of the pyrethrins have been recognised (Yamamoto *et al.*, 1971).

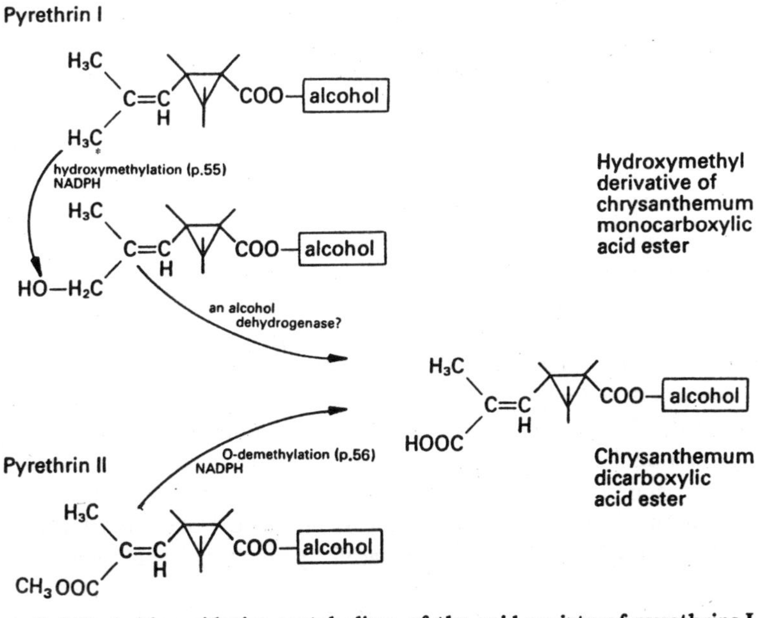

figure 7.4 Probable oxidative metabolism of the acid moiety of pyrethrins I and II

The rapid knock-down effect on flying insects possessed by all natural and some synthetic pyrethroids strongly suggest that they have a neurotoxic action. In view of this rapid action, Chadwick (1963) concluded that natural pyrethroids must elicit a response when they have penetrated no more than a micron (1 μm) below the surface of the cuticle. This statement implies that they need only to penetrate the thickness of the epicuticle, presumably triggering some more general response once they reach the peripheral nerve cells (figure 2.3). With passage of time, entry is progressively more complete, and insect ganglia are probably ultimate targets, although uncertainty still exists as to whether pyrethroids are best regarded as poisons of the peripheral or the central nervous system of insects.

When intact insects are treated with pyrethroids the behavioural and physiological manifestations of poisoning superficially resemble those caused by DDT (section 6.4). In particular, an initial period of sensory hyperexcitation leads successively to loss of coordination, ataxia, prostration, convulsions and death. Considerable quantitative differences have, however, been noted, the gradual onset of excitation characteristic of DDT poisoning contrasting with the rapid development of paralysis which usually occurs to insects poisoned with pyrethroids (Adams and Miller, 1980).

When pyrethroids are applied to isolated nerves or, *in vivo*, to longitudinal flight muscles of house flies, exaggerated repetitive discharges take place. These superficially resemble those caused by DDT (compare figure 6.3 with figure 7.5a). There are, however, several differences, some of which are described by Adams and Miller (1980). It is also of interest that the toxic effects of natural pyrethroids can, for a while after application, be reversed by washing excised preparations free from pyrethroid; this implies that, in the case of pyrethroids, the toxic action does not involve a strong chemical attachment of the poison to a neural receptor.

The work of Adams and Miller (1979) well illustrates the effect of pyrethroids on nerve–muscle preparations. Using **tetramethrin** as toxicant to flight motor units of the house fly, they confirmed earlier work which showed that trains of impulses occurred in poisoned preparations and concluded that motor unit discharges originate in peripheral motor nerve endings. Typical traces before and 10 min after poisoning are shown in figure 7.5a. Upon lowering the temperature of the nerve–muscle preparation, the muscle potentials evoked by giving shocks to poisoned nerve endings are shown in figure 7.5b. It will be seen that the duration of the discharge increases with decreasing temperature but the number of spikes in a discharge remains almost unaltered. Below about 19°C, similarly poisoned preparations show only a single firing upon receiving a single stimulus. Moreover, if the frequency of stimulation is increased from about 0.5 Hz to about 2 Hz, multiple discharges are replaced by single spikes at temperatures above 20°C.

Trains of discharges of the kind just described may well be the cause of the convulsions that poisoned insects suffer. It seems probable from the work on

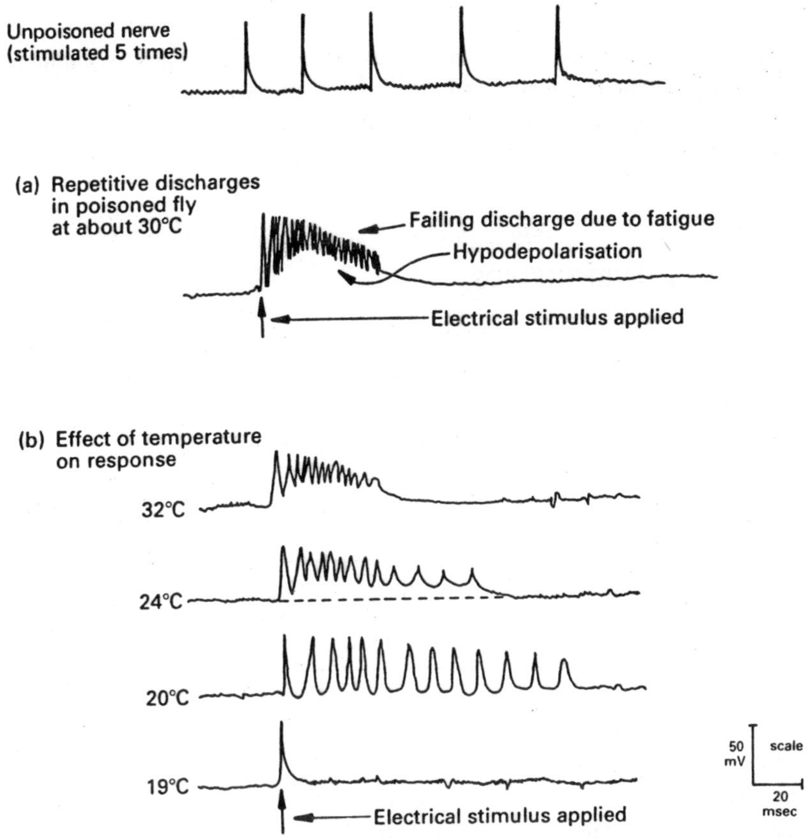

figure 7.5 Effect of a pyrethroid on flight motor activity of house flies. (a) The top recording shows flight motor activity before poisoning, the lower one is the response 10 min after applying the LD_{50} dose of tetramethrin. Stimulus applied by shocking. (b) Four recordings showing the muscle potentials evoked in a tetramethrin-poisoned fly at different temperatures. Above about 20°C, the number of repetitive discharges in a train is roughly constant, but the frequency increases and duration decreases with increasing temperature. Below about 19.5°C a poisoned nerve–muscle system gives only a single muscular discharge on shocking. (Idealised diagrams based on the work of Adams and Miller, 1979)

allethrins by Yamamoto (1970), Narahashi (1971) and Wang *et al.* (1972) that sodium and potassium ion conductances are altered when the nerve is poisoned. In particular, the potassium efflux stage (of recovery after passage of an impulse) lasts much longer in pyrethroid-poisoned than in normal axons.

There are, however, some difficulties in relating these physiological observatins to the lethal effect of pyrethroids on intact insects. It is not clear whether death is a consequence of a more intense manifestation of the same symptoms (i.e. whether intense convulsions lead to lethal paralysis). The reason for the difficulty is that the ***physiological events*** just described have a positive temperature coefficient whereas pyrethroids, like DDT, show a negative correlation between temperature and *insecticidal effectiveness.* This does not appear to be consistent with the notion that increased frequency of repetitive discharges is the direct cause of death.

7.5 SYNTHETIC PYRETHROIDS

The elucidation of the structure of the natural pyrethroids made possible the synthesis of related compounds which, while possessing similar or higher insecticidal activity than the natural compounds, are also more stable to light and to air. Some of the numerous compounds that have been tested or marketed are listed in table 7.2, together with some of their properties.

Table 7.2 Comparison of some properties of natural and synthetic pyrethroids

Compound	Knock-down[(1)] propensity	Relative[(2)] toxicity to house flies	Oral toxic[(3)] dose to rats mg/kg body wt	Relative[(1)] cost per cent
Pyrethrin	good	2	moderate	100
Bioallethrin	good	6	moderate	100
Allethrin	fair	3	moderate	60
Resmethrin	poor	42	high	155
Bioresmethrin	fair	100	very high	215
Tetramethrin	good	2	very high	65
Permethrin	poor	60	moderate	>100
Decamethrin	poor	1900	very low	>100
Fenvalerate	fair	38	low	>100

(1) Mostly from ITC (1976).
(2) After Elliott *et al.* (1978). These authors warn that their figures are for general comparison only.
(3) Pyrethrins are variously reported to have LD_{50}s of the order of 570–1100 mg/kg. This range of LD_{50}s is referred to as representing moderate toxicity. 'Low' implies that the LD_{50} is *below* this range and 'high' that it is *above* it. A substance for which the toxic dose is 'low', is, naturally, *highly* toxic.

The **allethrins** (figure 7.6a) are of the same order of toxicity to the house fly as are natural pyrethroids. A disadvantage, however, is that they have a weaker knock-down effect than natural pyrethroids. Their performance is improved when synergists are added, but in common with many other synthetic pyrethroids, the improvement is less marked than it is for the natural substances.

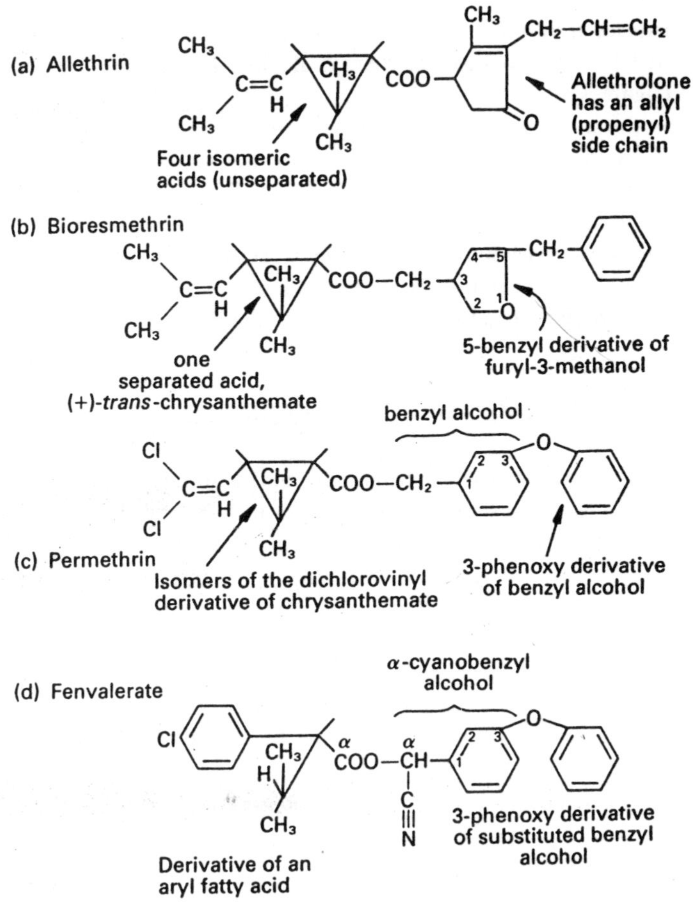

figure 7.6 Synthetic pyrethroids

An allethrin mixture is a brown liquid which is insoluble in water, soluble in oil, somewhat volatile and stable to heat.

Workers at Rothamsted Experimental Station synthesised a number of alcohols and attached them to (+) *trans* chrysanthemic acid. One of these synthetic pyrethroids, **bioresmethrin**, introduced in 1966, contains 5-benzylfuryl-3-methanol and is about fifty times as toxic to house flies as are the natural pyrethrins (figure 7.6b). In 1972, the same workers observed that, if appropriate alcohols were attached to chlorinated analogues of chrysanthemic acid, the resultant substance is much more stable to light and oxygen than are the natural pyrethroids. By replacing the two methyl groups of the *iso*butenyl moiety of the acid with chlorine atoms, and esterifying the product with a relatively simple alcohol, a photostable substance, **permethrin** (figure 7.6c) is formed.

In 1974, Japanese workers successfully prepared pyrethroid-like esters where, for the first time, the three-membered carbon ring of chrysanthemic acid was absent. Instead, they used a number of α-isopropyl arylacetic acid derivatives, often joined by α-cyano-3-phenoxybenzyl alcohol. **Fenvalerate** is a well-known example (figure 7.6d).

Perhaps the most noteworthy aspect of all these successes is that synthesis was based on the principle of trying to copy the *molecular geometry* of the natural materials rather than to mimic the structural chemistry of their molecules. Almost every part of the parent molecule can now be replaced by a unit of analogous structure without loss of insecticidal activity. Elliott and Janes (1978) divided the molecule of pyrethrin I into seven segments and discussed the variations in structure that are permissible in each of these segments.

7.6 OTHER INSECTICIDES OF PLANT ORIGIN

Nicotine and its near-relatives *nor*nicotine and anabasine (figure 7.7) are all viscous liquids. Nicotine darkens on exposure to light and should be stored in the dark. It boils at 247°C, has a vapour pressure of about 0.04 mm at 25°C and is miscible with water. It is a weak base ($pK_a = 8$), and forms salts with acids. The free base has been employed as a contact spray, and it is possible that much of it enters insects through the spiracles, possibly then acting in the vapour phase. The salts have also been used as insecticides. In earlier years a suitable balance between short and long term action (i.e. of fumigation as against persistence of a deposit) was obtained by formulating it as the salt of an acid of appropriate

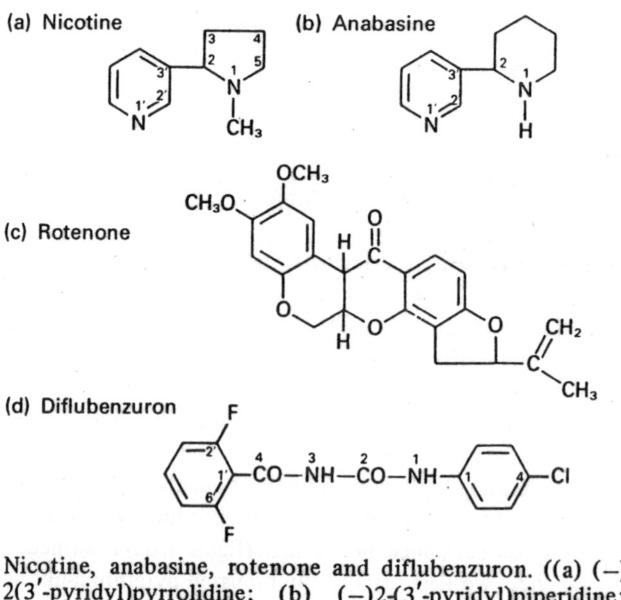

figure 7.7 Nicotine, anabasine, rotenone and diflubenzuron. ((a) (−)1-methyl-2(3′-pyridyl)pyrrolidine; (b) (−)2-(3′-pyridyl)piperidine; (d)1-(4-chlorophenyl)-3-(2′,6′-difluorobenzoyl)urea)

solubility and hydrolysability. For example, nicotine tannate is insoluble, and was used to achieve a stomach poison action.

Nicotine and **nornicotine** occur in certain members of the *Solanaceae* and especially in species of the genus *Nicotiana*. Anabasine occurs in *Anabasis aphylla*, a member of the *Chenopodiaceae*. Not much nicotine is used commercially now for it, and its analogues, are unpleasant toxic liquids and organophosphorus compounds are usually equally effective, especially for the control of aphids. Insofar as nicotine is needed (about 500 tons per year are used in the United States) it is extracted by steam distillation from the leaves and roots of the tobacco plant. In tobacco leaves it occurs at a concentration of between two and five per cent, usually as a salt of Krebs cycle acids (e.g. citric acid, malic acid).

Natural nicotine is the (–) isomer; the (+) isomer, prepared chemically, is of relatively low toxicity. Natural nicotine appears to possess a structure of the right dimensions and of the right basicity to enable it to bind on to acetylcholine receptors in the nervous system (Yamamoto *et al.*, 1962). It has been suggested, for example, that the distance between the two nitrogen atoms, namely 0.42 nm, mimics acetylcholine in that the esteratic oxygen atom and the positively charged nitrogen atom of acetylcholine are about 0.4 nm apart. According to Flattum and Sternburg (1970), there is evidence that the receptor that it attacks is in the neuromuscular junction of mammals but in synaptic ganglia of insects.

In both insects and vertebrates cotinine, a 5-keto derivative of nicotine, is usually the major metabolic product. It is probably formed by a NADPH-dependent hydroxylation reaction, the 5-hydroxy nicotine then being converted to cotinine by further dehydrogenation.

Rotenoids form another group of plant-derived insecticides which play a small but useful role in pest control in the home garden, the market garden and to control ectoparasites on animals. The insecticides are present in the roots of certain genera of the *Leguminosae*, especially *Derris* spp in the Far East and *Lonchocarpus* in South America. Long before they were first used as insecticides (about 1914) their toxic effect was exploited by local fishermen who used to macerate the plants and throw them into water to paralyse fish which then floated to the surface.

The active principles of **Derris** (of which there are about six, rotenone being the most active) can be extracted from the dried, ground-up roots by organic solvents. Alternatively, the still-popular 'derris dust' is made by mixing the ground roots with a clay diluent. Rotenone itself forms yellow laevorotatory crystals which dissolve in many organic solvents but are insoluble in water. Its structure is shown in figure 7.7c. It is of low but variable toxicity to mammals when given by mouth but it is extremely toxic to fish. It is destroyed by light and so is of limited persistence in the garden.

Rotenone does not appear to be a nerve poison but instead it brings about a slow paralysis. It probably inhibits electron transport in mitochondria, and this indirectly leads to the blocking of nerve conduction. In isolated mitochondria it does not affect succinate oxidation (which proceeds via succinate dehydrogenase, ubiquinone and the cytochromes). It is therefore presumed that it disrupts electron transport somewhere in the region of flavin mononucleotide, FMN (Horgan *et al.*, 1968).

7.7 STOMACH POISONS

Arsenic compounds, fluorides and borates are amongst the inorganic compounds still used for insect control. They are often ingested by the insects and so they are referred to as stomach poisons. Of these, only the arsenical compounds are sometimes sprayed on to crops, the fluorides and borates being used against such insects as cockroaches and ants. Arsenic compounds are not approved for use in crop protection in Britain.

Calcium and lead **arsenates** are the two commonest stomach poisons employed in agriculture. The primary function of the metal is to govern stability and solubility; these in turn determine the rate at which the substance decomposes to release soluble, toxic, arsenical compounds. If solubility is too high, or if hydrolysis is too rapid, soluble arsenates will both damage foliage and get washed off plants by rain. On the other hand, if the compound is too stable, it may not 'solubilise' during the brief time that it is within the insect gut; since it is initially insoluble, too great a stability implies that it would behave as an inert mineral and be voided almost unchanged.

Arsenates exist in a number of forms. The lead arsenate which is often used approximates to the composition $PbHAsO_4$ and contains some 20 per cent arsenic. It is employed primarily in orchards to control the pear slug, sawfly and caterpillars of codling, tortrix and winter moths. An advantage over most organic compounds is that its method of access to leaf-eaters implies a selective action. In particular, it does not harm predators of the red spider mite. In the USA, preparations containing calcium arsenate are often preferred to lead arsenate because of fears associated with environmental contamination with lead*. Calcium arsenate is, however, hydrolysed rather rapidly and can for this reason by phytotoxic. Calcium arsenate is used against the cotton boll weevil and against pests in orchards.

It is probable that arsenate uncouples oxidative phosphorylation. By masquerading as the phosphate ion, it forms high energy intermediates analogous to those formed by phosphate during oxidative and substrate phosphorylation (e.g. the diphosphoglycerate formed during glycolysis). The organo-arsenates, however, are much less stable than the corresponding phosphates and hydrolyse spontaneously, so squandering the energy of the linkage, whereas the phosphates only decompose in the presence of the appropriate enzymes, and thereby transfer to other compounds energy required to drive endogenous reactions. Another

***Professor Bryce-Smith, a colleague of the author, has campaigned vigorously against indiscriminate environmental pollution with lead. There is some evidence that children, in particular, may be vulnerable.**

view is that arsenates are first reduced to arsenites and that the latter then inhibit respiration by blocking the thiol groups of lipoic acid, an essential cofactor in the pyruvate dehydrogenase and α-ketoglutarate dehydrogenase reactions.

7.8 HYDROCARBON OILS

Various hydrocarbon oils are used as insecticides, ovicides, herbicides and as emulsifiable carriers of oil-soluble pesticides (figure 2.4). *Tar oils* (which contain phenols, amines and carboxylic acids as well as aromatic hydrocarbons) are very phytotoxic and are therefore not used on foliage except as non-selective herbicides. They are, however, used to control the eggs of mites and insects overwintering on dormant trees.

Unsaturated **petroleum oils** are also phytotoxic. They can be removed from crude petroleum by treating it with sulphuric acid. When this has been done, products of suitable boiling point and viscosity may be used on foliage. Often, the main reason for such use is as carriers of other pesticides – **fortified oils** are obtained by dissolving materials such as BHC or pentachlorophenol in diesel oils and then emulsifying these solutions in water. In such cases, the oil probably assists the main toxicant to penetrate into lipid surfaces since it is absorbed in, and thereby softens and disrupts, the waxy layers in leaves, insect cuticle or fungal spore walls. In this latter capacity the oils are also acting as penetrating and adhering agents. Refined oils also have special uses as selective herbicides (section 12.9).

It was said above that oils of suitable viscosity and boiling point may be used in the presence of foliage, for, over and above any of their uses as carriers or as spray supplements, alkane hydrocarbons can be toxic in their own right. On leaves, for example, if the average molecular weight is too low, the oils are innocuous both to leaves and to insects upon them. On the other hand, if the molecular weight (and therefore the viscosity and boiling point) is too high, both leaves and the insects upon them suffer damage. Somewhere between the two, some selectivity in favour of insect control is often feasible.

Hydrocarbon oils that are used for selective pest control on foliage are, therefore, carefully controlled boiling point fractions of petroleum and comprise a complex mixture of predominantly paraffinic hydrocarbons. This older name for the alkanes derives from the fact that the compounds are of low chemical reactivity (*'para affinis'*, without activity). For this reason the alkanes, together with many fluoroalkanes and chloroalkanes, are examples *par excellence* to use to illustrate the meaning of 'physical toxicity' (a term encountered in a more controversial context in relation to the mode of action of DDT in chapter 6). This follows from the fact that, quite apart from the thermodynamic considerations below, it is difficult to reconcile their chemical inertness with specific chemical attachment at a receptor or other hypothetical site of action.

The nature of 'physical toxicity' can best be explained and illustrated by considering what happens when alkanes related in homologous series are tested against some biological system. In brief, when alkanes (and certain other compounds of low chemical reactivity) are used in toxicity tests it is found that each member of the series causes an equal response when present in the external medium at about the same **relative saturation** even though the isotoxic external concentrations are very different. For the present purpose, relative saturation may be defined as the toxic concentration divided by the saturated concentration in the external medium (although for rigorous treatment various provisos must be attached).

It has long been recognised that the effective concentration of any poison at the internal site of action need bear no simple relation to the concentration in the external medium. This is particularly so when the external medium in which the poison is applied differs greatly in its physical chemistry from the 'biophase' where the toxic action is manifest. In consequence, whilst LD_{50}s and similar measures of toxicity are of profound practical importance in that they determine such things as cost of medical treatment or level of danger resulting from accidental ingestion of poisons, they may reveal little about what is going on at the site of action. In order to investigate structure–activity relationships it is necessary to remove the complicating factor of **phase distribution** which is superimposed upon the **intrinsic toxicity** of the substance. Very often, when a biologically active compound is chemically active, metabolism and excretion add further uncertainties which complicate the unravelling process.

The use of the relative saturations of closely related substances as measures of toxicity removes the complicating factor of phase distribution so long as the compounds under consideration are of low chemical reactivity and so long as a near-equilibrium condition between internal and external phases is rapidly achieved. Alkane hydrocarbons and chloroalkanes fulfil these requirements for physical toxicity. It can, perhaps, be postulated that all compounds could be physically toxic, but that chemically reactive substances cause irreversible damage at concentrations much lower than those which *would have been* physically toxic if such chemical reactivity had not existed.

Table 7.3 shows the concentrations of four hydrocarbons which induce an

Table 7.3 Iso-anaesthetic concentrations and relative saturations of alkanes applied as vapours to mice (Badger, 1946)

Substance	P_t toxic vapour pressure, mm. Hg	P_s saturation vapour pressure, mm. Hg	P_t/P_s
Pentane	85	794	0.11
Hexane	30	240	0.13
Heptane	12	81	0.15
Octane	6	28	0.20

equal degree of anaesthesia in mice (Badger, 1946), and illustrates the value of the concept of relative saturation as a device for disentangling intrinsic toxicity from phase distribution. Anaesthesia, being reversible, is a more reliable and sensitive index of toxicity than LD_{50}s to insects, phytotoxic effects, etc., but the conclusions reached below are probably more or less valid also for these less obviously quantifiable forms of toxicity.

The concentrations found to induce an equal level of anaesthesia are expressed as the vapour pressure, P_t, of the toxicant at the temperature (20°C) of the experiment. The toxic vapour pressure, P_t, is calculated from g/l and hence from mol/l on the assumption that the substance acts as a perfect gas (and remembering that in this circumstance one mole in 22.4 l exerts a pressure of 760 mm Hg at 273 K). P_s is the saturation vapour pressure of the homologue at the temperature of the experiment, and is obtained from reference books.

The use of the relative saturation to eliminate the contribution of phase distribution has been discussed by Ferguson (1939). Briefly,

(i) the relative saturation is an approximation to the thermodynamic activity, a, at which the stated toxic effect is manifest.

(ii) a, is, for any solute, the same in all contiguous phases once equilibrium has been attained, whether or not the solutions are ideal.

(iii) For non-ideal solutions, $a = \gamma C$ where C is the molar concentration of the toxicant in the phase under consideration and γ is the activity coefficient, a measure of the deviation from ideality of the dissolved solute in the solution.

(iv) Hence $a_{ext} = \gamma_{ext} C_{ext}$ for the external phase and $a_{int} = \gamma_{int} C_{int}$ for the internal phase so that, at equilibrium, $\gamma_{ext} C_{ext} = \gamma_{int} C_{int}$.

(v) Therefore, the ratio $\gamma_{ext}/\gamma_{int}$ is the ratio of the chemical concentrations, C, in the two phases – otherwise known as the partition coefficient.

Those who wish to pursue the thermodynamic aspects of the subject further will find both theory and illustrative data in articles by Ferguson (1939, 1951), by Brink and Posternak (1948) and by Hassall (1957). Even inert gases such as argon can show physical toxicity (Ferguson and Hawkins, 1949). For others, the principal points are illustrated by the data in table 7.3. Heptane, for example, appears in the Table to be about seven times as toxic as pentane when the external isotoxic concentrations are considered. Nevertheless, when seen from the point of view of the relative molar concentrations present at equilibrium in the internal phase, pentane is somewhat the more toxic of the two, for it acts at a lower relative saturation and hence at a lower molar concentration at the site of action.

The elimination of the complicating factor of phase distribution is a necessary preliminary to being able to 'recognise' physical toxicity. It is not, in itself, however, the point deserving attention here – phase distribution always has to be eliminated if one is studying in one phase an event which is actually occurring in another! The remarkable thing in the present instance is that, having eliminated phase distribution, there is a near constancy of the internal concentrations which have equitoxic effects (this conclusion is derived from the near-constancy of the P_t/P_s values on the likely assumption that lipid solutions of hydrocarbons do not vary greatly in ideality, i.e. γ_{int} is nearly constant for the four alkanes). There was, in advance, no reason for believing that having corrected for phase distribution, the internal concentrations should have been expected to be similar – they seldom are for chemically-active substances.

The significance of this near-constancy in biochemical terms is not entirely clear, but many people believe that equal numbers of molecules of all physical poisons sorbed on to a given area of lipoprotein surface distort it to roughly the same degree and so cause an equal level of abnormality of performance. Hydrocarbons, alkyl halides and ethers, are, of course, precisely the sorts of inert substances that would be expected to be absorbed into the lipid part of a lipoprotein membrane.

In various organisms, there are several types of membrane-located reactions, their relative importance depending not only on the species but on the age of the organism. Examples are neuronal activity in animals, photosynthesis in plants, oxidative phosphorylation and permeability of the plasmalemma. In addition, organisation is important during cell division, both in regard to chromosomal orientation and the behaviour of the nuclear membrane. Several workers have pointed out that candidate 'physical' poisons appear to affect various organised activities in different circumstances. For example, Gavaudan and Brebion (1945) observed that benzene and chlorobenzene have anaesthetic properties but they can also affect photosynthesis and cell division. Hassall (1961) suggested that carbamates could act as narcotics, inhibitors of the Hill reaction (section 13.2) or interfere with nuclear division. Holan (1969) similarly concluded that DDT could not only affect neuronal activity but also interfere with oxidative phosphorylation and with the Hill reaction. Perhaps that process which is of the greatest significance to an organism at the time of application is the one which responds most obviously to the presence of a physical poison.

7.9 DIFLUBENZURON

A substance currently attracting considerable attention is **diflubenzuron**, DFB (figure 7.7d). The reason for this interest is that it disrupts the metabolism of chitin, a polysaccharide of particular importance to arthropods. It thus appears to offer the prospect of a broad spectrum of insect control, with minimal environmental damage in most habitats. Moreover, so long as no untoward side effects appear, it could be expected to have a high safety margin for vertebrates. Its acute toxicity to mammals is, in fact, low. It has received conditional registration by the US Environmental Protection Agency for use in controlling the boll weevil, *Anthonomus grandis*, on cotton.

The first histological observation that DFB probably interfered with the deposition of cuticle was made by Mulder and Gijswijt (1973) and it was later demonstrated that this interference resulted from inhibition of the deposition of chitin within the integument. It now seems likely that the inhibition is a consequence of a rapid and complete blocking of **chitin synthetase** rather than to an activation of chitinase, an enzyme that breaks chitin down. Much of the work was done using larvae of *Pieris brassicae* (Deul *et al.*, 1978) but Hajjar and Casida (1979) obtained similar results using nymphs of the milkweed bug, *Oncopeltus fasciatus*. The latter workers showed that chitin formation was inhibited 50 per cent by 5.5×10^{-7}M DFB, which, under the conditions of the experiment, represented 0.2 μg per g of abdominal wall. Another manifestation of its action is that it prevents the hatching of eggs. In this instance, the embryonic development of eggs from treated female boll weevils apparently proceeds normally but stops short just before it is complete (Bull and Ivie, 1980).

Van Eck (1979) has recently developed an *in vitro* system which has facilitated investigation of chitin biosynthesis and its inhibition by DFB. Isolated body walls of housefly larvae convert tritium-labelled glucosamine to tritium-labelled chitin, a polysaccharide comprising N-acetyl glucosamine residues. Diflubenzuron inhibits the synthesis of chitin in this system and leads to an accumulation of tritium-labelled uridine diphospho-N-acetylglucosamine in the treated body wall – polysaccharides are formed by pre-existing primer molecules accepting sugar (or sugar-like) units from nucleoside diphosphosugars. Concomitantly there is, relative to untreated walls, a decrease in the incorporation of tritium into chitin, which suggests that the chitin polymerisation enzyme, chitin synthetase, is the target of diflubenzuron.

On the other hand, Cohen and Casida (1980) found that DFB did not affect the activity of chitin synthetase derived from the gut of *Tribolium castaneum* even though it was a powerful inhibitor of chitin polymerisation *in vivo*. One solution of this apparent paradox is that DFB may act by binding to some site vital in chitin polymerisation, but not a site on the enzyme itself. An alternative explanation, that DFB is converted *in vivo* to a more potent inhibitory substance, becomes less likely as a result of the work of Bull and Ivie (1980). These workers studied the effect of a variety of DFB derivatives on the hatchability of eggs of *Anthonomus grandis*. They concluded that aryl hydroxylated analogues of the sort likely to be formed by microsomal mono-oxygenases, were not intrinsically toxic when applied to adult female boll weevils, nor were they detectable within the eggs of adults treated with DFB itself.

The degradation of DFB by soil micro-organisms has been studied by Seuferer *et al.*, (1979) using species of *Fusarium*, *Cephalosporium*, *Penicillium* and *Rhodotorula*. Most of the identified metabolites were hydrolytic products or their acetylated derivatives. They included difluorobenzoic acid and chlorophenylurea (formed by hydrolysis of the amide linkage between atoms 3 and 4 in figure 7.7d) while chloroaniline is formed by hydrolysis of the amide linkage between atoms 1 and 2.

In higher animals, orally administered DFB does not accumulate significantly in body tissues. Much of it appears unchanged in the faeces but some is excreted in the urine as ring-hydroxylated metabolites and as hydrolysis products (Ivie, 1978). It is apparently not phytotoxic, nor does it inhibit the synthesis in animal cells of carbohydrates containing N-acetylglucosamine units (Bishai and Stoolmiller, 1979).

7.10 ACARICIDES

Mites are of considerable commercial importance as pests of fruit trees and of some vegetable crops. In addition they affect larders, barns and animal houses as well as stored food. Different genera of red mites attack fruit trees and greenhouse crops while white mites often thrive in places where grain is stored. Organophosphorus and carbamate compounds with acaricidal properties have been mentioned in previous chapters (pp. 72, 89, 110, 115) so this section is largely confined to some general considerations and brief discussion of a few additional compounds.

When first introduced, organic insecticides led in some districts to an alarming increase in the number of red spider mites after early-season foliage applications to fruit trees. This was partly a consequence of the insect predators of red spider mites being killed more readily than the mites themselves. Taken in conjunction with the very rapid life cycle of the red spider mites, this removal of biological control enabled the few surviving mites to generate an enormous population within a few weeks of applying a spray aimed at insect pests. In many cases, higher doses, which would theoretically have killed most of the mites, were financially uneconomic, were phytotoxic or were potentially hazardous to the consumer.

Some form of summer treatment is essential, for it is a practical impossibility to kill every over-wintering egg on a fruit tree by the use of a dormant season spray such as petroleum oil. Various insecticides were at first used against summer eggs, nymphal stages and adult mites. These compounds often proved unsatisfactory, not so much because of their limited acaricidal power but because of their toxicity to beneficial insects or to crop plants. In veterinary entomology, where these two problems do not arise, toxaphene and several carbamates, organophosphorus compounds and pyrethroids are quite effective acaricides. Much effort has therefore been put into a search for non-phytotoxic chemicals able to control phytophagous mites without doing unacceptable damage to their predators or competitors.

To this end, three types of approach proved valuable. The first was to try to design specifically acaridical chemicals. The second was to attempt to modify existing insecticidal structures to make them more acaridical but less insecticidal. The third line of approach arose from observations that several fungicides were able to control spider mites under field conditions. Attempts were therefore made to modify the molecular structure of certain fungicides and especially of those employed against the powdery mildews.

One line of approach to specific acaricides started from azobenzene (figure 7.8a), a substance used from about 1945 for the control of the glasshouse red spider mite. Early investigations by Eaton and his colleagues at East Malling Research Station (Eaton, 1949) showed that the azo group was not essential to toxicity, but that it could be replaced by a variety of other bridging groups (figure 7.8b).

An early discovery was diphenyl sulphone ($C_6H_5.SO_2.C_6H_5$) but it was not very successful because it lacked persistence. However, a tetrachlorinated derivative, **tetradifon** was introduced in 1954 (figure 7.8c). Tetradifon found considerable commercial application, the chlorine atoms conferring greater persistence as well as greater lipid solubility and chemical stability. Unfortunately, resistance has now limited its use. In fact, it has sometimes been found essential to use different acaricides in rotation, for the rapid life cycle of mites, coupled with the need to use rather persistent chemicals, provides the ideal conditions for the development of resistance.

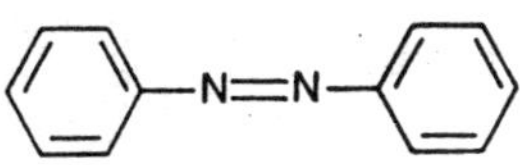

(a) Azobenzene

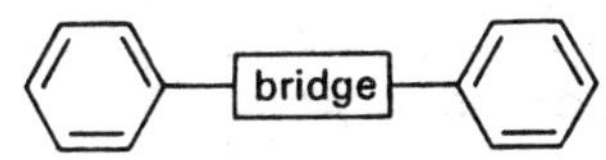

(b) Basic bridged-ring structure

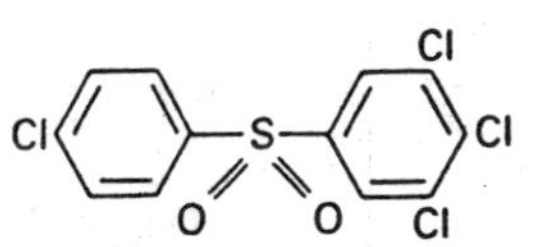

(c) Tetradifon

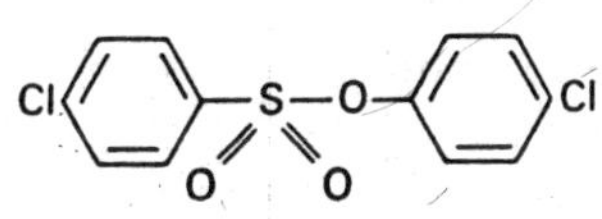

(d) Chlorfenson

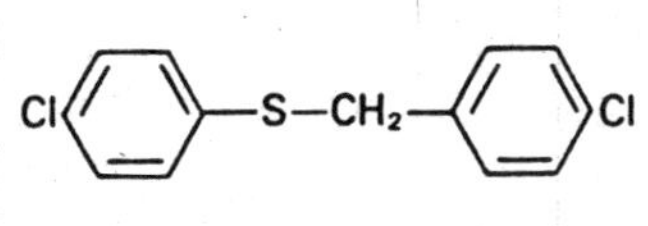

(e) Chlorbenside

figure 7.8 Bridged-ring acaricides. ((c) 4-chlorophenyl-2,4,5-trichlorophenyl sulphone; (d) 4-chlorophenyl-4-chlorobenzenesulphonate; (e) 4-chlorophenyl-4-chlorobenzyl sulphide)

Tetradifon acts against the eggs and nymphal stages of mites and it also makes the females sterile. In common with many other acaricides, it is not systemic because it is too insoluble in water but it is able to kill mites on both surfaces of a sprayed leaf. It is phytotoxic to cucumbers, but is used to control red spider mites on a variety of fruit trees and bushes and also on ornamental plants. McBride (1959) showed that it persisted two or three months in the peel of citrus fruits but that little of it was present in the pulp or juices.

Another possible bridge between phenyl or chlorinated phenyl rings is the sulphonate linkage. **Chlorfenson** (figure 7.8d) has been used commercially, but phytotoxic effects have been reported, including cracking of the epidermis of fruit and scorching of leaves. As with tetradifon, this compound is toxic to the hatching eggs and nymphal stages but is less effective against adult mites. Chlorfenson is of low mammalian toxicity (oral LD_{50} about 2000 mg/kg).

Chlorbenside has a sulphur atom in the bridging group (figure 7.8e) and was for many years a successful acaricide. Like the other acaricides mentioned above, it is a water-insoluble solid. It oxidises, after spraying, to the sulphoxide and the sulphone; this may ensure a longer persistence when the oxidation occurs after the substance has penetrated the leaf.

Several acaricides have been discovered which do not readily fit into other groups of pesticidal compounds. One of these, **chlordimeform** (figure 7.9) is a member of the formamidine family of pesticides, several of which are currently being tested for biological activity. It was developed as an acaricide but has since been found to be very effective against certain troublesome lepidopterous larvae. It has a broad spectrum of action against phytophagous mites, against eggs of some mites and against the cattle tick, *Boophilus microplus* (Zelenski *et al.*, 1978). In the case of the tobacco hornworm, *Manduca secta*, the adults are about a hundred times more susceptible than are the eggs or the larvae (Lund *et al.*, 1979). It is of interest that the plant protection afforded against these insects by chlordimeform appears to be via a non-lethal mechanism, for doses of about 10^{-7}M stimulated motor nerve activity and caused the insects to fall off the plants.

In rats, chlordimeform causes the accumulation of serotonin and noradrenalin in the brain, possibly because, at a concentration of 10^{-5}M, it inhibits monoamine oxidase, an enzyme that normally destroys these natural amines once they have fulfilled their function (Matsumura, 1975). On the other hand, Zelenski *et al.* (1978) observed that it antagonised the contractions induced by several vasoactive agents in aortic strips of the rabbit, the order of antagonism being

$$K^+ > \text{histamine} > \text{serotonin} > \text{noradrenaline}$$

A common metabolite of chlordimeform is 2-methyl-4-chloroaniline, formed by hydrolysis of the double bond (figure 7.9a). This metabolite has been shown by Hill *et al.* (1979) to be potentially carcinogenic, for it causes sarcomas when administered to rats in their diet. It becomes bound to protein, DNA and RNA in rat liver much more readily than to macromolecules in other tissues; the binding process involves an NADH dependent reaction.

Cycloprate is another substance which shows promise for use against phytophagous mites. It is the hexadecyl ester of cyclopropane carboxylic acid, CPCA (figure 7.9b). It appears to be rather specifically acaricidal and ovicidal, for eggs of insects, and motile stages of both mites and insects, are less sensitive to it than are the eggs of mites (Quistad *et al.*, 1979). In one recent trial (Schuster *et al.*, 1979) it controlled the two-spotted spider mite, *Tetranychus urticae*, on strawberry but (unlike other acaricides such as naphthoquinone and cyhexatin) it reduced the yield of the crop.

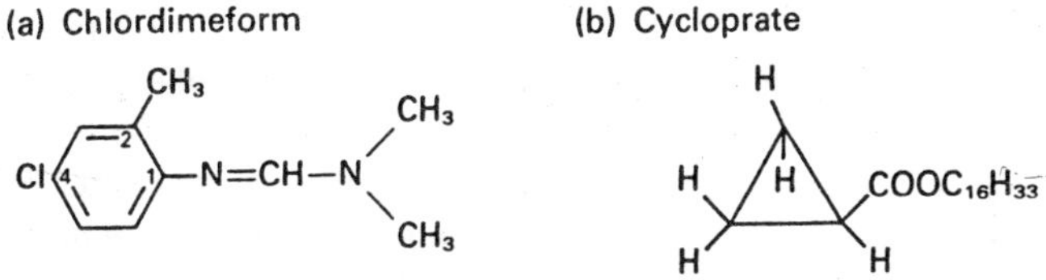

figure 7.9 Two other acaricides. ((a) N′-(4-chloro-2-methylphenyl)-N,N-dimethylformamidine; (b) Hexadecyl cyclopropane carboxylate)

According to Quistad *et al.* (1979), cycloprate may cause a lethal disruption of lipid metabolism. This may happen as a consequence of an interference with the transfer of fatty acids into mitochondria by blocking the normal fatty acid carrier, carnitine (compare Edwards and Hassall (1980), p. 197). In effect, there appears to be less free carnitine in mite eggs than in adult mites, and less free carnitine in mites than in insects. Evidence presented by these workers suggests that cycloprate is first metabolised to CPCA and this then conjugates with free carnitine, thereby decreasing its effectiveness as a carrier of fatty acids.

Another line of approach in the search for acaricides has been to modify the structure of established insecticidal substances. For example, **chlorbenzilate** and **dicofol** (figure 6.4) can both be regarded as DDT derivatives. It is of interest that dicofol is a metabolic product of DDT in certain insects (section 6.5). It would appear that the hydroxylation decreases insecticidal activity but simultaneously increases acaricidal effectiveness. In the case of chlorbenzilate the central part of the DDT molecule is so greatly modified that only a complex bridge is left between the chlorinated rings.

Of the ten **organophosphorus insecticides** for which specific mention of acaricidal uses occurs in the British List of Approved Products for Farmers and Growers (MAFF, 1980), all but two are dimethyl esters. In addition, all but two are mono- or di-thiophosphates, and the two exceptions have juxtaposed a phosphate oxygen atom and a double bond (–O–C=C–). The three common amides (dimethoate, vamidothion and phosphamidon) are also to be found in the list of acaricides. It is not possible to say at this time whether there is anything more than a coincidental link between any or all of these structural features and acaricidal activity.

Of various insecticides that have found use as acaricides outside the area of plant protection, a few are worthy of special mention. **Toxaphene** and **dioxathion** (a diethyl dithiophosphate) have been employed against economically important cattle ticks in Kenya (Crampton and Gichanga, 1979). Three of the major species have become resistant to toxaphene, but none has lost its susceptibility to dioxathion. On the other hand Meleney and Roberts (1979) reported that toxaphene was the most effective of seven established or candidate acaricides against the common scab mite, *Psoroptes ovis*, of cattle. Loomis *et al.* (1979) found that fenvalerate (figure 7.6) was a useful substance for the control of hen mites.

Amongst the acaricides that are also either fungicides or are related to fungicidal substances are **dinocap, binapacryl** and **dinitrocyclohexyl phenol.** All of these are based on the structure shown in figure 7.10a. It is of interest that dinocap (section 9.5) was first employed in 1945 as an acaricide. Later, its ability to control powdery mildews was recognised, and it is now mostly employed for the latter purpose. Another acaricide which has had considerable success in recent years is **cyhexatin,** which is tri(cyclohexyl) tin hydroxide (figure 7.10b). It was introduced in 1968 for use against phytophagous mites and is sprayed on apples, plums and rasberries. It has also been used against mites on animals; for example, Meleney and Roberts (1979) found that it was effective against the common scab of cattle if used by vat-dipping. It is not known whether the resemblance of cyhexatin to fentin acetate (section 8.6) is other than fortuitous.

(a) 2,4-dinitro-6-R-phenol derivatives

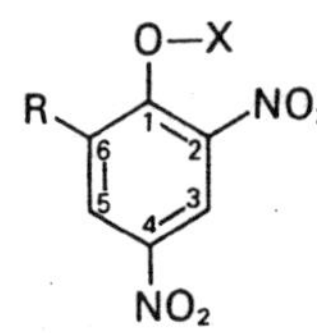

(b) Cyhexatin

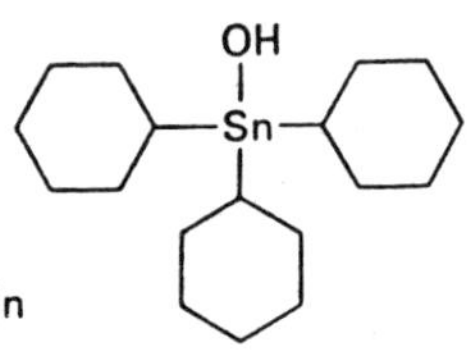

figure 7.10 Structures of some acaricides that are related to fungicides. (In dinitrocyclohexyl phenol, X = H and R = cyclohexyl. In dinocap, X = crotonate and R = *iso*-octyl (figure 9.7). In binapacryl, X = methylcrotonate and R = *sec*-butyl (figure 9.7). In dinobuton, X = *iso*propylcarbonate and R = *sec*-butyl). (Cyhexatin is tri(cyclohexyl) tin hydroxide)

7.11 NEMATICIDES AND MOLLUSCICIDES

There are numerous species of phytophagous nematodes; some live as ectoparasites on the roots of plants and others live within plants. In either case their feeding can result in stunted plant growth, poor yields and virus infection.

There are few reliable estimates of crop damage by nematodes but Strickland (1965) considered that 1 per cent of pest damage to cereals in Britain was caused by them. It is also considered likely that the potato root nematode can cause losses of up to 2 tons per acre in badly infected land.

It is usually necessary to integrate cultural, physical and chemical methods to achieve adequate nematode control. In addition it is highly desirable to be familiar with the life cycles and habits of the principal nematode species. For example, many species seasonally migrate to different depths in the soil, and effective control is more likely to be achieved by treatment during the phase of shallow occupation of the soil. Again, many eggs are covered by cysts impermeable to pesticides and some eggs only hatch when stimulated by substances secreted by the roots of the host plant species.

Many **nematicides** have a fumigant action. Such compounds are usually applied to the soil by injection or by sprinkling, the application often being followed by rotary cultivation. They are volatile liquids which quickly volatilise to fill the soil air spaces. Others are relatively non-volatile solids. **1,3-Dichloropropene** (figure 7.11a) and **DD mixture** (table 7.4) are used to control potato-root and root-knot nematodes; for best results the soil temperature should not be too low. Dichloropropene is normally applied to soil at a depth of 20–30 cm. **1,2-dibromo-3-chloropropane** has been used for soil treatment in citrus groves. Moje (1960) discussed the physicochemical characteristics likely to favour the nematicidal activity of halides and concluded that those containing nucleophilic centres were candidate nematicides.

Table 7.4 Some nematicides used in agriculture

Type	Substance
a. Volatile liquids or gases (with halogen)	methyl bromide ethylene dibromide DD mixture (1,2-dichloropropane plus 1,3-dichloropropene) chloropicrin 1,2-dibromo-3-chloropropane
b. Volatile liquids or gases (without halogen)	formaldehyde methyl isothiocyanate carbon disulphide
c. Relatively involatile solids	metham sodium dazomet thionazin aldicarb, oxamyl, carbofuran (chapter 6) parathion, demeton (chapter 4)

Some compounds, such as formaldehyde, carbon disulphide, chloropicrin and metham sodium can best be described as soil sterilants. They not only control nematodes, but a variety of soil insects and fungi as well. **Metham sodium** is applied to the soil as a drench and then breaks down to the volatile methyl isothiocyanate (section 9.2). **Thionazin** (figure 7.11b) is an organophosphorus compound which is often hand-applied to the furrow for potato nematode control. It is also used to control the bulb and stem nematodes to tulip and narcissus and the bud nematode of chrysanthemum. Other organophosphorus compounds employed to control stem, bulb and root nematodes include parathion and demeton-methyl (section 4.7). **Dazomet** (figure 7.11c) is of interest in that, after

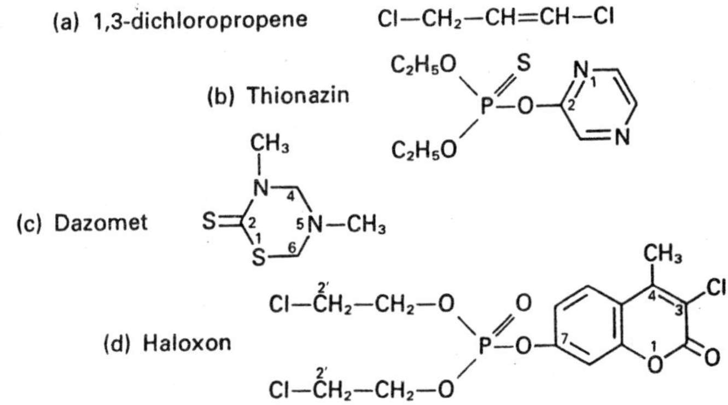

figure 7.11 Some common nematicides. ((b) Diethyl pyrazin-2-yl phosphorothionate; (c) tetrahydro-3,5-dimethyl-1,3,5-thiadiazine-2-thione; (d) di(2′-chloroethyl)-3-chloro-4-methylcoumarin-7-yl phosphate)

soil incorporation, it, too, breaks down to give methyl isothiocyanate. Several carbamate insecticides are also useful nematicides, chief amongst them being aldicarb, oxamyl, pirimicarb and carbofuran (table 5.2).

The compounds mentioned above are principally used in crop protection and are applied to the soil weeks, or in some cases months, before sowing or planting. Treatments involving volatile liquids are usually expensive. Nematicides have, however, agricultural uses outside crop protection, and several have been investigated for the control of helminth diseases of sheep and cattle. In addition to certain organophosphorus compounds already mentioned (section 4.10) **haloxon** (figure 7.11d) is used for this purpose.

The principal molluscs of importance in temperate zone agriculture are slugs and snails, and of the **molluscicides** mentioned elsewhere in this book, phorate (section 4.7) and methiocarb (section 5.5) have proved useful in crop protection. In addition, metaldehyde-bran baits are frequently used and can be very effective for small scale use in the home garden or in the market garden. Metaldehyde is a tetramer of acetaldehyde with the structure shown in figure 7.12a. It is said to have its effect by narcotising the slug or snail so that, in the absence of shelter, it desiccates.

In addition to their use to protect crops from slugs and snails, some molluscicides are of practical value in animal production, for several animal diseases are caused by organisms that have snails as alternate hosts. An important example is the liver fluke (*Fasciola hepatica*), an organism which flourishes even in temperate climates. Its alternate host is the land snail, *Limnaea spp*., which frequents meadows where sheep graze. The grass of such pastures is sometimes treated with **copper sulphate**, although a better nematode for this purpose is **N-triphenylmethyl morpholine** (figure 7.12c).

Finally, some molluscicides are highly toxic to certain water snails *in vitro* (although complications often occur in the natural environment). This aspect of

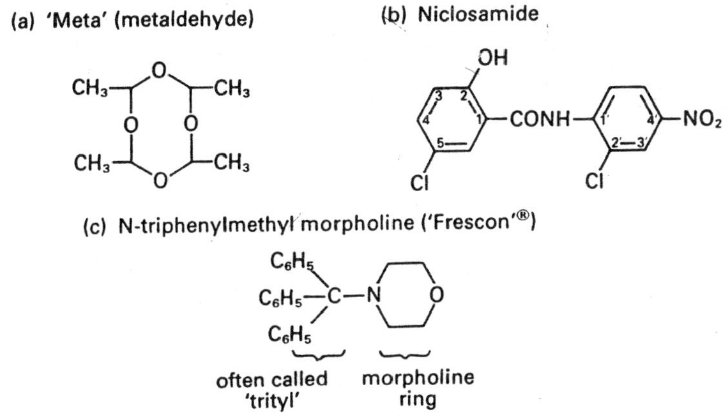

figure 7.12 Three molluscicides. ((b) 2′,5-dichloro-4′-nitrosalicylanilide)

molluscicidal action is of great importance, for *Schistosoma spp*., the organisms responsible for the debilitating disease of **bilharzia**, use species of aquatic snails as alternate hosts. Although this book is primarily concerned with the use of pesticides in agriculture, medical aspects of the use of pesticides cannot be totally divorced from our subject, for a weak and demoralised farmer is a poor farmer. This chapter therefore ends with a brief consideration of some chemical measures for the control of bilharzia (the simplest and cheapest of control measures is not to drink from, and excrete into, the same water).

Copper sulphate, at 2 ppm, can be quite effective in slow-moving non-alkaline water, especially if the vegetation and fish are expendable. It is, however, made ineffective by precipitation in alkaline waters and it is adsorbed on to clay particles. **Pentachlorophenate** is also sometimes used but it, too, is highly toxic to water plants, including algae, and so can destroy the food chain.

Niclosamide (figure 7.12b) is a safer substance to handle than is pentachlorophenate and it has a less devastating effect on water plants. It is, however, like many other anti-bilharzia molluscicides, unacceptably toxic to fish. Jobin (1979) regards it as the preferred molluscicide for the control of the snails that transmit *Schistosoma mansomi* and *S. haematobium*, and it is now in use for projects in Brazil, Puerto Rico, Egypt and Iran. The control of the snails is apparently satisfactory but the cost of the successful treatment varies greatly with the geographical situation. In particular, costs rise as the volume of a habitat increases and as the distance between such habitats widens. In common with most other organic molluscicides it is adsorbed on to organic matter, and the latter often abounds in the near-stagnant water that is the favourite habitat of the snails.

Another substance used in the control of vectors of *Schistosoma spp*. is **N-triphenylmethyl morpholine** (figure 7.12c). According to Brezden and Gardner (1980) it acts rather specifically on prosobranch and pulmonate aquatic snails, including *Biomphalaria glabrata*, the intermediate host of *Schistosoma spp*. The same workers reported that it initiates bursts of synaptic potentials. On the other hand, Plummer and Banna (1979), in a study of the effect of the molluscicide on the large neurons of *Archachatina*, found that some neurons were inhibited and others were excited by it. They concluded that it probably had a direct action on the nerve cells, possibly by altering the inward currents of sodium or of calcium ions.

A different approach to the control of *Biomphalaria glabrata* is illustrated by the work of Pereira *et al.* (1978). These workers made hexane extracts from the plant *Euphorbia cotinifolia* and found that as little as 2.5 ppm of the extracted and partly purified solid killed 90 per cent of adult snails under laboratory conditions. It was less effective against eggs, but 48 ppm killed 90 per cent of them. In the field, the extract was lethal for snails in concentrations of 20 ppm in stagnant water but lower concentrations than this were lethal to fish.

REFERENCES

Adams, M. E. and Miller, T. A. (1979). *Pestic. Biochem. Physiol.*, **11**, 218
Adams, M. E. and Miller, T. A. (1980). *Pestic. Biochem. Physiol.*, **13**, 137
Badger, G. M. (1946). *Nature, Lond.*, **158**, 585
Bishai, W. R. and Stoolmiller, A. C. (1979). *Pestic. Biochem. Physiol.*, **11**, 258
Brezden, B. L. and Gardner, D. R. (1980). *Pestic. Biochem. Physiol.*, **13**, 169
Brink, F. and Posternak, J. M. (1948). *J. cell. comp. Physiol.*, **32**, 211
Bull, D. L. and Ivie, G. W. (1980). *Pestic. Biochem. Physiol.*, **13**, 41
Casida, J. E. and Yamamoto, I. (1967). Abstr. 6th Internat. Congr. Crop Protn., Vienna, 1966, p. 233
Chadwick, P. R. (1963). *Pyrethrum Post*, **7**, 25
Cohen, E. and Casida, J. E. (1980). *Pestic. Biochem. Physiol.*, **13**, 129
Crampton, P. L. and Gichanga, M. M. (1979). *Bull. ent. Res.*, **69**, 427
Deul, D. H., de Jong, B. J. and Kortenbach, J. A. M. (1978). *Pestic. Biochem. Physiol.*, **8**, 98
Eaton, J. K. (1949). *Proc. 2nd Internat. Congress Crop Protn.*, p. 119
Edwards, N. A. and Hassall, K. A. (1980). *Biochemistry and Physiology of the Cell*, 2nd edn, McGraw-Hill, Maidenhead
Elliott, M., Farnham, A. W., Janes, N. F., Needham, P. H., Pulman, D. A. and Stevenson, J. H. (1973). *Nature, Lond.*, **246**, 169
Elliott, M. and Janes, N. F. (1978). *Chem. Soc. Revs.*, **7**, 473
Elliott, M., Janes, N. F. and Potter, C. (1978). *Ann. Rev. Ent.*, **23**, 443
Ferguson, J. (1939). *Proc. R. Soc.*, **127B**, 387
Ferguson, J. (1951). 26 Colloques Internationaux du Centre National de la Recherche Scientifique, p. 25
Ferguson, J. and Hawkins, S. W. (1949). *Nature, Lond.*, **164**, 963
Flattum, R. F. and Sternburg, J. G. (1970). *J. econ. Ent.*, **63**, 62
Gavaudan, P. and Brebion, G. (1945). *Méml Servs chim. Etat.*, **32**, 410
Hajjar, N. P. and Casida, J. E. (1979). *Pestic. Biochem. Physiol.*, **11**, 33
Hassall, K. A. (1957). *J. Sci. Fd Agric.*, **7**, 415
Hassall, K. A. (1961). *J. exp. Bot.*, **12**, 47
Hill, D. L., Shih, T.-W. and Struck, R. F. (1979). *Cancer Res.*, **39**, 2528
Holan, G. (1969). *Nature, Lond.*, **221**, 1025

Horgan, D. J., Ohno, H., Singer, T. P. and Casida, J. E. (1968). *J. biol. Chem.*, **243**, 5967

ITC (1976). International Trade Centre UNCTAD/GATT. *Pyrethrum: a Natural Insecticide with Growth Potential*, Geneva, 193 pp.

Ivie, G. W. (1978). *J. agric. Fd Chem.*, **26**, 81

Jobin, W. R. (1979). *Am. J. trop. Med. Hyg.*, **28**, 142

Loomis, E. C., Bramhall, E. L. and Dunning, L. L. (1979). *J. econ. Ent.*, **72**, 856

Lund, A. E., Hollingworth, R. M. and Shankland, D. L. (1979). *Pestic. Biochem. Physiol.*, **11**, 117

MAFF (1980). *Ministry of Agriculture, Fisheries and Food, List of Approved Products for Farmers and Growers*, H.M.S.O., London

Matsumura, F. (1975). *Toxicology of Insecticides*, Plenum Press, New York

McBride, J. J. (1959). *J. agric. Fd Chem.*, **7**, 255

Meleney, W. P. and Roberts, I. H. (1979). *J. med. Ent.*, **16**, 52

Moje, W. (1960). *Adv. Pest Control Res.*, **3**, 181

Mulder, R. and Gijswijt, M. J. (1973). *Pestic. Sci.*, **4**, 737

Narahashi, T. (1971). *Adv. Insect Physiol.*, **8**, 1

Pereira, J. P., De Souza, C. P. and Mendes, N. M. (1978). *Revta bras. Pesqui med. Biol.*, **11**, 345 (Portuguese; quoted from Abstract)

Plummer, J. M. and Banna, H. B. (1979). *Comp. Biochem. Physiol. C. Comp. Pharmac.*, **62**, 9

Quistad, G. B., Staiger, L. E., Schooley, D. A., Sparks, T. C. and Hammock, B. D. (1979). *Pestic. Biochem. Physiol.*, **11**, 159

Schuster, D. J., Price, J. F., Howard, C. M. and Albregts, E. E. (1979). *J. econ. Ent.*, **72**, 360

Seuferer, S. L., Braymer, H. D. and Dunn, J. J. (1979). *Pestic. Biochem. Physiol.*, **10**, 174

Strickland, A. H. (1965). *Jl R. Soc. Arts* No. 5102, **113**, 62

van Eck, W. H. (1979). *Insect Biochem.*, **9**, 295

Wang, C. M., Narahashi, T. and Scuka, M. (1972). *J. Pharmac. exp. Ther.*, **182**, 442

Yamamoto, I. (1970). *Ann. Rev. Ent.*, **15**, 257

Yamamoto, I., Kamimura, H., Yamamoto, P., Sakai, S. and Goda, M. (1962). *Agric. Biol. Chem.*, **26**, 709

Yamamoto, I., Elliott, M. and Casida, J. E. (1971). *Bull. Wld Hlth Org.*, **44**, 347

Zelenski, J. D., Robinson, C. P. and Pento, J. T. (1978). *Pestic. Biochem. Physiol.*, **8**, 278

8 Fungicides: General Principles; Inorganic and Heavy Metal Fungicides

8.1 FUNGI AND FUNGICIDES

Problems associated with the chemical control of fungi are in some ways different from those that arise in relation to the control of insects. In particular, fungal mycelium has an almost limitless propensity to regenerate from a few surviving hyphal strands. Consequently, most fungicides are primarily protective in action, for a fungus which is already well established is often difficult to eradicate. Systemic fungicides have eradicant properties, but even for pathogens for which a suitable systemic compound exists, eradication may not be worthwhile, for a well-established fungus may have set back plant growth irreparably.

A fungicide which kills on contact can check the growth of mycelium and limit the production of reproductive structures. Such action delays or prevents the disease from spreading from infected plants to healthy ones nearby, but the majority of successful non-systemic fungicides are most effective when applied prior to the arrival of the infection. Non-systemic fungicides must therefore normally possess some degree of persistence and the majority of such fungicides are, in fact, insoluble in water. Some fungi are ectoparasites, and these are more vulnerable to contact sprays than are endoparasites. However, even ectoparasites produce nutritive branches (haustoria) which penetrate plant cuticle and enter plant cells where they may survive when the external mycelium is destroyed.

In addition to airborne pathogens, fungi may attack plants through the soil or be transmitted on or in the seeds. Such fungi are frequently controlled by soil or seed treatment, often by the use of fungicides specifically developed for these purposes. Clearly, a knowledge of the habits and life cycles of such fungi is an essential pre-requisite for effective chemical control.

In 1971, fungicides manufactured in the USA fell into three main groups (table 1.1). The tonnage of sulphur fungicides (68,000 tonnes) produced at that time was ten times that of copper compounds, and all the organic fungicides together represented little more than half the tonnage of sulphur products. Such figures reflect the relatively much higher innate toxicity of organic compounds. It has been estimated (Anon, 1976) that, in terms of money spent by US growers on fungicides in 1974, the dithiocarbamates led the market ($406 m), the phthalimides coming second ($337 m) and copper compounds third ($150 m).

A strictly taxonomic classification of fungi is possibly of less immediate practical value in the present context than is an approach based on a scheme used by Barnes (1977). In table 8.1 the pathogenic fungi are first classified according to their life style and then fungicides of use against pathogens in each group are inserted.

First, pathogenic fungi may be divided according to whether the principal route of infection is via air, seeds or soil. Those that are airborne vary in relation

Table 8.1 Classification of fungi according to their life style, with some fungicides effective against them

Airborne infections	
Group 1	Infect plants *via* aqueous phase (Inter alia, *Oomycetes*, e.g. downy mildews)
Fungicides:	heavy metals; dithiocarbamates, phthalimides, dichlofluanid (most systemic fungicides are ineffective)
Group 2	Not directly dependent on water to infect the plant; initially, at least, live under the cuticle (*Inter alia*, some *Ascomycetes*, e.g. powdery mildews)
Fungicides:	dinitrophenols; some systemic fungicides, e.g. 2-amino pyrimidines, triforine, pyrazophos, benzimidazoles
Group 3	Spread under wet conditions but live beneath the cuticle; mycelium is hydrophobic (*Inter alia*, some *Ascomycetes* and *Fungi Imperfecti*, e.g. *Botrytis* (grey mould); *Venturia* (apple scab); *Pseudocercosporella* (eyespot of wheat); *Sclerotinia* (brown rot of fruit); *Septoria* (leaf spot of wheat))
Fungicides:	phthalimides; dichlofluanid Some systemic fungicides, e.g. benzimidazoles
Seedborne infections	
	Many types of fungi; some live on seeds, others inside them
Fungicides	(as seed treatments or occasionally soil treatment near the seeds): thiram, quintozene; dichlone; cuprous oxide; organomercury compounds Some systemic fungicides e.g. carboxin, triforine, benomyl
Soilborne infections	
	Many types of fungi
Fungicides	(as granules or as a surface spray, or put in furrows, or worked into soil: nabam, quintozene, dichloran, chloroneb, copper compounds Some systemic fungicides, e.g. drenches of thiophanate-methyl; thiabendazole, ethirimol (note: *seed* or *soil* applications of some systemic fungicides can be used against certain *airborne* infections

to their requirement for water. Those in Group 1 enter plant cells by aqueous transmission and infection is favoured by damp conditions, often because access is gained by free-swimming zoospores. Examples are *Oomycetes* of which the downy mildews merit special mention, as does *Phytophthora infestans*, the cause of late blight of potatoes. Some fungi from other taxonomic groups also fall into Group 1 because, although the spores are not motile, they nevertheless gain access as a result of the movement of dew and rain droplets. *Septoria*, which is responsible for leaf spot and glume blotch disease of wheat, is a specific example.

The airborne fungi of Group 2 (table 8.1) are different in that they prefer dry conditions at the time of infection. The mycelium grows on the leaf surface and is frequently hydrophobic, as are the conidiophores. Pathogenic *Ascomycetes* of the Order *Erysiphales* belong to this group and are commonly known as the powdery mildews.

The airborne fungi of Group 3 infect and spread when the plants are moist. Once established, the hydrophobic mycelium lives in the surface layers of the plant. The pathogens of this group come from several taxonomic groups although an important representative, *Botrytis* grey mould, is a member of the *Fungi Imperfecti*. Group 3 also includes a number of rots and scabs of great commercial importance.

The fungi that infect via the seeds also belong to several different taxonomic groups. The *Basidiomycetes* are represented by the smuts and bunts, some of which, because they can lay waste to cereal crops, are of fundamental importance to mankind. Some *Ascomycetes*, such as *Pyrenophora avenae*, an important pathogen of grasses and cereals, also infect plants via their seeds.

Soilborne infections also involve fungi from different taxonomic groups. *Pythium* is one of several genera which cause damping off; it is a member of the *Oomycetes*. Other soilborne fungi belong to the *Basidiomycetes* (e.g. the onion smut, *Urocystis cepulae*) or to the *Fungi Imperfecti* (e.g. *Fusarium* and *Verticillium*).

The numerous fungicides in use today are not equally effective against all pathogenic fungi and it is found that fungi from different habitats or belonging to different taxonomic groups often respond to different types of fungicides. The heavy metal and dithiocarbamate fungicides control many of the airborne pathogens of Group 1. An example is *Phytophthora infestans* which is controlled by the use of copper fungicides*, although fentin acetate and dithiocarbamates such as zineb and maneb have to some extent replaced copper for this purpose. There are still very few systemic fungicides marketed for the control of diseases

* In 1842–1848 the course of history was changed by the Irish potato famine brought about by potato blight. Vast numbers of people died or emigrated to the United States. Someone wrote long ago (ashamedly, I've lost the reference), that, in tragic irony, the mountains of Snowdonia, on a clear day visible in Ireland, were scarred by unsightly piles of copper slag from the copper workings there. At that time neither the cause of the disease, nor its cure were, of course, known.

caused by *Oomycetes* although certain acetamide derivatives and acylalanines have recently been developed for this purpose and phosetyl-Al is widely employed in France to control *Plasmopara viticola*, the cause of downy mildew of vine.

Powdery mildews in Group 2 are frequently controlled by dinitrophenols such as dinocap and binapacryl. Many systemic compounds are now also used for this purpose, an example being the pyrimidine, bupirimate, which is used to control powdery mildew of roses. Similarly, dimethirimol is used against powdery mildew on cucumber and melon. Group 3 fungi are often controlled by the phthalimides (e.g. captan) and by dichlofluanid. Systemic fungicides of use against such organisms as *Botrytis* and *Venturia* include benzimidazoles such as benomyl and thiophanate-methyl.

Various seedborne infections are controlled by a variety of contact fungicides, including seed dressings containing thiram, quintozene or drazoxolon. Organomercury seed dressings are often used to protect cereal seeds against smut fungi (*Ustilaginales*). Some systemic fungicides are also effective against smut fungi (e.g. carboxin is used against loose smut of barley and wheat). They are particularly useful against fungi which live in the embryo and so cannot be reached by superficial poisons. The oxathiin fungicide, carboxin, has been used for this purpose (e.g. against loose smut of barley).

Various soilborne infections are controlled by contact fungicides such as nabam, dichloran, quintozene, chloroneb and copper compounds. Certain systemic fungicides (e.g. thiophanate-methyl) are used to control Fungi Imperfecti and related pathogens. In addition soil sterilants such as methyl bromide and chloropicrin, which are used (especially in glass houses) to control nematodes and weed seeds, also control soilborne fungi.

8.2 PENETRATION OF FUNGICIDES INTO FUNGI

Fungal cell walls are negatively charged and comprise fibres of chitin and cellulose, together with some protein, polysaccharides other than cellulose (e.g. glucans), and lipids (Bartnicki-Garciá, 1968). In the case of *Mucor*, chitosan (30 per cent) and phosphate are present and in some fungi callose and nucleic acid derivatives are also present (Aronson, 1965). It is probable that most non-cationic fungicides readily pass through the cell walls of most fungi, the main barrier to entry being the lipoprotein membrane within it. The walls of some fungal spores may be less permeable and more rigid than those of the general mycelium, their lipid and callose components possibly providing a physical barrier to penetration. However, spores usually become more susceptible to fungicides at the time of emergence of the germ tube. Electron microscope studies of the conidia of *Botrytis* show that the wall is double, only the outer one of which ruptures when the germ tube emerges.

Charged particles have difficulty in penetrating the lipoprotein plasmalemma and the entry of such substances may be facilitated by their temporary conversion to non-ionic forms. Thus, weak acids and bases usually only enter in the undissociated form, weak cations may form *covalent salts* with fatty acids (which therefore act as carriers) or *metal chelates* may be formed with amino or keto acids (compare section 8.3). The relative rates of entry of closely-related substances is frequently related to the sizes of the oil–water partition coefficients. Entry may, in fact, comprise a large number of steps in which the substance partitions between aqueous and lipid phases (a similar phenomenon is the basis of some forms of chromatography). Within the cell, fungicides are probably transported through cell or syncytium by protoplasmic streaming.

Internal distribution between subcellular components varies greatly with the type of fungus and fungicide as is shown by the work of Owens and Miller (1957). They discovered that 18 to 48 per cent of a dose of various metals remained attached to spore wall components of *Aspergillus niger* but that less than 4 per cent was held back by the walls of *Neurospora sitophila*. Most of the metal that penetrated the walls of *A. niger* became attached to mitochondrial components, whereas the bulk of that which entered *N. sitophila* was present in the 12 000 g supernatant fluid (i.e. the endoplasmic reticulum plus the cytosol). In the case of dichlone (section 9.6), 75–78 per cent remained in the cytosol of the spores of both fungi. Cycloheximide binds to the ribosomes of *Saccharomyces* (Siegel and Sisler, 1965) and this may account for its fungitoxic action.

One of the most striking aspects of the influx of many fungicides is the rapidity of the process – it sometimes leads rapidly to the virtual removal of all toxicant from a dilute circumambient solution. Superficially, such rapid entry may seem to militate against the possibility of diffusion being the main force at work, for it is not unusual for the 'concentration' in the spore (defined as the weight of fungicide per unit volume of spore) to be ten thousand times that in the surrounding aqueous phase. Moreover, entry is often 60 to 90 per cent complete within 1 min of exposure to the external aqueous solution (Brown and Sisler, 1960; Somers, 1963).

These data do not necessarily imply that influx occurs 'against the concentration gradient', for not only may diffusion be associated with partition into lipids but, in addition, the ionic or molecular species may be different on either side of the cell membrane. The second possibility is illustrated by a simple analogy; imagine a suitable vessel divided by a partition which allows only copper ions to pass through, and where one compartment contains copper sulphate solution and the other a dilute solution of hydrogen sulphide. Soon, almost all the copper ions would end up as insoluble copper sulphide on one side of the partition, but simple diffusion down a concentration gradient, followed by immobilisation of the diffused copper, is still the driving force.

In contrast to the rapidity with which fungicides enter fungi, any efflux is usually gradual. This efflux is, in the case of fenarimol (section 10.4) energy-dependent and is inhibited by low temperature, by anaerobiosis and by respiratory inhibitors (de Waard and van Nistelrooy, 1980). The net effect of influx and efflux at any moment is known as the *uptake* of the fungicide. As will be seen later, resistance is often ascribed to different rates of destruction of a poison by susceptible and resistant strains. It is therefore of interest that these authors attribute fenarimol resistance of *Aspergillus nidulans* to a more active energy-dependent efflux mechanism in resistant strains of this mould.

8.3 COPPER FUNGICIDES: THEIR PREPARATION, PROPERTIES AND USES

Copper ions in solution are toxic to all forms of life, so copper fungicides used on foliage are necessarily insoluble in water. A variety of copper compounds has been used for this purpose. Some, such as freshly-prepared **Bordeaux mixture**, are gelatinous and have the advantage that the copper is thereby bound to the leaf surface. Moreover, gels dry slowly and this increases the probability of the spore being killed by the fungicide, for spores usually germinate under damp conditions. To others, such as **copper oxychloride**, a sticking agent such as starch is added to increase adhesion after spraying. Most copper fungicides (table 8.2) are formulated either as dusts or as wettable powders.

Copper fungicides are used to control potato blight but they are also active against a wide variety of fungi. In this respect they can be contrasted with many organic fungicides, for the latter often have a narrow biological spectrum.

Table 8.2 Some copper fungicides, past and present

Substance	Approximate formula	Comments
Bordeaux mixture	$CuSO_4.3Cu(OH)_2$	Made from $CuSO_4$ and lime. Used as a spray
Copper oxychloride	$CuCl_2.3Cu(OH)_2$	Made from NH_4OH and $CuCl_2$. Used as a spray and a seed dip
Basic copper carbonate	$Cu(OH)_2.3CuCO_3$	Used as a dust
Cuprous oxide	Cu_2O	Yellow or red copper oxide. Used for seed treatment and as a spray
Burgundy mixture	$CuSO_4.3CuCO_3$	Obsolete; made from $CuSO_4$ and Na_2CO_3
Cheshunt compound	$CuSO_4 + NH_4.HCO_3$	A soluble copper fungicide used as a soil drench against damping off
Copper naphthenates	aromatic carboxylic acids	Too toxic for use on plants. Used to treat timber in barns, fences

Copper dusts are used to control the downy mildew of hop, and copper sprays are used *inter alia* against cane and leaf spots, peach leaf curl and tomato leaf mould. In addition, a few formulations (e.g. the water-soluble Cheshunt compound) are applied to soil to control damping-off diseases. **Cuprous oxide** has been employed for seed treatment and it is also widely used as a spray, especially in the tropics. Copper derivatives of **aromatic carboxylic acids** (naphthenic acids) are too toxic for foliage application but are useful for the treatment of timber in barns, fences and greenhouses.

Bordeaux mixture is an insoluble precipitate formed by adding, at room temperature, a concentrated solution of copper sulphate to a slight excess of lime suspended in the rest of the water. The precise method of preparation influences such physical properties as the fineness of the precipitate and the ease of dispersion. It also influences the speed of conversion of excess lime to carbonate after spraying, and this, in turn, affects the retention of the deposit by foliage. Similarly, the fungicidal efficiency of Bordeaux precipitate decreases on storage, probably because of changes in the extent of aggregation and variation, with time, in the crystalline form. For this reason, Bordeaux mixture is one of the few pesticides which can, with advantage, be prepared by the grower himself.

Bordeaux mixture is not formed, as one might imagine, from equimolar amounts of lime and copper sulphate. Such proportions, by double decomposition, would give a mixture of insoluble copper hydroxide and sparingly soluble calcium sulphate. However, if an aqueous suspension of authentic copper hydroxide is boiled, an intramolecular shift occurs in the copper hydroxide to give black, hydrated cupric oxide. When Bordeaux suspension is similarly heated, it stays green. Instead, copper sulphate and calcium hydroxide react in a molar ratio of 4 to 3:

$$4CuSO_4 + 3Ca(OH)_2 \longrightarrow CuSO_4.3Cu(OH)_2 + 3CaSO_4 \tag{8.1}$$

The main component of Bordeaux mixture is thus a basic copper sulphate which is a molecular complex containing all the copper and none of the calcium. After deposition on the leaf, the precipitate acts as a reservoir from which copper ions can be mobilised, while the calcium sulphate and any excess lime help to increase the binding of the precipitate to the leaf surface.

Mobilisation of the copper from the Bordeaux mixture involves the formation of complexes rather than of cupric salts. It is fairly generally accepted that solubilisation is mainly brought about by exudates from the germinating fungal spores themselves; to use a well-worn expression, the fungal spores become their own executioner. No doubt similar exudates can originate from the host plant, but fungal spores can be killed by Bordeaux precipitate on microscope slides so the presence of the host plant is not essential.

It is known that small but measurable quantities of various metabolic intermediates are exuded by germinating fungal spores. Amongst these are amino- and keto-acids, and copper readily forms chelate complexes with such substances.

The amino acid, *glycine*, aminoacetic acid, is quantitatively the most important component of most exudates, and possible bidentate and tridentate ligands with copper ions are shown in figure 8.1.

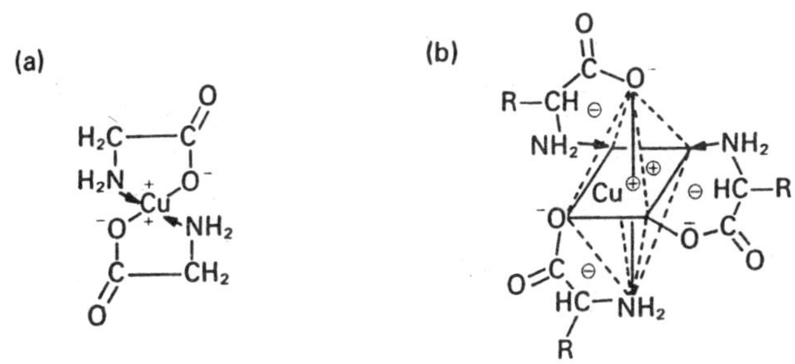

figure 8.1 Possible chelate complexes between copper ions and glycine. (a) A bidentate ligand (electrically neutral); (b) an octahedral tridentate ligand (electrically negative)

The exudate hypothesis was first introduced by Wain and has been discussed by him (Wain, 1959), but the 'suicide theory' has its critics. Tröger (1958), could not confirm that amino acids were exuded by fungal spores in the absence of copper, and concluded that the production of significant amounts of such substances in its presence was an abnormal toxic response to the metal rather than a cause of the metal's mobilisation.

Another interesting and potentially important aspect of complexes relates to their level of stability. Tröger (1960) observed that certain copper complexes were not toxic to the spores of *Fusarium* and it seems possible that, so far as fungitoxicity is concerned, the stability requirements of an organometal complex may be rather precariously balanced. If the complex is sufficiently stable it will chelate and transport the copper to the fungal spore; once there, it will lose the metal to some vital internal component capable of forming an even stabler complex with it. However, if it is very stable, it will mop up any free copper ions in the vicinity but may not deposit them at the potential site of action. In this second case the complex would be acting not so much as a carrier of a lethal agent but as an inactivating mechanism. This may apply not only to copper, but to other heavy metals (section 8.4).

For historic reasons, **Bordeaux mixture** has been chosen as the example to illustrate the general chemistry of copper fungicides – its fungitoxic properties were accidentally discovered because a lime and copper sulphate mixture was applied to roadside vines to deter the wayfarer from sampling the grapes. Nowadays, **copper oxychloride** is probably the most widely employed copper fungicide. It is a basic copper chloride with a structure analogous to that of the main component of Bordeaux mixture (table 8.2). It can be purchased both as a wettable powder and as a colloidal liquid (which contains very small particles permanently suspended in a dispersing agent). **Burgundy mixture** is a basic copper compound made by adding a small excess of sodium carbonate to a solution of copper sulphate. It was often phytotoxic because of excess alkalinity and is seldom used nowadays. **Cuprous oxide** on the other hand, is used both for seed treatment and as a foliage spray to control potato blight and other diseases. It is made by the electrolytic oxidation of metallic copper. There is usually little to choose between these various products in terms of field efficiency, some being marginally preferable to others in particular localities or on particular crops. All can occasionally be phytotoxic, typical symptoms being leaf scorch and russetting, especially on fruit trees and bushes.

8.4 COPPER FUNGICIDES: THEIR MODE OF ACTION

The mode of fungitoxic action of copper is uncertain, although it is initially fungistatic rather than fungicidal (i.e. it stops growth, rather than kills). This important observation is nearly as old as the chemistry of fungicides (an interesting review of this early work has been given by McBrien (1964)). At the turn of the century, Herzeberg, Tubeuf and Volkart independently observed that spores of bunt or smut fungi treated with copper sulphate do not germinate if washed and placed in water, but do germinate normally if placed on gelatin or in soil. The probable explanation is that these media contain complexing agents with a greater affinity for copper than have the binding sites within the spores. Hecke later showed that most of the copper could be washed out of poisoned spores by treatment with dilute acid and that the spores were then able to germinate normally. He also observed that spores had the ability to accumulate large amounts of copper even from external solutions of very low copper concentration. Up to some limiting value, the factor deciding the amount of copper absorbed from solutions of copper salts is not the concentration as such, but the total amount available from the medium. Rate of mixing determines the time it takes for uptake to be complete when the concentration is very low and the volume large.

Factors which influence the amount of copper taken up by fungal spores may throw some light on the mode of action of copper ions, or at least on their availability at the site of action. As could be anticipated from Hecke's results, the amount of copper absorbed becomes less as the hydrogen ion concentration of the external medium increases. Secondly, reagents which form stable complexes with copper may reduce both the uptake and the toxic effect of copper. Thus EDTA (ethylenediamine tetra-acetate) reverses the inhibition of germination of *Alternaria* spores pre-treated with copper. Various sulphydryl compounds, including glutathione, cysteine and British Anti-Lewisite (BAL) have a similar effect.

It will have been noted that the probable importance of chelate complexes has now been mentioned in two related but different contexts – the solubilisation of copper from an insoluble fungicidal residue (section 8.3) and the removal of deposited copper already sorbed on or in poisoned but still potentially viable fungal spores. A picture thus seems to be emerging, not of the importance of a chelate complex *per se*, but of the relative stabilities of different complexes, present at each stage along the line from the components of a fungal exudate to the toxic complex at the ultimate site of action within the spore. This is worthy of further examination.

Many of the most fungitoxic heavy metals are transition elements (e.g. Cu, Ag, Ni). For those that are not transition elements, those of larger atomic weight and of variable valency are often the more toxic (e.g. Zn is less toxic than Cd, which is less toxic than Hg). These general relationships, imperfect though they are, suggest that fungitoxicity (when defined in terms of the magnitude of the applied dose) is a function of the ease with which atoms gain, lose or share electrons. A physicochemical term often used to express the tendency of an atom to attract electrons is *electronegativity*. As used by Pauling, it is the difference between the actual bond energies, as determined experimentally, and the values which can be calculated (from dissociation energies) on the assumption that these same bonds were purely covalent. In effect, the electronegativity is thus a measure of the *ionic character of a covalent bond**.

Bond polarities greatly influence both physical and chemical properties. For example, they affect melting point and solubility and determine the reactivity of a chemical linkage. Consequently, for related elements, many chemical properties roughly parallel electronegativity. Amongst these is the ease of formation of chelate complexes and the subsequent stability of such complexes.

We thus come back to our main theme. The fungitoxicities of a number of metal ions were compiled by Horsfall (1956) and, *inter alia*, the relative sizes of these figures have been compared to Mellor and Maley's (1948) data on the relative stability of chelate complexes and also to relative electronegativity values. Remembering that the nature of the metal is known but the precise nature of the complex(es) it forms *en route* to its site of toxic action is not, a quite striking similarity exists between the order of fungitoxicity of the different cations and the ease with which the metals form covalent or coordinate bonds in chelate-complexes (table 8.3).

It is still not clear whether copper exerts its ultimate toxic effect in the cuprous or the cupric form. In unpublished work, the present writer found that copper absorbed on to algal cells was replaced by (univalent) silver ions almost quantitatively on an equivalent-for-equivalent basis if the copper was assumed to be divalent. This demonstrates that, in algae, the vast majority of copper ions absorbed form cupric complexes but it does not exclude the possibility that a small but important percentage could form cuprous complexes at a relatively few vital sites.

Somers (1963) observed that spores of *Neurospora* absorb 25 per cent more

* The reader who needs to brush up bond chemistry is referred to a helpful little book by G. I. Brown (1961).

Table 8.3 Partial correlation of fungitoxicity with chemical properties of metal ions and metal chelate complexes

Fungitoxicity	Electronegativity[1]	Chelate stability[2]
{ Ag (highest)	Cu (highest – about 2.0)	Hg (highest)
{ Hg	Hg	Cu
Cu	Ag	Ni
Cd	{ Ni	Pb
Ni	{ Co	{ Co
Pb	Pb	{ Zn
Co	Zn	Cd
Zn (lowest)	Mn (lowest – about 1.4)	Mn (lowest)

[1] Haissinsky, as quoted by Somers (1961). A somewhat different plot is given by Danielli and Davies (1951).
[2] Mellor and Maley (1948).

copper under anaerobic conditions than when air is present. Similarly, McBrien and Hassall (1967) showed that cells of *Chlorella vulgaris* exposed to copper under aerobic conditions suffer drastic respiratory collapse if, after removing excess copper, the cells are then exposed to a period of anaerobiosis. These and other data suggest that fungi, algae and some bacteria are more vulnerable to the toxic action of copper when air is excluded, but the reason for this increased susceptibility remains obscure.

Most heavy metals, including copper, are able to precipitate proteins if the effective concentration is high enough. Using Somer's (1963) data, which showed a 100-fold concentration of copper in fungal spores compared to that in the immediate environment, Corbett (1974) calculated that the internal concentration of copper could rise to 1.5 mM. This is possibly sufficient to bring about non-specific protein precipitation. However, much of the copper may be attached to negative sites on wall components, in which case, as was seen earlier, calculations based on weight of copper per unit volume of cells could be misleading.

8.5 INORGANIC AND ORGANIC MERCURY COMPOUNDS

Several inorganic mercury compounds have been used for seed and soil treatment, but their use has decreased, largely because of concern for operator safety and for environmental reasons. **Mercuric oxide** finds some use in Britain to control apple canker and as a seal for pruning cuts or other injuries to trees and shrubs.

Mercurous chloride (calomel) is phytotoxic and cannot be employed as a fungicide on foliage other than that of grass. It is often formulated as a 4 per cent dust for soil application to control clubroot (*Plasmodiophora*) of brassicas and white rot of onions, but a suspension of it in water is sometimes used as a root dip. It has insecticidal and nematicidal properties especially when mixed into the soil. When used as a fungicide, or against clubroot, it can give incidental suppression of cabbage root fly and onion fly (MAFF, 1980). Another use of calomel is to control moss in turf.

The mode of action of inorganic mercury compounds remains uncertain, but many of the comments made in the previous section about copper are also relevant here. The toxic action of mercuric chloride when added to spores can be reversed by thiol reagents such as cysteine and it is generally believed that it reacts with thiol groups in proteins. Many dehydrogenases have active thiol groups and Webb (1966) referred to more than 40 such enzymes which were partly inhibited by concentrations of less than 10 μM mercuric ions.

Organomercury compounds were first used for seed treatment some sixty years ago. This use continues, but for reasons of safety they are now seldom employed as foliage sprays. The general formula is

$$R{-}Hg{-}X$$

where R nowadays is an aryl or alkoxyalkyl group and X is frequently derived from an acid. The bond Hg–X is said to be salt-like (the cation being $R.Hg^{+}$) but this is often obscured by the fact that acids, HX, are so chosen that the total complex is highly insoluble.

Phenylmercury chloride (PMC) was introduced in 1942 as a **foliage spray** for the control of *Nectria galligena* (which causes apple canker) and *Venturia inequalis* (which causes apple scab). The acetate (PMA) and the nitrate (PMN) can also be used against these organisms. However, so far as Europe is concerned, an EEC Directive (79/117) effectively restricts organomercury foliage sprays to use in Northern Ireland on culinary apple trees in autumn to control canker. Swinburn *et al.* (1977) reported that PMN was at that time the fungicide most widely used for this purpose in that province.

Although phenylmercury acetate is used in **seed treatments**, many of the organomercury preparations currently employed for this purpose contain somewhat more complex active ingredients, **methoxyethyl mercury silicate** (MEMS) and the corresponding acetate being two of them (figure 8.2).

Alkyl mercury compounds are seldom used now because of their virulent and traumatic neurotoxic action. Particularly serious incidents occurred in Scandinavia (Hook *et al.*, 1954) and in Iraq (FAO, 1974). The clinical picture of poisoning is described in the latter reference; mild symptoms include headache and paraesthesia of lips, fingers and toes. More severe poisoning leads to tremors and loss of side vision. In severe cases, irreversible mental deterioration may set in.

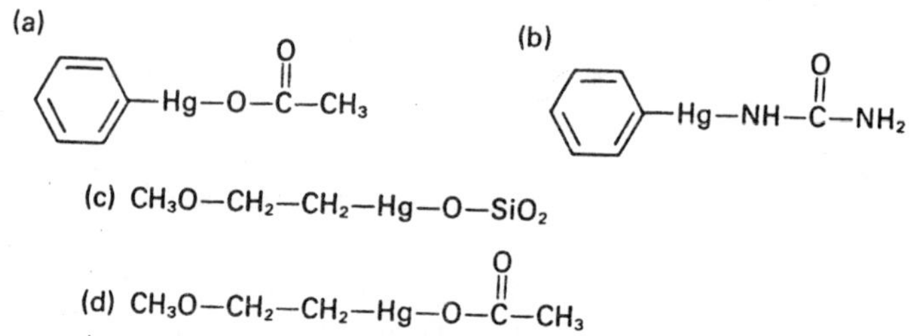

figure 8.2 Some organomercury fungicides. ((a) Phenyl mercury acetate, PMA; (b) phenyl mercury urea; (c) methoxyethyl mercury silicate, MEMS; (d) methoxyethyl mercury acetate)

Of the remaining substances, some of those of lower molecular weight are somewhat soluble in water (PMA, PMN) or are appreciably volatile (PMC), but most of the aryl and alkoxyalkyl compounds are associated with groups such as silicic acid or urea and are almost insoluble. The relative eradicant vis-à-vis protective action probably reflects differences in solubility, volatility and chemical stability. Methoxyalkyl derivatives are much less stable to hydrolysis than are aryl compounds and are therefore less likely to remain long in the environment. No episode of poisoning has been reported that has involved the consumption of seed treated with aryl or alkoxyalkyl mercury compounds (FAO, 1974).

Organomercury **seed treatments** have been widely used to protect cereal seeds from a variety of diseases (table 8.4). The seeds are coated with a suspension of the poison or dusted with the dry powder. In the latter case the seeds are first coated with an adhesive to assist the toxic material to stick to their surface.

Table 8.4 Some important diseases of cereals that are controlled by organomercury seed treatment[1]

Crop	Disease	Pathogen
Wheat	Bunt	*Tilletia caries*/foetida
	Seedling blight	*Fusarium* spp
	Snow mould	*Fusarium nivale*
Barley	Covered smut	*Ustilago hordei*
	Leaf stripe	*Helminthosporium gramineum*
	Seedling blight	*Fusarium* spp
Oats	Loose smut	*Ustilago avenae*
	Covered smut	*Ustilago levis*
	Leaf spot	*Helminthosporium avenae*
Maize	Leaf spot	*Helminthosporium* spp
Sorghum	Seedling blight	*Fusarium* spp

[1] Data selected from Table 1 of FAO Agricultural Studies Report No. 95 (FAO, 1974).

They are also used to control seedborne diseases of sugar and fodder beet. The mercury content of seed dressings is usually between 0.5 and 2 per cent.

The need for caution in the handling and use of organomercury compounds should not disguise the fact that the method of seed treatment has considerable advantages over foliage application for controlling specified diseases. It is a method of use which enables an expensive poison to be applied very economically (grain treated with mercury seed dressing adds about 1 g/ha of mercury to the soil). Moreover, seed treatment in many industrialised countries is now done by the seed merchant rather than by the grower. This is a highly desirable development, for reputable merchants have expertise, qualified staff and appropriate equipment for carrying out the operation safely. When seed treatment is possible, the grower need therefore not handle dangerous concentrates personally. On the other hand, *all treated grain must, of necessity, be sown; it must never be employed to prepare food for man or livestock.* Similarly, treated grain must be stored securely prior to sowing, so that it is inaccessible to children.

To reduce the risk of treated grain being consumed by accident, it is conventional to colour the seed dressing, the colour both acting as a warning as a code to indicate the nature of the fungicide or insecticide in the dressing. Unfortunately, this safety precaution has, on occasion, misfired with disastrous consequences. On more than one occasion, a starving community has deliberately washed the seed-treated grain, assuming that the removal of the colour was a demonstration that the poison had also been removed. Since serious symptoms may be delayed for weeks, a false sense of security may compound the error beyond the point of no return (FAO, 1974).

There is one serious problem associated with the use of treated seed. When grain is treated with **alkylmercury** fungicides or with some **organochlorine** insecticides, it can be a hazard to birds. In springtime especially, when alternative food is scarce, the more voracious devourers of the farmers' newly-sown seed may pay dearly for their predations. This is an unfortunate complication for it has brought into disrepute a method which, when applicable, is one of the safest and most efficient of all methods of pest and disease control. The reason for the greater safety is that a seed dressing is present on a seed below ground whereas the harvested part of many crops is above ground; that harvest, moreover, is separated by the length of a growing season from the time of application of poison. It should also be stressed that so far as mercury is concerned, serious damage to seed-eating birds and to birds of prey has always been a consequence of the consumption of seed treated with *alkyl*mercury seed dressings; aryl and alkoxyalkyl compounds are not known to have this devastating effect (FAO, 1974). Since the use of alkylmercury compounds is now extremely restricted (see below), the problem is no longer an acute one.

If the farmer should wish personally to apply dry organomercury dressings to seeds (in Britain, liquid dressings are not made available to him) it must be remembered that most organomercury compounds are skin irritants and some are sufficiently volatile to be toxic when inhaled. Protective clothing and a respirator should therefore always be worn.

In view of the potential hazard associated with the use of organomercury seed dressings, attempts have been made to restrict the use of such compounds and wherever possible to employ alternatives. The joint FAO/WHO Meeting in Rome in March, 1974, further recommended that alkylmercury compounds should only be employed for the treatment of nuclear stocks of seed. Organomercury compounds are very effective against many important diseases of staple food crops (table 8.4) and have a broad spectrum of attack, whereas, unfortunately, many of the replacement substances have a restricted range or a limited effectiveness. In addition, some promote the development of resistance or are very expensive to employ. **Hexachlorobenzene** is effective against wheat bunt except where resistant strains occur, but it is scarcely less hazardous to man than are the organomercurials. **Quintozene** has also been used for the treatment of various kinds of seeds in some countries. **Benomyl** and **thiabendazole** are costly and are ineffective against barley leaf strip. **Dithiocarbamates** such as maneb and thiram are moderately broad spectrum compounds and often prove highly successful. **Carboxin** is highly effective against diseases borne within the seed (e.g. loose smut of barley) but is very expensive for general use.

The use of organomercury compounds on foliage is less important than their use as seed dressings. They are, however, still used in some countries to control canker of apple. Some tend to be phytotoxic in wet seasons and some commercially prized varieties of apple, such as Cox's Orange, are particularly susceptible to damage. To avoid residue problems, the last application should occur at least six weeks before harvesting.

The mode of action of organomercury compounds remains obscure. Possibly, they break down to inorganic mercury at the site of action, the organic complex merely acting as a vehicle to transport toxic mercury ions into the organism (compare copper in organic complexes, section 8.3). If this is the case, the comments about thiol groups of dehydrogenases made in respect to inorganic mercury may also apply here. However, Ulfvarson (1969) concluded that compounds of the type R–Hg–X probably all broke down to give ions of the type $R{-}Hg^+$. These may then be toxic in their own right. This certainly seems to be the case for methyl mercury ions, $CH_3{-}Hg^+$, for they are more toxic than are equal numbers of free mercuric ions. In addition, mitotic abnormalities have been observed in the root tips of *Allium cepa* upon treatment with concentrations of PMA as low as 0.0001 per cent. At low concentrations the spindle action was inhibited with increasing time of contact, while higher concentrations inhibited cytokinesis, leading to binucleate cells with diploid or even tetraploid nuclei. At concentrations of 0.01 per cent, c-mitotic effects were noticed after several hours contact (Bielecki, 1977).

In view of the fact that inorganic and organic mercury are both used in crop protection, it is perhaps worth noting that it has recently become evident that methylmercury ions can be formed in biological systems from inorganic mercury ions (Vallee and Ulmer, 1972; Beijer and Jernelov, 1979).

8.6 ORGANOTIN COMPOUNDS

In 1950, the Institute for Organic Chemistry in Utrecht commenced investigations into the possible agricultural uses of organotin compounds. It was found that alkylated tetravalent tin compounds of the types $RSnX_3$, R_2SnX_2, R_3SnX became more toxic to fungi as the number of alkyl groups increased to three, but that the nature of the group X (chloride, hydroxide, acetate) was less important. Compounds of the type R_4Sn were less toxic than those with the general formula R_3SnX.

Of the alkyl compounds, *bis*(tri-n-butyltin)oxide has been used to proof fabrics against rot and recently it and the corresponding fluoride have been used as molluscicides and as mosquito larvicides (Cardarelli, 1978). However, trialkyltin compounds are unsuitable for use on crops because they are too phytotoxic. On the other hand, certain triphenyltin derivatives are well tolerated by some plants. For example, Hatzios and Penner (1978) found that young soybean and rice plants suffered no respiratory or photosynthetic damage, nor did electron microscopy reveal any damage to cell ultrastructure, when they were treated with **triphenyltin acetate** at up to 2.2 kg/ha. Triphenyltin (fentin) acetate and triphenyltin hydroxide are currently recommended for use on ornamental plants in Britain as well as for the control of potato blight. Fentin hydroxide has also been reported to be second only to sulphur in efficacy in the control of *Erysiphe betae*, the cause of powdery mildew disease of sugar beet (Byford, 1978). Triphenyltin compounds often control fungi at doses of not much more than one-tenth of those at which copper is fungitoxic.

Fentin acetate is slightly soluble in water (20 ppm) but it is readily hydrolysed to fentin hydroxide. The mode of action is uncertain. Rose and Aldridge (1972) demonstrated that it inhibited oxidative phosphorylation in mammalian mitochondria but the work does not seem to have been extended to fungi. More recently, Mottley (1978) has shown that photophosphorylation in pea chloroplasts is inhibited as a result of interference with the chloroplastic electron transport system. He also provided indirect evidence that organotin compounds may inhibit calcium-dependent ATPase in *Chlamydomonas*. On the other hand, Leow *et al.* (1979) have shown that triphenyl tin compounds inhibit adenylate cyclase activity in brain homogenates at a concentration of only 1 μM. It is possible that an attack on membrane integrity is a common denominator in these different manifestations of toxicity, but the cause of specific fungitoxic action remains obscure.

In view of the uses of trialkyltin compounds mentioned earlier, it is of interest that Seinen *et al.* (1979) have demonstrated that di-n-butyltin and related compounds have antilymphocytic properties and marked immunosuppressive action.

8.7 SULPHUR AND LIME SULPHUR

In 1971 the amount of sulphur used by United States farmers was 2.5 times as great as that of other fungicides. Yet, despite intensive research, we are little nearer understanding the mechanism of its fungicidal activity than was Forsyth who, in 1803, first described the use of a sulphur spray on fruit trees. Horsfall (1956) has summarised in an amusing and scholarly way the vacillations of the views of the scientific fraternity on the mode of action of sulphur fungicides.

Briefly, it was suggested as early as 1879 that sulphur acted in the elemental state. After nearly a hundred years of research (most workers had given up by about 1965!), and after the suggested involvement of nearly every combined form of the element imaginable, the majority view now is that sulphur probably acts in the element state. Admittedly, Mach's idea in 1879 was that the toxicity of sulphur was 'physical', the crystals acting as lenses that focussed light on the fungal tissue. In contrast, the modern idea is 'chemically' based, the sulphur being thought to be an antimetabolite of oxygen. Excluding Mach, the main proponent of the *direct action* hypothesis was Sempio (1932).

On the other hand, the *oxidation* hypothesis suggested that sulphur dioxide and sulphur trioxide are the actual toxicants, but it was eventually shown that sulphur in this form is no more toxic than would be expected from the hydrogen ion concentration of their solutions in water. The *hydrolysis* theory suggested that sulphur is hydrolysed by water to active polysulphides. This hypothesis was championed by Hubert Martin, one of the pioneers of the subject of chemical crop protection. The *reduction* theory suggested that hydrogen sulphide is the actual toxicant. This possibility was apparently disproved in 1953 when it was shown that colloidal sulphur was from five to fifty times more toxic than was hydrogen sulphide. (This is perhaps not totally conclusive evidence, for it would be possible to imagine a cyclic catalytic effect which liberated a sulphide or polysulphide at the site of action; an analogy can be found in the cyclic catalytic effect of the quaternary ammonium weed killers, section 12.4.) Earlier, Yarwood had tested the efficacy of lime sulphur against a powdery mildew with and without the addition of the powerful oxidising agent, $KMnO_4$. Unfortunately for this particular reduction theory, the permanganate prevented hydrogen sulphide production but the mildew was nevertheless controlled.

Spores of some fungi have a remarkable capacity to reduce sulphur – some spores can take up nearly 2 per cent of their own weight of sulphur and then convert most of it into hydrogen sulphide within two hours. It thus seems likely that the biochemical processes leading to sulphide formation cannot be far removed from those implicated in the fungitoxicity of elemental sulphur. Several workers have suggested that sulphur competes with oxygen for electrons and thereby fouls the machinery of intermediary metabolism. For example, Owens (1960) considered that the S_6 molecule of sulphur is converted to a free radical which interferes with the conversion of acetate to citrate by *Neurospora*. Tweedy (1964), working with *Monilinia*, provided evidence that sulphur competes with mitochondrial cytochromes for electrons.

The trouble with all such theories which attribute a very highly specific form of toxicity (in this case, fungitoxicity) to interference with an all but ubiquitous biochemical process, is that it is necessary to postulate that in the one case the toxicant does get to its site of action and that in all the others it does not. It is virtually impossible to counter such an argument, yet scarcely ever is the evidence to support it unequivocal. Sulphur is an excellent illustration of this dilemma, because in this case it must be assumed that the factors of access and distribution are all-important, for sulphur is powerfully fungicidal yet it is normally harmless on plants. Moreover, of all plant protection chemicals (table 1.2) it is probably the only one which is apparently non-toxic to mammals – short of causing death by choking. Yet work with radioactive sulphur applied to plants has shown that internal sulphur compounds become rapidly labelled, and the main labelled end-product is hydrogen sulphide (McCallan and Miller, 1957). Surely this is evidence that sulphur does indeed get into plants and does entangle with the metabolic machinery, yet most plants are not readily damaged by neutral sulphur preparations.

Sulphur is sometimes used as a fumigant, a colloidal smoke being made by vaporising it in lamps. In addition it is applied to plants in three formulations – as wettable powders, as colloidal sulphur and as lime sulphur. *Wettable powders* in Britain contain not less than 70 per cent w/w of sulphur and the particle size range is also officially specified. Not less than 40 per cent of the sulphur should be in particles of 6 μm or less in diameter. In practice, most of the particles are within the range 1–10 μm. A dispersing agent and a diluent such as kaolin or zinc oxide are usually also present and these account for the remaining 30 per cent of the wettable powder preparation.

The composition of *colloidal sulphur* liquid formulations (suspension concentrates) is similarly subject to official specification in Britain. A suspension concentrate must contain 40 per cent or more of sulphur and not less than 90 per cent of the particles should be less than 6 μm diameter. In practice, most of the particles are within the range of 0.3–3.0 μm.

Lime sulphur is an orange-coloured liquid which should contain at least 24 per cent w/v of polysulphide sulphur. It is made by boiling a suspension of lime with an excess of flowers of sulphur. In effect, alkali-catalysed hydrolysis of sulphur occurs, the calcium salts of the acids hydrogen sulphide and hydrogen sulphite being formed. Excess sulphur then reacts with each of these to give calcium polysulphide and calcium thiosulphate. Split into the individual steps, the process can thus be represented as follows:

$$3S + 3H_2O \rightleftharpoons 2H_2S + H_2SO_3 \qquad (8.2)$$

$$3Ca(OH)_2 + 2H_2S + H_2SO_3 \longrightarrow 2CaS + CaSO_3 + 6H_2O$$

$$CaS \xrightarrow{xS} CaS.Sx \qquad\qquad CaSO_3 \xrightarrow{S} CaS_2O_3$$

The polysulphide sulphur, S_x, can be regarded as a solution of sulphur, for the complex, $CaS.S_x$, rapidly decomposes (compare KI_3, formed when iodine dissolves in potassium iodide). Experience has shown that it is the polysulphide sulphur content of lime sulphur, rather than the total sulphur, which provides the best index of fungicidal efficiency in the field.

Compared with organic fungicides, a very high rate of application of inorganic sulphur fungicides is necessary. On the other hand, sulphur has the great advantage of being cheap and does not present a hazard to man. A further advantage is that it has a fairly wide biological spectrum – in sharp contrast to the insecticide situation, selective control of different fungi present on plants is seldom necessary or desirable.

Lime sulphur is often a more efficient fungicide than sulphur suspensions of equal sulphur content. This probably reflects the high availability of sulphur when applied in true solution, the efficiency possibly being augmented by the mild alkalinity. On the other hand, many plants are damaged by it, including prized varieties of apples such as Cox's Orange Pippin, Rival and Beauty of Bath. Over-spraying with lime sulphur on warm days (above 26°C) is particularly liable to cause problems. Acute damage is visible as marginal leaf burns and chronic damage includes the formation of premature abscission layers leading to early drop of leaves and fruit.

Sulphur sprays are used against apple scab, for which a full spray programme in the United Kingdom classically required the successive application of four sprays; an early one of higher concentration at the 'green cluster' stage of growth; a second, less concentrated, at the 'pink bud' stage; a third, dilute, at petal fall; and a fourth at the fruitlet stage. One or more of the later applications may be replaced by sprays containing organic fungicides such as captan or dodine (especially on *sulphur shy* varieties). Other uses of sulphur sprays include the control of peach leaf curl and of powdery mildews. For example, Byford (1978) found that two sprays of sulphur, the first applied at the onset of the attack, suppressed severe infections of sugar beet by the powdery mildew, *Erysiphe betae*, and increased the yield of sugar by 13 per cent. Sulphur sprays are also used against the gall mite which causes 'big bud' of blackcurrants; this is an instance of an interesting occurrence which will be encountered elsewhere, namely, that those fungicides that are particularly effective against the *Erysiphales* are also very often quite useful acaricides as well.

At first sight, lime sulphur does not seem to conform to the general rule that a persistent material must be both stable and insoluble. However, it is sulphur and not lime sulphur which is the stable, insoluble and persistent substance. Instead, lime sulphur illustrates another generalisation – that materials which undergo non-destructive cyclic chemical changes after application may thereby become more firmly bound to the leaf surface. The principal changes occurring in lime sulphur when its components come into contact with water, air and carbon dioxide are given by the following equations:

Calcium thiosulphate

(a) $2CaS_2O_3 + O_2 \longrightarrow 2CaSO_4 + 2S$

Calcium polysulphide (O_2)

(b) $CaS.Sx + O_2 \longrightarrow CaS_2O_3 + (x-1)S$ (8.3)

(c) $CaS.Sx + 2H_2O \longrightarrow Ca(OH)_2 + H_2S + Sx$

(d) $CaS.Sx + H_2CO_3 \longrightarrow CaCO_3 + H_2S + Sx$

The precipitation is probably a cyclic process, for the alkali produced in equation (c) may help to re-dissolve any of the precipitated sulphur. In consequence, the deposit may become more and more dispersed, and comprise finer and finer particles. The calcium carbonate produced in equation (d) will help to bind the sulphur firmly to the leaf surface, and the thiosulphate produced in equation (b) will become the reactant for equation (a). As stated earlier, the hydrogen sulphide may contribute to the fungitoxic action, as may hydrogen polysulphides formed as intermediate hydrolysis products.

REFERENCES

Anon (1976). *Fm Chem.*, Sept., 1976 (quoted from Barnes, 1977)

Aronson, J. M. (1965). *The Fungi, an Advanced Treatise, Vol. 1* (eds. G. C. Ainsworth and A. S. Sussman), Academic Press, New York, p. 47

Barnes, G. (1977). *Herbicides and Fungicides: Factors affecting their Activity* (ed. N. R. McFarlane), Chemical Society Special Publication No. 29. Burlington House, London, p. 35

Bartnicki-Garciá, S. (1968). *Ann. Rev. Microbiol.*, 22, 87

Beijer, K. and Jernelov, A. (1979). *Topics in Environmental Health, Vol. 3: The Biogeochemistry of Mercury in the Environment* (ed. J. O. Nriagu), Elsevier/North-Holland Biomedical Press, Amsterdam

Bielecki, E. (1977). *Acta biol. cracov. Sér. bot.*, **20**, 67

Brown, G. I. (1961). *A Simple Guide to Modern Valency Theory*, Longmans, London

Brown, I. F. and Sisler, H. D. (1960). *Phytopathology*, **50**, 830

Byford, W. J. (1978). *Ann. appl. Biol.*, **88**, 377

Cardarelli, N. F. (1978). *Mosquito News*, **38**, 328

Corbett, J. R. (1974). *The Biochemical Mode of Action of Pesticides*, Academic Press, London

Danielli, J. F. and Davies, J. T. (1951). *Adv. Enzymol.*, **11**, 35

FAO (1974). *The Use of Mercury and Alternative Compounds as Seed Dressings*, Report of a Joint FAO/WHO Meeting, Geneva, March 1974, FAO Agricultural Studies No. 95, FAO, Rome

Hatzios, K. K. and Penner, D. (1978). *Pestic. Biochem. Physiol.*, **9**, 70

Hook, O., Lundgren, K. D. and Swensson, A. (1954). *Acta med. scand.*, **150**, 131

Horsfall, J. G. (1956). *Principles of Fungicidal Action*, Chronica Botanica Co., Waltham, USA

Leow, A. C. T., Towns, K. M. and Leaver, D. D. (1979). *Chem.-Biol. Interact.*, **27**, 125

MAFF (1980). *Ministry of Agriculture, Fisheries and Food. Agricultural Chemicals Approval Scheme: List of Approved Products for Farmers and Growers*, H.M.S.O., London

McBrien, D. C. H. (1964). Aspects of the Uptake of Copper and its Physiological Effects upon Chlorella vulgaris, Ph.D. Thesis, University of Reading, England

McBrien, D. C. H. and Hassall, K. A. (1967). *Physiologia Pl.*, **20**, 113

McCallan, S. E. A. and Miller, L. P. (1957). *Contr. Boyce Thompson Inst. Pl. Res.*, **18**, 497
Mellor, D. P. and Maley, L. (1948). *Nature, Lond.*, **161**, 436
Mottley, J. (1978). *Pestic. Biochem. Physiol.*, **9**, 340
Owens, R. G. (1960). *Devs ind. Microbiol.*, **1**, 187
Owens, R. G. and Miller, L. P. (1957–8). *Contr. Boyce Thompson Inst.*, **19**, 177
Rose, M. S. and Aldridge, W. N. (1972). *Biochem. J.*, **127**, 51
Seinen, W., Vos, J. G., Brands, R. and Hooykaas, H. (1979). *Immunopharmacology*, **1**, 343
Sempio, C. (1932). *Memorie R. Accad. Ital. Classe di scienze fisiche matematiche e naturali, 3, Biol.*, **2**, 1 (cited from Horsfall, 1956)
Siegel, M. R. and Sisler, H. D. (1965). *Biochim. biophys. Acta*, **103**, 558
Somers, E. (1961). *Ann. appl. Biol.*, **49**, 246
Somers, E. (1963). *Meded. LandbHoogesch. OpzoekStns Gent*, 28, 580
Swinburne, T. R., Cartwright, J., Flack, N. J. and Brown, A. E. (1977). *Rec. agric. Res. (Belfast)*, **25**, 57
Tröger, R. (1958). *Arch. Mikrobiol.*, **29**, 430
Tröger, R. (1960). *Phytopath. Z.*, **40**, 91
Tweedy, B. G. (1964). *Phytopathology*, **54**, 910
Ulfvarson, U. (1969). *Fungicides, an Advanced Treatise*, Vol. 2, (ed. D. C. Torgeson), Academic Press, New York, p. 303
Vallee, B. L. and Ulmer, D. D. (1972). *Ann. Rev. Biochem.*, **41**, 91
de Waard, M. A. and van Nistelrooy, J. G. M. (1980). *Pestic. Biochem. Physiol.*, **13**, 255
Wain, R. L. (1959). *Some Chemical Aspects of Plant Disease Control*, R. Inst. Chem. Lectures, Monographs and Reports, No. 3, Russell Square, London
Webb, J. L. (1966). *Enzyme and Metabolic Inhibitors*, Vol. 2, Academic Press, New York

9 Non-systemic Organic Fungicides

9.1 CLASSIFICATION

Organic substances with fungicidal properties are classified in table 9.1. Of the groups listed the organometallic compounds were dealt with in the previous chapter, systemic compounds are considered in Chapter 10, and the remainder provide the contents of this chapter.

Table 9.1 Classification of organic fungicides

A. Organometallic compounds (chapter 8)	
1. Organomercury	e.g. phenylmercury acetate
2. Organotin	e.g. fentin acetate
B. Organosulphur compounds	
1. Dithiocarbamates	e.g. thiram, ziram, maneb
2. Phthalimides	e.g. captan, captafol, folpet
C. Dinitrophenol derivatives	e.g. dinocap, binapacryl
D. Chlorinated derivatives of benzene and naphthalene	
1. Chloronitrobenzenes, chloronitroaniline	e.g. quintozene, dicloran
2. Chloronitriles	e.g. chlorothalonil
3. Chloroquinones	e.g. dichlone
4. Chlorophenols	e.g. pentachlorophenol*
E. Cationic surfactants	
1. Guanidine derivatives	e.g. dodine acetate
2. Imidazoline derivatives	e.g. glyodin acetate*
F. Other non-systemic compounds	
1. Dicarboximides	e.g. iprodione*, procymidone*
2. Dithianon*	
G. Systemic antifungal compounds (chapter 10)	
These are classified in section 10.1	

Compounds in groups A–F inclusive have a protective, non-systemic action, although a few may eradicate minor infections. Those marked with an asterisk are not considered in this book.

9.2 DITHIOCARBAMATES: CHEMISTRY, UPTAKE AND USES

In terms of tonnage employed, the dithiocarbamates are the most used organic fungicides (table 1.1). They fall into two main groups (table 9.2), the most noteworthy difference between them being that the molecules of substances in Group 2 have a hydrogen atom attached to the carbamate nitrogen (figure 9.1).

The fungicides in Group 1 are all derivatives of dimethyl dithiocarbamic acid, although two sorts of derivatives are in use. Most are metallic complexes (except in the case of sodium, they are not salts) but one, thiram, is a disulphide oxidation product (figure 9.2b). Thiram is the only disulphide used as a fungicide but its diethyl analogue is used medically to treat alcoholics. All forms of one substance may be biologically inter-convertible since disulphide forms can be reduced to thiol forms and metal complexes can dissociate.

The dithiocarbamates of both groups probably enter fungi as un-ionised molecules of a weak acid, of a disulphide derivative, or of a covalent complex. Nabam, for example, which is soluble in water and highly ionised, is of low toxicity to spores of *Monilia* if environmental conditions do not favour oxidation. In contrast, the water-insoluble and lipophilic zinc complex, ziram, is toxic whether or not oxidising conditions exist. For toxicity at the site of action, however, it may be essential for an ionic form to be readily assumed, no matter in which form the substance penetrates the fungus.

An early theory was that conversion of active ingredient to dithiocarbamate ions occurs outside the fungus, a complex then being formed with *extracellular* copper ions. According to this theory, a copper complex is the actual migratory species. However, it now seems more probable that the union of dithiocarbamates with *endogenous* copper ions is an integral part of their mode of action within the fungus (section 9.3).

Soluble dithiocarbamates are used for soil treatment while the insoluble heavy metal complexes and disulphides are valuable for foliage spraying, seed treatment and for post-harvest protection. Some of the more important uses of these and other non-systemic fungicides appear in table 9.3.

Metham-sodium is not only a fungicide but a general soil sterilant. At about 100 kg/ha it destroys soil arthropods (e.g. millipedes and woodlice), nematodes (e.g. the potato cyst nematode) and weed seedlings (e.g. *Oxalis*, chickweed) as well as soilborne fungi. For reasons of cost, its use is normally restricted to glasshouse soil and to land to be used for high value crops. It decomposes to give methyl isothiocyanate, CH_3.NCS, which probably acts as a fumigant. It is essential that both metham-sodium and isothiocyanate should have disappeared before a crop is introduced into the soil.

Thiram was originally introduced to control *Botrytis* although it is used nowadays for seed treatment, especially as a protection against damping-off in beans, peas and maize and against *Verticillium* wilt of lucerne. The report of Ellis and Paschal (1979) is one of several recent illustrations of the use of thiram for seed treatment. They found that thiram (and also captan) moved into the

Table 9.2 Classification of dithiocarbamate fungicides

Group, and its main feature	Chemical designation	Type of derivative	Examples
Group 1. There is no hydrogen atom attached to the nitrogen atom	Dimethyl dithiocarbamates (figures 9.1a, 9.2)	(i) metal derivatives	
		(a) sodium salt	(no common name)
		(b) zinc complex	ziram
		(c) ferric complex	ferbam
		(ii) oxidation product	thiram
Group 2. There is one hydrogen atom attached to the nitrogen atom	(a) Bis-dithiocarbamates (figure 9.1b)	metal derivatives	
		(a) sodium salt	nabam
		(b) zinc complex	zineb
		(c) manganese complex	maneb
		(d) mixed complex	mancozeb
	(b) Monomethyl dithiocarbamate (CH_3.NH.CS.SH)	(a) sodium salt	metham-sodium

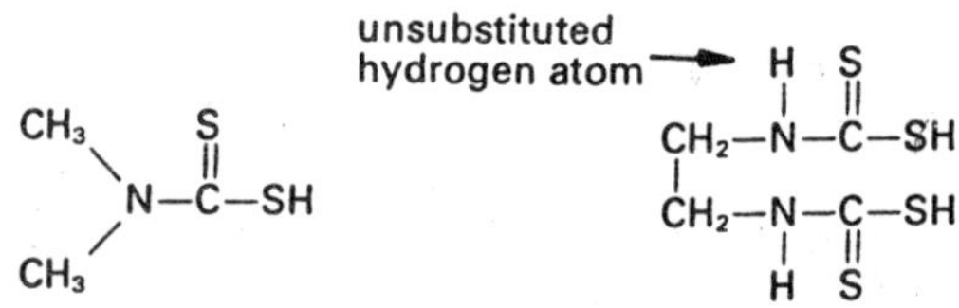

(a) Dimethyldithiocarbamate (Group 1) (examples of fungicidal derivatives appear in fig. 9.2)

(b) Ethylene *bis*-dithiocarbamate (Group 2a) (Zineb and maneb are fungicidal derivatives)

figure 9.1 Structure of (a) a di-substituted and (b) a mono-substituted dithiocarbamate

tissues of the seed coat of the pigeon pea (*Cajanus cajan*) but did not penetrate to the embryo. The treatment proved effective against seven pathogens that are internally seedborne on this crop, for none normally is located deeper than the testa.

Warm chemical soak treatments have been shown by Maude *et al.* (1969) to be effective against many deep-seated pathogens. The principle underlying this procedure is that seeds soaked for about 24 hours at 30°C become part-imbibed with water and in consequence they absorb sufficient amounts of slightly soluble fungicides to eliminate specific pathogens. For example, a suspension of 0.2 per cent thiram, the solubility of which is about 30 ppm, has been used to protect celery against leaf blight (*Septoria apiicola*), peas against leaf and pod spot (*Ascochyta pisi*) and carrots against black rot (*Stemphylium radicinum*). More recently, Soteros (1979) found that, of several methods investigated, a warm thiram soak proved to be the most effective means of controlling *Alternaria* spp. on carrot seeds. Many aspects of anti-fungal treatment of vegetable seeds have been usefully summarised by Maude (1978).

Ferbam has been used successfully against apple scab, but it gives poor control of powdery mildews. It is therefore sometimes employed with elemental sulphur as a double spray for foliage. A disadvantage is that it leaves an unsightly black deposit which can only be washed off picked fruit with difficulty.

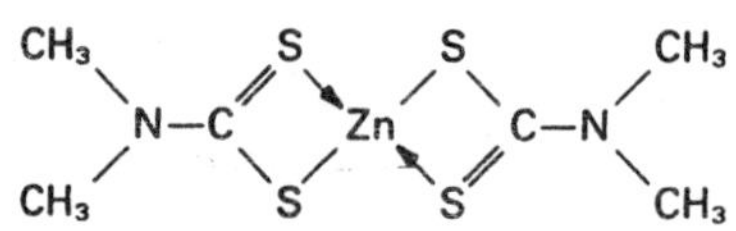

(a) Chelate metal complex (ziram)

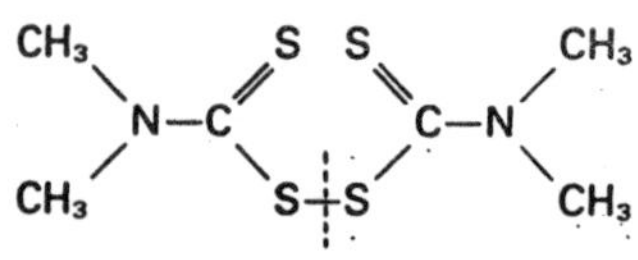

(b) Disulphide oxidation product (thiram)

figure 9.2 Two types of derivatives of dimethyl dithiocarbamate

Maneb, **zineb** and **mancozeb** (a complex containing both manganese and zinc) are the most important dithiocarbamate sprays for foliage application. Much of their importance rests in the fact that they have partly replaced copper fungicides for the control of potato and tomato blights. **Mancozeb** is tending to supersede maneb for the control of potato blight – half of the potato crop in the United Kingdom is currently being sprayed with a mixture of mancozeb and metalaxyl (section 10.6). Dithiocarbamates generally persist on sprayed leaves for a shorter time than copper fungicides, a property which can be disadvantageous in seasons when conditions favour potato blight, for more spray applications are then necessary. **Zineb** is, in addition, employed against downy mildew of hop, lettuce and onion, and against bubble and cobweb of mushrooms. It is less frequently applied to top fruit for, like thiram, it can taint produce or cause the corrosion of tin cans. For vegetables, it has been partly replaced by maneb and mancozeb, since these often provide superior control. **Maneb** is used against black spot of roses and against tulip fire (MAFF, 1980). Both maneb and mancozeb are components of cereal sprays in Europe, giving some control of rusts and of some other diseases.

Various dithiocarbamates have also found use for *post-harvest protection* of foodstuffs. Banana hands are often dipped in a suspension of **maneb**; **thiram**, applied as a post-harvest dip, has been used to control strawberry rot. Such uses are feasible because of the low toxicity to higher animals of insoluble dithiocarbamates (Fishbein, 1977). Post-harvest uses of dithiocarbamates and other fungicides have been reviewed by Eckert (1967).

The development of serious resistance under field conditions is far less common when protective fungicides are used than it is for systemic compounds (section 10.1) but nevertheless under laboratory conditions resistance can be readily induced. An example is provided by the work of Anilkumar and Sastry (1979). Starting with a sensitive strain of *Rhizoctonia bataticola* that just tolerated 25 μg/ml of thiram the strain was subjected to selection pressure on a growth medium containing increasing concentrations of thiram. After six serial passages through such media it was able to grow on a medium containing 500 μg/ml of thiram.

9.3 DITHIOCARBAMATES: MODE OF ACTION AND METABOLISM

Despite their chemical similarity, there is evidence that dialkyl dithiocarbamates and bis-dithiocarbamates do not act in the same way. This difference is often attributed to reactivity associated with the N–H bond present in Group 2 compounds. Whether or not this is so, the biological spectrum of action of compounds in the two groups is undoubtedly different. Moreover, histidine usually reduces the fungicidal effectiveness of compounds in Group 1 but has no effect on the fungitoxicity of the monoalkyl compounds of Group 2.

Several dimethyl dithiocarbamate derivatives, including both the sodium salt and the disulphide form, exhibit a remarkable and rare phenomenon termed a

Table 9.3 Some of the more important uses of protective (non-systemic) fungicides

Fungicide	Uses (foliage-applied unless otherwise stated)
Metham-Na	Soil sterilant – soil fungi, nematodes, millipedes
Thiram	Botrytis, cane spot, rusts; post-harvest dip for strawberries; seed treatment against damping off, *Verticillium* wilt, *Fusarium* spp. and root rots
Maneb, mancozeb	Potato, tomato blight; tulip fire; black spot of roses; cereal rusts
Zineb	Potato, tomato blight; downy mildew on hop, lettuce; tomato leaf mould and root rot; celery leaf spot
Captan	Apple, pear scab; *Botrytis* on strawberries, lettuce; black spot of roses; stem rot of tomato
Folpet	Black spot on roses; some other diseases of ornamentals; grape downy mildew
Captafol	Potato blight; *Rhynchosporium* leaf blotch of barley
Dichlofluanid	*Botrytis* on several bush and cane fruits and on strawberries; septoria glume blotch of wheat
Dinocap	Powdery mildews and red spider mites on apples, strawberries, gooseberries, roses
Binapacryl	Powdery mildew and red spider mites on apples
Quintozene	Seed treatment of potato tubers against dry rot; inhibits potato sprouting; a smoke for *Botrytis* on greenhouse plants
Dicloran	Dust for *Botrytis* on tomatoes, lettuce, cyclamen
Chlorothalonil	Several diseases on bush and cane fruit and on soy beans; also on strawberries
Dichlone	Seed treatment, especially for legumes
Dodine	Apple scab, especially when infection incidence high

bimodal response curve. As increasing amounts of dithiocarbamate are added to mould cultures growing in the presence of a little copper, it is found that toxicity first increases with dithiocarbamate concentration, then diminishes, and finally increases again (figure 9.3). In other words, over a restricted concentration range, increasing the external concentration of fungicide elicits a decreasing response from the fungus.

The existence of such a bimodal curve probably gives a clue either about the way the fungicide gets into the fungus or more likely, how it brings about its toxic action once it has reached its site of action. Earlier theories (e.g. Kaars Sijpesteijn and Janssen, 1958) favoured the first of these and more recent theories favour the second. The location of various molecular species and the significance of their presence thus remains debatable but it is commonly agreed that the bimodal curve is probably a reflection of the relative abundance at

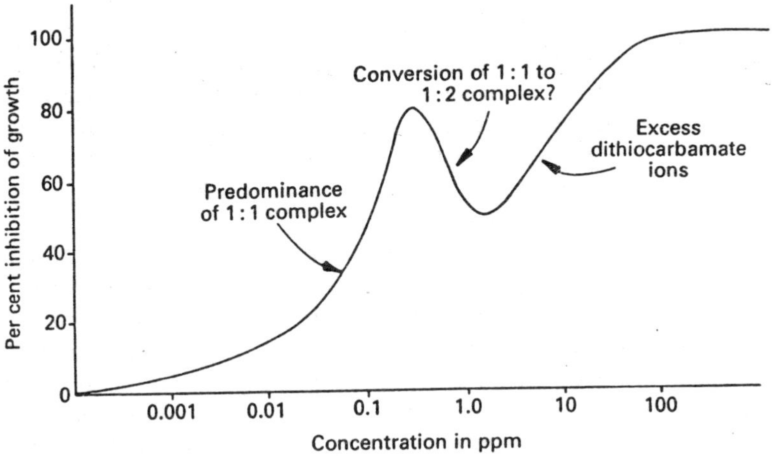

figure 9.3 Bimodal dosage-response curve of *Aspergillus* to sodium dimethyl dithiocarbamate

the site of action of different molecular species. Alternatively it could reflect the absolute abundance of one molecular species, as the external level of dithiocarbamate increases.

The proponents of the theory that bimodality reflects mode of action consider that free dithiocarbamate ions are formed within the cell, irrespective of which complex or oxidation product is applied externally. Once inside, it is assumed that the ions form a complex with internal heavy metals, of which copper is an obvious candidate. At low dithiocarbamate concentrations a resonating complex is believed to be formed comprising one metal ion united to one dithiocarbamate molecule (figure 9.4a). It is this complex which is believed to be toxic to those fungi which show a bimodal curve, and which would then be responsible for the left hand rising portion of the curve. There is spectrophotometric evidence for the existence of this 1:1 complex, for on addition of a solution of copper sulphate to dimethyl dithiocarbamate, a characteristic absorption peak occurs at 380 nm. This peak decreases and eventually disappears as more dimethyl dithiocarbamate is added. This is attributed to the formation of a more stable and less toxic 2:1 complex, analogous to that shown for zinc in figure 9.2a. When all the internal heavy metal has been converted to 2:1 dithiocarbamate complex, excess dithiocarbamate accumulates and is apparently toxic in its own right, perhaps by inactivating sulphydryl enzymes.

An important clue to the mode of action of bis-dithiocarbamates (Group 2) may be provided by the observation that spores, the germination of which was inhibited by nabam, commenced to germinate normally when cysteine was added to the growth medium. Presumably a sulphydryl enzyme is the target, although which of the many possible candidates is involved remains undetermined. Data on enzymes which can be inhibited by various dithiocarbamates has been tabulated by Owens (1969, p. 243).

The nature of the active species (if it is not the applied bis-dithiocarbamate) is also uncertain, for the labile hydrogen atom referred to above renders these compounds very unstable. At least fourteen breakdown products are feasible, any one of which might be, or might give rise to, the actual toxicant. Of the two principal theories, one implicates an isothiocyanate as the active species while the other suggests it is ethylenethiuram disulphide (figure 9.4b).

The isothiocyanate theory is based upon the fact that the soil fumigation action of **metham-sodium** (table 9.1) arises from the formation of a volatile isothiocyanate. However, Owens (1969) failed to confirm any similarity in the (abnormal) internal balance of metabolic acids after treating *Rhizoctonia* with authentic methyl isothiocyanate and with nabam. The work of Morehart and Crossan (1965) revealed rather that there was a similarity in the pattern of free amino acid distribution in *Colletotrichum* after the mycelium had been treated with maneb and with ethylene thiuram disulphide (figure 9.4). The latter can

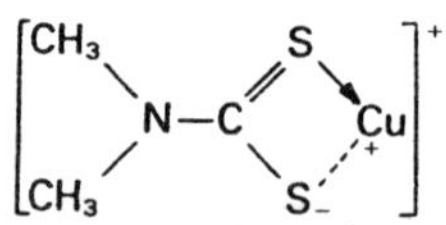

(a) The charged, resonating 1 : 1 complex formed between copper ions and dimethyl dithiocarbamate

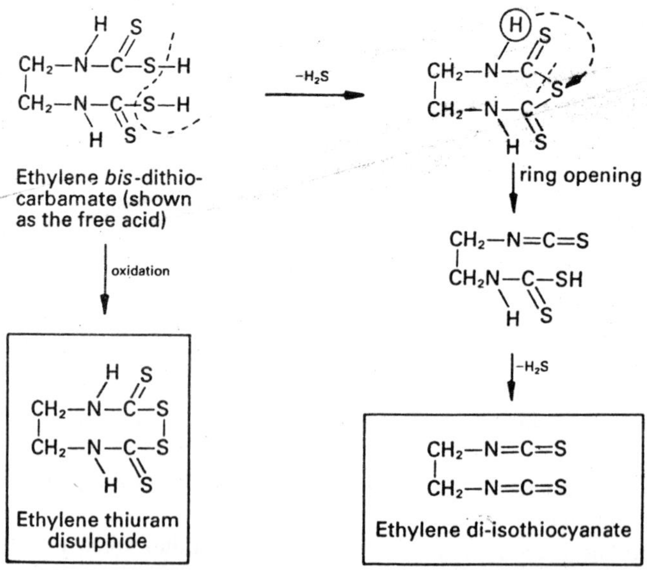

(b) Two derivatives of ethylene bis-dithiocarbamate that have been implicated in its fungitoxic action

figure 9.4 Possible fungitoxic derivatives of (a) dimethyl dithiocarbamate and (b) ethylene bis-dithiocarbamate

probably be formed from all the common fungicides in Group 2, and it has therefore been proposed that this disulphide is the actual toxic agent.

All dithiocarbamate fungicides, but especially the ethylene derivatives, are unstable in most plants, breaking down chemically and photochemically (Woodcock, 1977) as well as in the presence of enzymes. Common breakdown products of various dithiocarbamates include sulphur, carbon disulphide, isothiocyanates, thiuram disulphide (or analogues) and thiourea derivatives. Ethylene thiourea is a major metabolite of zineb, but also formed from it are ethylene urea and 2-imidazoline. Secondary metabolism of dimethyl dithiocarbamate (Group 1) results, *inter alia*, in conjugates in which this substance is joined to glucose and to alanine.

In contrast to what happens in plants, some bacteria, including *Pseudomonas*, break **thiram** down with the formation of carbon disulphide and dimethylamine. For different reasons, a similar situation arises in rats fed **ziram** (Vekshtein and Khitsenko, 1971). No unchanged ziram (Group 1, figure 9.2) was found in the urine, although thiram and a derivative of dimethyl dithiocarbamic acid were detected. It is probable that the acidic gastric juices contributed to the breakdown, because carbon disulphide and dimethylamine were present in the exhaled air and these two substances are also formed when ziram is added to gastric juices *in vitro*.

Although the acute toxicity of dithiocarbamates to vertebrates is moderate (table 1.2), there is evidence that they or their breakdown products may be a chronic hazard to some organisms. Zdzienicka *et al.* (1979) tested the mutagenic activity of thiram on four histidine-requiring strains of *Salmonella typhimurium*. Mutations were observed in all four strains, two with and two without prior metabolic activation of the thiram. Sulphydryl compounds abolished the mutagenic activity in all four strains.

Chernoff *et al.* (1979) observed that rather large doses of **maneb** caused hydrocephalus in foetuses from pregnant females to which it was fed. More importantly, ethylene thiourea, a common metabolite of maneb, caused teratogenic effects including defects of the digits of foetuses when administered at only 40 mg/kg/day. However, such abnormalities were not observed in the young of mice, hamsters or guineapigs fed doses much higher than this. Ethylene thiourea is also reported to have weakly carcinogenic properties.

High residues of dithiocarbamates have occasionally been observed in vegetable crops protected with these fungicides. The fear of possible mutagenic, teratogenic or carcinogenic effects, arising from observations of the sort illustrated above, has led to the use of dithiocarbamates being restricted in Canada and in some other countries.

9.4 THE PHTHALIMIDE GROUP

This group comprises **captan, folpet, captafol** and **dichlofluanid** (figure 9.5).

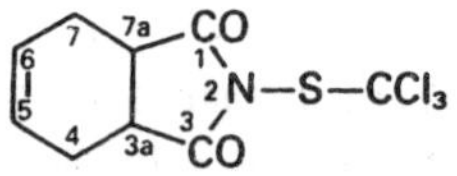

(a) Captan

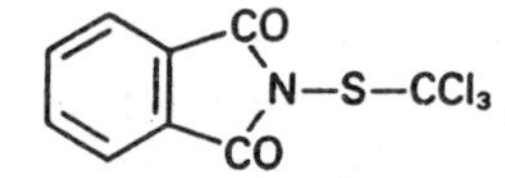

(b) Folpet

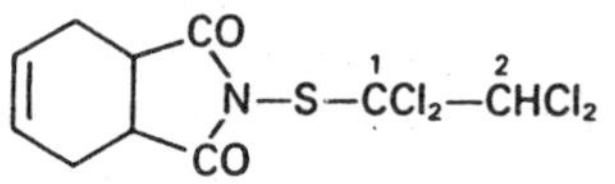

(c) Captafol

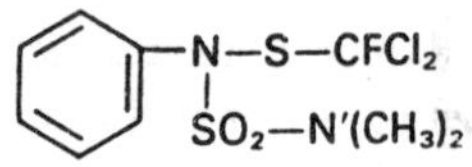

(d) Dichlofluanid

figure 9.5 Fungicides of the phthalimide group. ((a) 3a,4,7,7a-tetrahydro-N-(trichloromethylthio)-phthalimide; (b) N-(trichloromethylthio)-phthalimide; (c) 3a,4,7,7a-tetrahydro-N-(1,1,2,2-tetrachloroethylthio)-phthalimide; (d) N-(dichlorofluoromethylthio)-N′,N′-(dimethyl)-phenylsulphamide)

After the dithiocarbamates they are, on a tonnage basis, the most used organic fungicides (table 1.1). Some of their uses are summarised in table 9.3.

Captan was introduced as a broad spectrum protective fungicide in 1952. It is insoluble in water, not very soluble in oil, and is formulated (as are many insoluble substances) in the form of a wettable powder. It is primarily employed against apple and pear scab, and compared to a normal sulphur spray programme, it often decreases fruit damage, increases yield, and improves the appearance of fruit skins. A full spraying schedule may comprise up to eight fortnightly applications to orchard trees; phytotoxicity under British conditions is usually low. However, it does not control powdery mildews, so if these are present in a mixed infection, a double spray of captan and dinocap (or one of its analogues) is usually essential. It also controls *Botrytis* mould on strawberries and lettuces, potato and tomato blight, and *Didymella* stem rot of tomatoes. It is also quite useful for the control of bitter rot of apples, a disease caused by *Gloeosporium* spp. which affects apples in storage.

Folpet is a white insoluble solid similar in properties to captan, but it is somewhat more phytotoxic to top fruit. In consequence, it is used more frequently for disease control on ornamentals, and especially to control black spot of roses. **Captafol** is superior to captan for the control of potato blight but it is more expensive. It is recommended in Britain (MAFF, 1980) for use against *Rhynchosporium* leaf blotch of barley, against Septoria glume blotch of wheat and against white tip of leeks. Northover (1978) found that a pre- of a post-leaf fall application of captafol reduced the incidence of peach leaf curl caused by *Taphrina deformans* in the following spring, and that it was more effective than captan under conditions which favoured severe infestation. Captafol has also been shown to be more effective than several other fungicides at inhibiting the recolonisation by *Fusarium oxysporum* of freshly steamed soil; a 200 μg/ml solution kept the soil pathogen-free for three weeks (Rowe and Farley, 1978). These authors concluded that the technique provided an economic method for the control of *Fusarium* crown and root rot of tomatoes. **Dichlo-**

fluanid is recommended (MAFF, 1980) for the control of *Botrytis* on a number of bush and cane fruits, as well as for *Botrytis* and for leaf mound on glasshouse tomatoes, and for downy mildew on cauliflower seedlings. In Europe it is widely used for the control of *Botrytis* on grape vines.

The entry of **captan** into fungal spores from an external medium occurs very rapidly, the rate increasing with rising temperature. The latter is often taken to mean that influx involves a chemical mechanism rather than physical sorption, but when molecular species undergo chemical attachment to cellular components after entry it is questionable whether a simplistic application of the laws of adsorption physics leads to very reliable conclusions about the mechanism of absorption.

Richmond and Somers (1963) observed a correlation between the amount of **captan** absorbed and the thiol content of the spores. They concluded that captan was taken up by thiol sites, but they were unable to show any correlation between thiol content of spores of different fungi and the susceptibility of these fungi to captan. They also observed that when spores were pre-treated with thiol-inactivating agents the treated spores absorbed less captan than untreated controls but nevertheless the treated and untreated spores were equally susceptible to the fungicide. Richmond and Somers concluded that captan attaches itself indiscriminately to thiol groups. In consequence, they argued, much of it is immobilised by less vital thiol groups and thus is unable to reach vital thiol groups at the site of action. An analogy is detectable between the role of immobilising thiol groups here and the competition between general body lipids and vital nerve cell lipids in relation to the pharmacodynamics of organochlorine insecticides.

A common criterion of toxicity is the amount of toxicant that needs to be absorbed per gram of spore substance in order to effect a response. On this basis, captan cannot be regarded as a very toxic substance, for whereas penicillin or organophosphorus compounds can exert their toxic effects on appropriate organisms when only 0.1 to 10 μg are present per gram of cell or body weight, the toxic dose of captan is often of the order of 10^3 μg/g spore weight (Somers, 1962). It is the ability of fungi to absorb captan (and several other organic fungicides) from large volumes of very dilute external solutions which gives the impression that these substances are intrinsically very toxic. Their low inherent toxicity and limited persistence probably account, at least in part, for the low incidence of fungal resistance so far observed under field conditions.

It is unlikely that intact molecules of captan or its analogues are responsible for the fungitoxic effect. Whether the thiol is located at the primary site of action, or elsewhere, reaction with fungicide probably involves the participation of two thiol groups and leads to the formation of phthalimide, a disulphide, HCl and thiophosgene (figure 9.6). The latter is both unstable and volatile and at most sites it soon disappears. At the site of toxic action the same reaction probably occurs, but the conversion of thiol to disulphide may in this case provide the mechanism which blocks enzyme activity. In addition, it is also possible

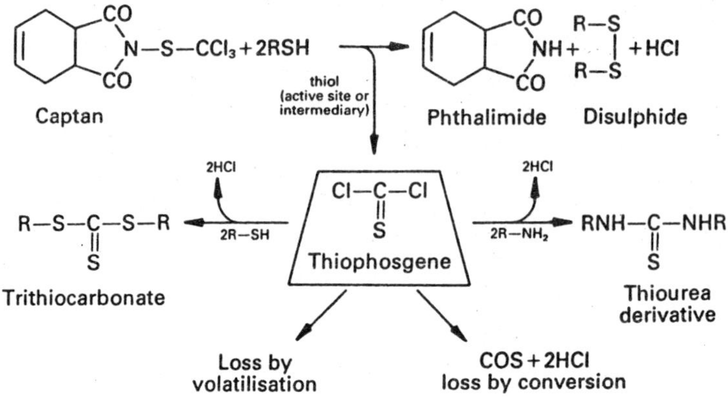

figure 9.6 Reaction of captan with thiol groups and possible subsequent reactions of thiophosgene

that at such sites the very reactive **thiophosgene** may attack more thiol groups to form thiocarbonates or may attack amino acids to form thiourea derivatives.

Using a marine fungus, *Lagenidium callinectes*, Ruch and Bland (1979) have studied the cytological changes that occur at lethal doses of captan. Ultrastructural changes in zoospores exposed to 3.2 ppm captan included the 'washing out' of mitochondrial matrix and disappearance of the mitochondrial cristae. They concluded that the main effect of the fungicide is on mitochondria, but they also observed the disappearance of the nuclear matrix and the swelling of the cisternae of the endoplasmic reticulum.

Fungicides of the captan group are toxic to fish but the oral toxicity to mammals is very low. This low toxicity probably reflects hydrolysis of the fungicide by acid gastric juices, for the LD_{50} by intraperitoneal injection is less than 100 mg/kg. Chronic tests in which captan was fed to dogs for 66 weeks showed that the no-effect level was at least 300 mg/kg per day. Nevertheless, the captan group of fungicides has been exhaustively tested, for these substances are distantly related to the teratogenic drug, phthalidomide. Specific teratogenic effects have, in fact, been observed in chick embryos and much detailed evidence has been collected by Fishbein (1977). More recently, Martin and Lewis (1979) observed that when 12 ppm captan was injected into chick embryos *in vivo*, within six days the rate of incorporation of thymidine mononucleotide into DNA had greatly decreased.

Mutagenicity can be investigated in well-established tests (e.g. the *Ames test*) using bacteria such as *Escherichia coli* or *Salmonella typhimurium*, and in such investigations, captan, captafol and folpet can give positive reactions. Moriya *et al.* (1978) contrasted such results with the absence of symptoms in chronic tests on mammals and demonstrated that the results of bacterial tests can be altered if the medium contains sulphydryl compounds or drug metabolising enzymes. They concluded that these, and blood, may in practice protect most mammals from small doses of potential mutagens. It is therefore of some interest that Dalvi and Ashley (1979) found that 10^{-4}M captan inactivated rat liver cytochrome P_{450} (section 3.4) but that 5 x 10^{-4}M glutathione afforded the enzyme almost complete protection.

Captan and related compounds present an unusual residue problem, for they accelerate the corrosion of tin cans. This is able to cause tainting of canned produce, but Klayder (1963) reported that if canning is accompanied by the customary heat processing, most of the captan deposited on fruit is destroyed. Nutahara and Yamamoto (1978) found that captan, captafol and folpet were all rapidly decomposed in homogenised cucumber fruit, 1 ppm captafol disappearing within ten minutes. Decomposition was also fast in cabbage and potato tuber but was much slower (half-life, 60 min) in tomato and apple homogenates. The rate of decomposition was roughly correlated with the pH, decomposition being non-enzymic and it took place most rapidly at high pH.

9.5 DINITROPHENOL DERIVATIVES

Various dinitrophenol derivatives have found uses as insecticides, fungicides and herbicides. One member of the group was employed for moth-proofing as early as 1892, the first herbicide was patented in 1932 and the first fungicidal dinitrophenol derivative was discovered in 1949.

It was in that year that Sprague observed that an esterified dinitrophenol, **dinocap** (figure 9.7) controlled both mites and powdery mildews (*Erysiphales*). It differs from the herbicidal members of the dinitrophenol group in being both insoluble in water and non-ionic. It is particularly fortunate that dinocap and its analogue binapacryl are able to control powdery mildews, for it will be recalled that the fungi which cause these diseases are not readily controlled by most of the non-systemic organic fungicides currently available.

Dinocap is a mixture of at least two isomeric substances (figure 9.7a, b). Dinocap-4 is probably more fungitoxic than is dinocap-6, but the latter is more acaricidal. In this respect dinocap is somewhat atypical, because most dinitrophenols employed in crop protection are 2,4-dinitro-6-alkyl phenol derivatives (compare binapacryl and the herbicide, dinoseb, figure 9.7c, d).

Dinocap and **binapacryl** are crotonic and methyl-crotonic acid esters respectively. Both are hydrolysed by alkalis to dinitro alkyl phenates. Formulation is usually as a wettable powder or, for low volume application, as an emulsifiable (or miscible) liquid. They have some eradicant as well as protective action, the eradicant properties being accentuated by factors which favour increased penetration. Such factors include the presence of much spreading agent or the use of an oily emulsion, but dinocap is liable to be phytotoxic if applied in conjunction with oil.

The principal use of **dinocap** is for the control of powdery mildews on apples,

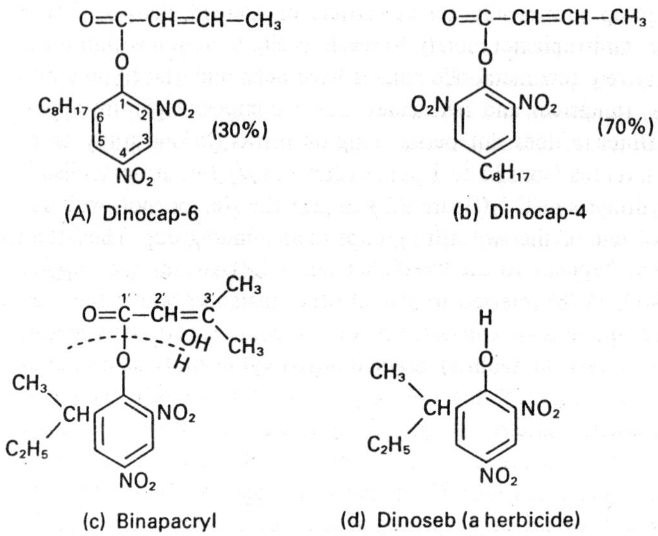

figure 9.7 Dinitroalkyl phenol derivatives. ((a) 2,4-dinitro-6-*sec*-octyl phenyl crotonate; (b) 2,6-dinitro-4-*sec*-octyl phenyl crotonate; (c) 2,4-dinitro-6-*sec*-butylphenyl-3′-methylcrotonate; (d) 2,4-dinitro-6-*sec*-butyl phenol (this is formed when the fungicide, binapacryl, is hydrolysed))

strawberries, gooseberries and roses. A combined spray of dinocap with captan or some similar fungicide is a valuable method of controlling mixed fungal infections, especially in apple orchards and on hops. **Binapacryl** is mainly employed to control powdery mildew and red spider mites on apples. Addition of pyrazophos (section 10.6) improves the control of powdery mildew (MAFF, 1980). In Britain, seven days should elapse between spraying and harvesting when either of these chemicals is used on an apple crop. Additionally, after spraying with binapacryl, livestock should be excluded from the treated area for a month. Both substances are very toxic to fish.

It is well known that simple dinitrophenates uncouple mitochondrial oxidation from phosphorylation – biochemists have used the parent substance, 2,4-dinitrophenol, specifically to study this important activity. (One of the best general reviews on this subject is probably still that of Simon (1953), but it does not deal specifically with fungicides.) Both dinocap and binapacryl are probably hydrolysed by enzymes in fungi to liberate free dinitrophenols which are then fungitoxic. Curiously, this logical assumption appears to be largely unsubstantiated because of the absence of suitable experimentation. However, if the assumption proves to be correct, the hydrolysis is another example of *lethal metabolism*, for the fungus becomes its own executioner (analogous examples in various branches of crop protection are to be found on pp. 57, 82, 265). The C-6 alkyl groups (figure 9.7) and the unsaturated acids attached to the phenolic hydroxyl group presumably act as carriers or 'shaped charges' that assist the 'toxophoric' dinitrophenol moiety to reach its site of action within the fungus.

Comparatively few metabolic studies have been undertaken on dinitrophenol insecticides, fungicides and herbicides, but a composite picture appears to be emerging. Dinocap does not persist long on plants (in one study on cucumber plants, the level fell from 15 to 1 ppm within 7 days). Initial metabolism probably involves hydrolysis at C-1 (figure 9.7) to give the free phenol, with subsequent reduction of one of the two nitro groups to an amino group. The latter is analogous to what happens to the weedkiller dinoseb (Hawkins and Saggers, 1974). Menzie (1969, 1978) referred to several other instances where the nitro groups of dinitro compounds are converted to amino groups (in studies variously carried out on locusts, rats and rabbits). Some dinitroalkyl phenols are also known to be degraded by oxidation of the C-6 alkyl group; for example, the *sec*-butyl side chain of dinoseb is oxidised to an isobutyric acid derivative when this herbicide is sprayed on apples (Hawkins and Saggers, 1974) and when it is fed to rats and rabbits (Ernst and Bar, 1964). No similar work appears to have been reported on dinocap, but it would be surprising if the octyl group on C-6 could not be hydroxylated by microsomal systems (chapter 3).

In view of the universal importance of oxidative phosphorylation, it is perhaps surprising that neither dinocap nor binapacryl is very toxic to mammals. The oral LD_{50} of dinocap for rats is about 1000 mg/kg body weight, whereas the corresponding figure for DNOC is about 30 mg/kg. The difference is probably related to the fact that the free phenol is the actual toxicant. The work of Bough *et al.* (1965) probably illustrates this, for it happens that the hydrolysis of binapacryl produces the herbicide dinoseb (this is why this substance has been included in figure 9.7) and dinoseb is very toxic to animals. They showed that, on feeding *binapacryl* to guinea pigs, death of the animals occurred when the level of *dinoseb* in the blood was the same as that which led to death when *dinoseb itself* was fed to the animals.

9.6 CHLORINE-SUBSTITUTED AROMATIC FUNGICIDES

Individual members of this diverse group of protective fungicides usually have limited but nevertheless locally important uses (table 9.3). In addition, pentachlorophenol, PCP, is used to preserve wood.

Quintozene (PCNB) and **tecnazene** (TCNB) are fungicidal nitrobenzenes (figure 9.8a). Quintozene is principally employed to control soil and seedborne fungal diseases, including damping off of brassicas and *Rhizoctonia* on cucumber, lettuce and flower crops. It can be used as a seed dressing or worked into the soil before planting. It is formulated as a dust and as a wettable powder.

Quintozene is insoluble in water and persists for some time in aerated soils. Thus, Wang and Broadbent (1972) found that its half-life in three types of soil varied from 4.7 to 9.9 months. However, in water-saturated unsterilised soil

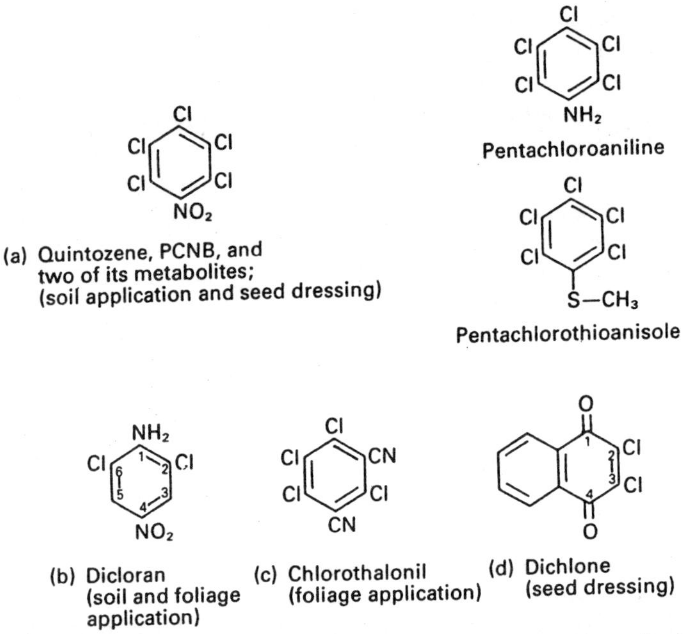

figure 9.8 Chloroaromatic protective fungicides. ((a) Pentachloronitrobenzene; (b) 2,6-dichloro-4-nitroaniline; (c) tetrachloro-*iso*-phthalonitrile; (d) 2,3-dichloro-1,4-naphthoquinone)

Ko and Farley (1969) found that it was largely converted *inter alia*, to pentachloroaniline within a few weeks. Pentachloroaniline was also one of several metabolites found in potatoes grown in quintozene-treated soil (Gorbach and Wagner, 1967) and in greenhouse soil samples regularly treated with quintozene (de Vos *et al.*, 1974). Clearly, the reduction of the nitro group to amino is commonly brought about by enzymes in soil organisms (compare section 9.4). It is of interest that several groups of workers have also found pentachlorothioanisole (figure 9.8a) amongst the metabolites of quintozene. The origin of the methylthio group of this substance is uncertain, although it could come from methionine or from a glutathione conjugate. Nakanishi and Oku (1969) found that sensitive strains of *Rhizoctonia* absorbed more quintozene than did resistant strains, but that the latter produced greater quantities of pentachloroaniline and pentachlorothioanisole (anisole is $C_6H_5.O.CH_3$).

Macris and Georgopoulos (1969) observed that quintozene is not effective against species of fungi which possess little chitin in their cell walls (e.g. *Phytophthora*). Conversely, sensitive fungi such as *Neurospora*, when analysed after treatment with quintozene, contained less hexosamine in their cell walls than did untreated specimens of the same fungus. It thus seems possible that quintozene interferes with chitin synthesis.

Tecnazene (tetrachloronitrobenzene, TCNB) was introduced in 1947 for use against dry rot in seed potatoes. Its half-life in soils is shorter than that of quintozene, probably because it is more volatile. It was soon evident, however, that it also inhibits sprouting of potato tubers in storage and it is now marketed as a dust for this purpose. It is less safe than quintozene on lettuces for it can delay hearting and stunt growth. Smoke formulations are used to control *Botrytis* on greenhouse plants, including tomatoes and chrysanthemums (MAFF, 1980). Since tecnazene is mixed with soil, it is of interest that it has been reported to have nematicidal properties (Rodriguez-Kabana *et al.*, 1978). At a rate of 0.025 g/kg soil a significant decrease in numbers occurred for several species of nematodes, including *Meloidogyne* spp., *Pratylenchus* spp. and *Tylenchorhynchus* spp.

Tecnazene is metabolised in pigeons and in rabbits to give tetrachloroaniline, a tetrachloro-aminophenol and a mercapturic acid derivative (section 3.6). Thus, nitro-group reduction seems to occur in the same way as for quintozene and the single aromatic C–H bond, can in addition, be hydroxylated. The mercapturic acid derivative is of interest in view of the thio-metabolites of quintozene (figure 9.8a). Literature upon tecnazene metabolism has been collected by Menzie (1969, 1974). The mode of action of tecnazene is uncertain, but since strains of fungi resistant to quintozene show cross-resistance to tecnazene, it is possible that tecnazene, too, interferes with some aspect of chitin metabolism.

Dicloran is a dichlorinated nitroaniline (figure 9.8b). It is used both as a soil-applied fungicide and as a foliar spray. Being insoluble in water and strongly adsorbed by clay particles, it is rather persistent in soils, in which its half-life is often upwards of a year). It is used as a foliar spray on lettuces and tomatoes to control *Botrytis*; in view of its persistence, a three weeks safety period should elapse between the last application and harvesting, even although its mammalian toxicity is low.

Dicloran is taken up by plant roots, and moves in the transpiration stream to the leaves, where it is soon destroyed enzymatically. Primary metabolism in plants and animals probably occurs by the same major route, namely the reduction of the nitro group to amino and the conversion of the (newly-formed) amino group to hydroxyl. In animals, the hydroxyl group may then conjugate with sulphate or with glucuronic acid (Menzie, 1969, p. 154) in the manner outlined in chapter 3.

The mode of action of dicloran is uncertain but it may affect fungal growth by inhibiting protein synthesis. This possibility is supported by the work of Weber and Ogawa (1965), who found that it reduced the rate at which radioactive leucine was incorporated into the proteins of *Rhizopus*.

Chlorothalonil (figure 9.8c) is a tetrachlorinated dinitrile of benzene. It controls a wide range of crop pathogens, especially on fruit canes and bushes, and shows particular promise for the control, by foliar application, of *Botrytis*, *Phytophthora* and *Rhizoctonia*. It has been reported to be the most promising of several substances tested, as foliar sprays, for the control of *Ascochyta*, a blight of faba beans (Kharbanda and Bernier, 1979). It reacts rather indiscriminately with thiol groups in cells, a reaction which possibly accounts for its biological activity, for exogenous thiol compounds reverse its fungitoxic action (Vincent and Sisler, 1968). However, glutathione is a substance frequently involved in detoxication (section 3.5) and is itself a thiol, so this same reactivity of chlorothalonil with thiols may be one mechanism whereby it is detoxified.

Dichlone (figure 9.8d) is a dichlorinated naphthoquinone. In the USA it is occasionally used as a seed dressing for legumes because it has a somewhat similar biological spectrum to organomercury compounds (section 8.5) and yet it is of much lower toxicity to man. It is not very useful as a foliar spray because it is decomposed by light. Dichlone is taken up rapidly by fungal spores and, on the basis of weight absorbed per gram of spores, it is amongst the most effective of protectant organic fungicides. Typically, the dose of dichlone causing 50 per cent inhibition of spore germination (I_{50}) is of the order of 500 μg/g, whereas the I_{50} of most organic fungicides is many thousands of μg per gram. It must, however, be remembered that the high figure for many fungicides (e.g. copper, captan) often reflects attachment to inert or less vital constituents of cell or cell wall. Dichlone inhibits the oxidation of both acetate and pyruvate and there is some evidence that this could result from its reaction with the thiol group of coenzyme A. It can, however, react with thiol groups in other compounds and also with amino groups, so its mode of action remains uncertain.

9.7 CATIONIC DETERGENTS

Various cationic imidazolines and amidines have been marketed as foliage-applied fungicides, but the most important is probably **dodine**, n-dodecyl guanidine acetate:

$$C_{12}H_{25}{-}NH{-}C(NH){-}NH_3^{+}\text{.........}^{-}OOC.CH_3 \qquad (9.1)$$

It was introduced in 1956 and has been used to control apple scab in seasons when the incidence of the disease is high. Byrde *et al.* (1962) found it to be no more effective than captan in seasons when the incidence was low. In addition to its protective function there is some evidence that it can eradicate recent (but not well-established) scab infections. It may be used on any growing edible crop without danger to the consumer and apples can be picked within one day of the final application.

Dodine, being a cationic detergent, is related to some of the spray supplements mentioned in chapter 2 and it is incompatible with anionic spreading agents. The base is moderately strong, so that the acetate is largely ionised within the pH range of biological importance, but a varying degree of hydrolysis will occur at different pHs. This probably explains why the toxicity of dodine to fungi varies with the pH of the external solution – in one instance, the substance was found to be twenty times more toxic at pH 7.8 that it was at pH 5.1.

The entry of dodine into spores is very rapid, at least 85 per cent of the amount finally present internally being taken up from the external medium within one minute of contact (compare copper cations, section 8.3). Such entry, from external solutions of appropriate concentration, can lead to an uptake of two per cent of the fungal cell weight. It appears to be rather firmly attached to anionic cell constituents such as (immobile) phosphate and carboxyl groups, for whereas captan can be washed out of captan-treated spores, those treated with dodine acetate retain all the detergent they have absorbed (Somers, 1966).

Dodine probably attacks lipoprotein membranes, disrupting vital processes which occur there. It may, in particular, inhibit mitochondrial electron transport (Pressman, 1963). In common with other detergents, it probably causes lysis of membranes by masquerading as membrane phospholipid, the more hydrophilic part of the molecule dissolving in membrane lipid and the more hydrophobic portion altering the attachment of the membrane protein. It is of interest that recent work with positive ion and non-ionic detergents has made it possible to separate membrane-bound components of hepatic microsomes and then to reconstitute functional mono-oxygenase (section 3.4) by putting the components together again.

REFERENCES

Anilkumar, T. B. and Sastry, M. N. L. (1979). *Phytopath. Z.*, **94**, 126

Bough, R. G., Cliffe, E. E. and Lessel, B. (1965). *Toxic. appl. Pharmac.*, **7**, 353

Byrde, R. J. W., Clarke, G. M. and Harper, G. W. (1962). *Rep. Long Ashton Res. Stn Univ. Bristol* (for 1961)

Chernoff, N., Kavlock, R. J., Rogers, E. H., Carver, B. D. and Murray, S. (1979). *J. Toxic. envir. Hlth*, **5**, 821

Dalvi, R. R. and Ashley, W. M. (1979). *Drug Chem. Toxic.*, **2**, 245

Eckert, J. W. (1967). *Fungicides: an Advanced Treatise*, Vol. 1 (ed. D. C. Torgeson), Academic Press, New York, p. 287

Ellis, M. A. and Paschal, E. H. (1979). *Seed Sci. Technol.*, **7**, 75

Ernst, W. and Bar, F. (1964). *Arzneimittel Forsch.*, **14**, 81

Fishbein, L. (1977). *Antifungal Compounds* (ed. M. R. Siegel and H. D. Sisler), Marcel Dekker, New York, Chapter 14

Gorbach, S. and Wagner, U. (1967). *J. agric. Fd Chem.*, **15**, 654

Hawkins, D. R. and Saggers, V. H. (1974). *Pestic. Sci.*, **5**, 497

Kaars Sijpesteijn, A. and Janssen, M. J. (1958). *Nature, Lond.*, **182**, 1313

Kharbanda, P. D. and Bernier, C. C. (1979). *Can. J. Pl. Sci.*, **59**, 661

Klayder, T. J. (1963). *J. Ass. off. agric. Chem.*, **46**, 241

Ko, W. H. and Farley, J. D. (1969). *Phytopathology*, **59**, 64

Macris, B. and Georgopoulos, S. G. (1969). *Phytopathology*, **59**, 879

MAFF (1980). *List of Approved Products for Farmers and Growers*; for details see p. 334

Martin, D. H. and Lewis, R. A. (1979). *Xenobiotica*, **9**, 523

Maude, R. B., Vizor, A. S. and Shuring, C. G. (1969). *Ann. appl. Biol.*, **64**, 245

Maude, R. B. (1978). C.I.P.A.C. Monograph 2 (ed. K. Jeffs), *Seed Treatment*, Chapter 9

Menzie, C. M. (1969, 1974, 1978). *Metabolism of Pesticides*, with Updates. For details, see references, chapter 5

Morehart, A. L. and Crossan, D. F. (1965). *Delaware Univ. Agric. Exptl Sta. Bull.*, 357, 26pp.
Moriya, M., Kato, K. and Shirasu, Y. (1978). *Mutat. Res.*, **57**, 259
Nakanishi, T. and Oku, H. (1969). *Phytopathology*, **59**, 1761
Northover, J. (1978). *Pl. Dis. Reptr.*, **62**, 706
Nutahara, M. and Yamamoto, M. (1978). *J. Pestic. Sci. Jap.* (Nihon Noyakugaku Kaishu), **3**, 101 (Japanese: reported from Abstract only)
Owens, R. G. (1969). *Fungicides: An Advanced Treatise*, Vol. 2 (ed. D. C. Torgeson), Academic Press, New York, Chapter 5
Pressman, B. C. (1963). *J. biol. Chem.*, **238**, 401
Richmond, D. V. and Somers, E. (1963). *Ann. appl. Biol.*, **52**, 327
Rodriguez-Kabana, R., King, P. S. and Adams, J. R. (1978). *Nematropica*, **8**, 69
Rowe, R. C. and Farley, J. D. (1978). *Phytopathology*, **68**, 1221
Ruch, D. G. and Bland, C. E. (1979). *Can. J. Bot.*, **57**, 2116
Simon, E. W. (1953). *Biol. Rev.*, **28**, 453
Somers, E. (1962). *Sci. Prog., Lond.*, **50**, 218
Somers, E. (1966). *Colston Pap.*, **18**, 299 (18th Symp. Colston Res. Soc., Bristol)
Soteros, J. J. (1979). *N.Z. Jl agric. Res.*, **22**, 191
Vekshtein, M. Sh. and Khitsenko, I. I. (1971). *Gig. Sanit.*, **36**, 23
Vincent, P. G. and Sisler, H. D. (1968). *PhysiologialPl.*, **21**, 1249
de Vos, R. H., ten Noever de Brauw, M. C. and Olthof, P. D. A. (1974). *Bull. envir. Contam. Toxic.*, **11**, 567
Wang, C. H. and Broadbent, F. E. (1972). *Proc. Soil Sci. Soc. Amer.*, **36**, 742
Weber, D. J. and Ogawa, J. M. (1965). *Phytopathology*, **55**, 159
Woodcock, D. (1977). *Antifungal Compounds*, Vol. 2 (ed. M. A. Siegel and H. Sisler), Marcel Dekker, New York, p. 209
Zdzienicka, M., Zielenska, M., Tudek, B. and Szgmczyk, T. (1979). *Mutat. Res.*, **68**, 9

10 Systemic Fungicides

10.1 CLASSIFICATION; ADVANTAGES AND DISADVANTAGES OF SYSTEMIC FUNGICIDES

Following the success of systemic insecticides and translocated herbicides, strenuous efforts made to develop systemic fungicides were increasingly rewarded from about 1965. The work was based on the premise that a systemic fungicide might well be able to attack internal mycelium and penetrating haustoria and thereby overcome the problem (section 8.1) that arises from the fact that even small pieces of surviving mycelium can regenerate. It was also hoped that systemic fungicides would reduce dependence upon external protective poisons, which suffer from the double disadvantage that they are subject to weathering and that they cannot give protection to plant growth initiated after spraying.

In table 10.1, systemic fungicides are divided into seven groups. Their principal uses are also summarised in the table and it will be seen that while the benzimidazoles have a rather wide antifungal spectrum most of the others are somewhat specific in their action. Members of most of the main botanical divisions of plant pathogens are attacked by members of one or other of the seven groups, but until quite recently the *Oomycetes* were an exception. However, Edgington *et al.* (1980), in a recent review of systemic fungicides, referred to several compounds which show great promise for the control of important pathogens belonging to this group.

When considering systemic fungicides, a little caution is necessary about the precise meaning of 'systemic'. It was mentioned elsewhere that this term does not necessarily imply highly efficient uptake through plant *leaves.* Several systemic fungicides (e.g. benomyl) can indeed be effectively applied to leaves, which they may enter to some extent. This entry is followed by translocation across leaves and by acropetal movement towards the leaf tips and margins. There is little translocation out of sprayed leaves, so the action cannot be regarded as truly systemic. However, many fungicides that are little more than quasisystemic (compare section 4.3) when applied to foliage are nevertheless absorbed excellently when they enter plants through the roots (after soil or seed application), or through the testa (following seed treatment).

Most systemic fungicides appear to attack very specific processes in fungi and activities at many of these sites of action seem to be influenced by single gene mutation. Probable or possible specific actions of various systemic fungicides

Table 10.1 Some systemic fungicides; their classification and uses

Group	Examples	Examples of uses
Antibiotics	streptomycin	bacterial diseases (stone and pome fruit)
	kasugamycin	*Pyricularia* (rice blast fungus)
Benzimidazoles	benomyl	wide spectrum, e.g. *Botrytis*, *Gloeosporium*; powdery mildews; scab (apples); *Rhynchosporium*; *Cercospora* leaf spots (bananas, sugar beet); eye-spot in cereals
	thiophanate-methyl	wide spectrum, similar to benomyl
	thiabendazole	reduction of gangrene, dry rot and other diseases (stored potatoes)
Pyrimidines	ethirimol	seed treatment and sprays against powdery mildew (barley)
	bupirimate	powdery mildew (apples and roses)
Piperazines	triforine	powdery mildew (spring barley); rusts (cereals)
Morpholines	tridemorph	mildew (barley, oats, winter wheat)
	dodemorph	powdery mildew (roses)
Oxathiins	carboxin	for seed treatment – rust, smut, bunt (cereals); Sclerotium (sugar beet); seed borne cotton diseases
	oxycarboxin	mainly a foliage spray on cereals, ornamentals
Miscellaneous	pyrazophos	powdery mildew (hops)
	chloroneb	for soil-action – against *Rhizoctonia*, *Pythium* and a few other *Oomycetes*
	metalaxyl	*Phytophthora* and some other *Oomycetes* by soil application and seed treatment
	IBP	*Pyricularia* (rice blast fungus)
	phosetyl-Al	*Phytophthora* (downy mildew of vines)

include interference with nucleotide base synthesis, with polynucleotide or protein formation and with the synthesis of steroids and of components of lipoprotein membranes. It is perhaps because of their highly specific effects that they select out existing mutants or even favour rapid gene mutation. In consequence, their use has led to a serious complication. In the decade or so since they were introduced, more *resistance* has appeared than in the previous fifty years when only protective fungicides were employed. Furthermore, resistance of fungi engendered by systemic fungicides tends to be long-lasting – that is to say, it persists for many generations after use of the compound has ceased – whereas resistance caused by protective fungicides is sometimes rather ephemeral (although resistance to dodine and organomercury compounds has been rather stable). The genetics of fungicide resistance and the biochemical mechanisms which are responsible for it have been reviewed by Georgopoulos (1977). This author has listed 41 instances of resistance developing to just one systemic compound – thiabendazole – in 23 fungi during the period 1970 to 1976. Similarly Dekker (1977) listed 17 species that were reported between 1969 and 1976 to have developed resistance to benomyl.

The use of systemic fungicides is increasing and the potential seriousness of the resistance problem cannot be over-emphasised. Alternation of the use of systemic and other fungicides could be expected to reduce the rate at which resistance develops, as also in some cases, may the use of synergists. However, the semi-permanence of resistance to systemic fungicides, and the frequency of occurrence of cross-resistance, are factors that signal caution in their use and may yet militate against their long-term success.

10.2 ANTIBIOTICS

The medically-important **streptomycin** has been used in crop protection as a bacteriostatic and antifungal agent. Its principal use is to control bacterial diseases of stone and pome fruits, but it has also been used for early spraying of downy mildew on hops. It is soluble in water and is readily taken up by the roots of plants. Streptomycin is somewhat phytotoxic, causing chlorosis by interfering with the synthesis of chlorophyll. It probably attacks prokariotes by attaching itself to ribosomes, one molecule of streptomycin uniting with one 30-S ribosomal subunit. It appears to prevent the ready movement of peptidyl-t-RNA from the P site to the A site* (it is to the peptidyl-t-RNA that the new amino acid is brought on its own t-RNA, and it is to this that it is attached during protein synthesis). Streptomycin has a much lower affinity for non-bacterial ribosomes than it has for those in prokariotes, and it does not affect

*** Further details are given in Edwards and Hassall (1980).**

protein synthesis in eukaryotic cytoplasm. Some yeasts and *Oomycetes* are sensitive to it and it probably affects *mitochondrial* protein synthesis in such organisms (Schatz and Mason, 1974).

Cycloheximide, another antibiotic produced by *Streptomyces griseus*, is a complex glutarimide derivative. It is too toxic for use in medicine and its use in crop protection also is often limited by phytotoxicity. It is more lipophilic than streptomycin and this is possibly the reason why it is more antifungal and less antibacterial than is streptomycin. At concentrations of 2–5 ppm it has been used to eradicate cherry leaf spot and wheat stem rust. In Britain, it has been accepted for use in forest nurseries to control needle blight of red cedar (Martin and Worthing, 1976).

Cycloheximide was shown in 1958 to be a potent inhibitor of protein synthesis and later work has indicated that it probably does this by preventing the transfer of amino acids from t-RNA to the growing peptide chain at ribosomal sites (Cremlyn, 1978). It certainly acts late in the incorporation process for it neither interferes with the activation of amino acids nor with their transfer to t-RNA. There is evidence that its action may be a consequence of its binding to the 60-S subunit of eukaryotic ribosomes (Roa and Grollman, 1967). In contrast to streptomycin, it inhibits protein synthesis in eukaryotic cytoplasm, but not in mitochondria.

Blasticidin S, a complex pyrimidine derivative isolated from *Streptomyces griseochromogenes*, is effective against a range of bacteria. It is much less toxic to most fungi, an important exception being *Pyricularia oryzae*, the organism responsible for paddy blast of rice. It has been marketed to control this disease, but it damages rice at concentrations not greatly above those needed to control rice blast disease.

At five ppm, blasticidin S inhibits the incorporation of glutamic acid into the protein of the rice blast fungus. This probably happens as a result of it attaching itself to the A site receptor of the 60-S subunit of the ribosome which aminoacyl-t-RNA should occupy. This effectively blocks the peptidyltransferase reaction and in consequence prevents peptide chain elongation (Corbett, 1974).

Kasugamycin is another antibiotic used against the rice blast fungus. It is an aminoglycoside and was isolated from *Streptomyces kasugaensis*. It is now used more widely in Japan than is blasticidin S, principally because the phytotoxicity is much lower. In addition, it is degraded in plants much less rapidly than is blasticidin. Although it can enter rice plant roots, it is used as a foliar spray since it is not economical to add it in large quantities to paddy water.

It is of interest that it is eradicant but not protective in action (contrast section 8.1). The reason is that it has little effect on *Pyricularia oryzae* before the fungus penetrates the host plant. On the other hand, it prevents the spread of the organism within the plant and it inhibits sporulation. The mode of action of kasugamycin appears to be similar to that of blasticidin, for it selectively inhibits the initiation of bacterial protein synthesis. It has no effect on nucleic acid metabolism.

10.3 BENZIMIDAZOLES

Benzimidazole (figure 10.1a) is the parent substance of a family of systemic fungicides, including benomyl, thiophanate-methyl and thiabendazole. If one of the hydrogen atoms of the amino group of carbamic acid is replaced by the benzimidazole radical, benzimidazolyl carbamate is formed, the methyl ester of which is called MBC or carbendazim and is shown in figure 10.1b. Benomyl is a n-butylcarbamoyl derivative of carbendazim, the carbamoyl substitution being on N-1 of benzimidazole (figure 10.1c). Note that in compounds such as *butyl-carbamate*, one of the amino nitrogen atoms is replaced by the substituent, whereas in *carbamoyl* compounds the carboxylic acid group is attached, with loss of water, to a base (in this case, to a substituted benzimidazole).

Benomyl is a wide-spectrum systemic fungicide, introduced in 1967. A few of its uses are listed in table 10.1, but it has been used for foliar application, for seed and soil treatment, as a root and bulb dip, and for post-harvest protection of produce. Despite its wide spectrum of action, the *Peronosporales* (pathogenic *Oomycetes*), some *Basidiomycetes* and dark-spored *Deuteromycetes* are insensitive to it, both *in vitro* and in the field (Erwin, 1973). Thus, amongst the diseases that it does *not* control are downy mildew of vine, potato blight, and damping off by *Pythium* (but *Fusarium* is sensitive).

Benomyl is applied to *foliage* to control a variety of powdery mildews (*Erysiphales*). As will be seen shortly, rather little actually enters the leaves by

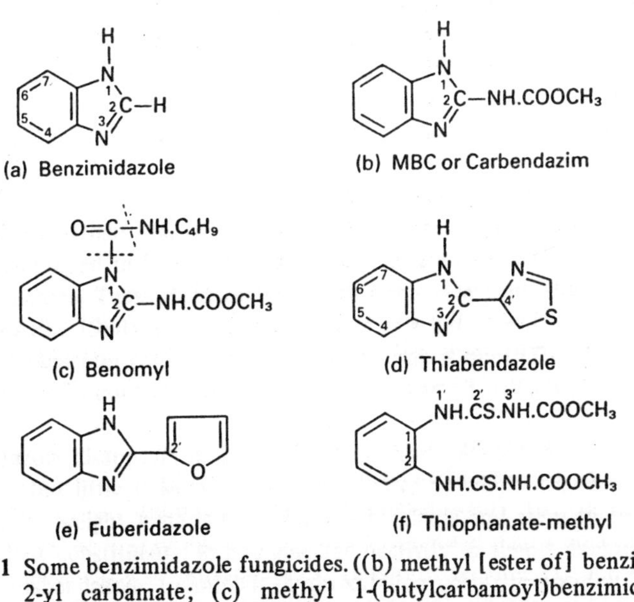

figure 10.1 Some benzimidazole fungicides. ((b) methyl [ester of] benzimidazol-2-yl carbamate; (c) methyl 1-(butylcarbamoyl)benzimidazol-2yl carbamate; (d) 2-(thiazol-4′-yl) benzimidazole; (e) fuberidazole; (f) 1,2-di(3′-methoxycarbonyl-2′-thioureido) benzene)

this route of application, most of the deposit actually acting as a superficial protectant. Backman *et al.* (1979) found that benomyl was more effective than fentin hydroxide against *Septoria* and *Colletotrichum* on soybeans, being particularly effective when damp conditions between flower blossom and pod fill favoured infection. In such cases yield was often 10–15 per cent more than that obtained from untreated plants. It is very effective against brown rot of stone fruit. When it is used in apple orchards prior to leaf fall it prevents the formation of perithecia during wintertime. Rather similarly, attempts have been made to decrease spoilage during storage by application at flowering time. An example is provided by Mason and Dennis (1978), who applied it to raspberries to decrease berry contamination by *Botrytis*. Such pre-harvest sprays decreased the percentage of berries with *Botrytis* mycelium developing in the proximal row of drupelets (i.e. arising from infection of the flower parts) but the fungicide did not translocate or persist sufficiently to prevent mycelium developing elsewhere on the berry.

Benomyl has been used as a bulb or corm *dip* to control, *inter alia*, various rots caused by species of *Botrytis*, *Fusarium* and *Sclerotinia*. It is also employed as a *seed treatment* to control loose smut of winter wheat (Martin and Worthing, 1976) as well as *Ascochyta* infections of pea seeds and, mixed with captan, to control anthracnose disease (*Colletotrichum*) of white beans (Edgington *et al.*, 1980). *Post-harvest protection* can often be afforded by dipping or dusting; an example is the treatment of banana hands to prevent the spread of *Gloeosporium musarum*. Ricci *et al.* (1979) reported the use of 10 minute dips no later than three days after harvest to control post-harvest rot in stored tubers of the cush-cush yam (*Dioscorea trifida*). This is caused by *Penicillium oxalicum*. Other examples of post-harvest protection have been provided by Eckert (1969).

Soil treatment with benomyl and other systemic compounds can be a very effective, if sometimes expensive, approach to disease control. It is particularly useful for *high value crops* such as strawberries, for which even vascular wilts can sometimes be economically controlled. The main problem is that high treatment levels are needed to maintain a toxic dose for uptake by roots. However, it is often financially feasible to use it in greenhouses and sometimes also in the field. It can be applied either as a soil drench or by soil mixing. Examples are the control of *Verticillium* wilt of tobacco by applying benomyl as a pre-plant soil treatment and of *Verticillium* and *Fusarium* wilts of tomato by using it as a soil drench (Edgington *et al.*, 1980).

Benomyl is rapidly hydrolysed within plants, the butylcarbamoyl side chain being removed to give **carbendazim**, the methyl ester of benzimidazole carbamic acid (figure 10.1b). This substance is fungitoxic and is indeed marketed as a systemic fungicide which is used as a spray on European cereals, especially to control eye-spot (*Pseudocercosporella*). Carbendazim is formed in plants from benomyl and is probably the active substance at the site of action. This view is supported by the fact that, if an aqueous suspension of benomyl is hydrolysed by boiling, the fungicidal effectiveness in plants decreases whereas the antifungal activity *in vitro* increases (Fuchs *et al.*, 1972).

The rate of breakdown of carbendazim (including carbendazim formed from benomyl) appears to differ considerably in different plants. Ayanaba *et al.*, (1973) found more than 75 per cent of a ^{14}C label still present in carbendazim 50 days after root uptake by peas, but in other plants both side chains of benomyl can be removed to give considerable amounts of 2-aminobenzimidazole. There is general agreement that the benzimidazole ring is not readily ruptured.

In soils, the half-life of total benzimidazole derivatives is often six months or more after addition of benomyl or carbendazim to the soil. These total residues do, however, include 2-aminobenzimidazole, which is not fungicidal. Since 1972, when Helwig reported the isolation from soil of micro-organisms capable of using benomyl as a sole source of carbon, several similar reports have appeared. This phenomenon has been investigatged by Fuchs and de Vries (1978), using strains of *Pseudomonas*. These workers were unable to detect growth when benzimidazoles other than benomyl were employed. They concluded that the probable explanation was that the n-butylamine moiety was removed from N-1 (figure 10.1a) and then, after deamination, used as an energy source. A report that carbendazim can act as a (poor) carbon source can be explained similarly, the methyl formate side chain probably being dislodged and then metabolised.

Axness and Flecker (1979) have studied the breakdown of labelled benomyl in rats following oral ingestion. Within 24 h of feeding, 40 per cent of the ^{14}C and 55 per cent of the tritium originally in the labelled benomyl appeared in the urine. One major metabolite comprised butylamine attached to glucuronic acid, so in this case hydrolysis presumably occurred at the CO–NH linkage indicated in figure 10.1c. In addition, butylcarbamoyl derivatives of cysteine and acetylcysteine were recovered from the urine. This possibly implies that transfer to glutathione of a butylcarbamoyl group occurred by rupture of the $C–N_1$ linkage indicated in figure 10.1c, followed by further metabolism of the sort described in section 3.6.

Thiabendazole (figure 10.1d) is a non-carbamate member of the group. It is principally employed in Britain to reduce the damage done by various diseases to stored potatoes. These diseases include dry rot, skin spot, silver scurf and gangrene (MAFF, 1980). It has also been used as a bulb dip to protect certain bulbs against *Fusarium* rot, as a seed treatment against bunt (*Tilletia*) of wheat and as a dressing to control blue and green mould on citrus fruit (see Cremlyn, 1978).

Thiabendazole disappears more rapidly from both plants and soil than does carbendazim, but little is known about the routes by which it is metabolised. High molecular weight materials are formed in plants and seem to be firmly bound to plant constituents; extractable radioactivity is almost exclusively in unchanged thiabendazole. In animals, hydroxylation occurs in position 7 in the benzene moiety (figure 10.1d) and this, after the formation of sulphuric or glucuronic acid conjugates, is excreted in the urine. Metabolism in animals is of some importance for thiabendazole is also employed as an anthelmintic. One of several recent investigations into its use for this purpose led to the conclusion that at 100 mg/kg body weight it is effective in removing trichostrongyloid larvae in sheep (Ogunsusi, 1978).

Fuberidazole (figure 10.1e) is closely related to thiabendazole but has a furane ring attached to benzimidazole. It is used in Europe for seed treatment. It was early shown to be valuable as a seed treatment against *Fusarium*, and both *F. culmorum* of peas and *F. nivale* of rye can be controlled by it. Good control of the snow mould, *Calonectria nivalis* and some reduction in the incidence of infections by *Puccinia triticini* have also been reported (see Woodcock, 1977). In common with other benzimidazoles it inhibits mycelial growth rather than the germination of spores, probably because it interferes with DNA synthesis (see below).

Thiophanate-methyl (figure 10.1f) is another fungicide with a wide biological spectrum. It is used to control *Botrytis*, *Gloeosporium* and powdery mildews. In addition, it controls scab of apples as well as *Sclerotinia* on several crops and *Cercospora* on sugar beet. It is used as a paint for pruning cuts and as a seed treatment for onions. It is included in the benzimidazole group because it undergoes decomposition and rearrangement within the crop plant to give rise to carbendazim. As in the case of benomyl, this is probably the actual toxicant at the site of action, a fact which helps to explain why benomyl, carbendazim and thiophanate-methyl have similar biological spectra.

Benomyl, thiabendazole, fuberidazole and thiophanate-methyl all act on fungi at concentrations ranging from one to 100 ppm. Within this range, lower concentrations prevent elongation of germ tubes of sensitive species while higher concentrations reduce spore germination. Bourgois *et al.* (1977) observed that benomyl alters the ultrastructure of the microconidia of *Fusarium*, causing a marked thickening of the cell walls and the development of lysosome-like vesicles in the cytoplasm. In contrast to the relatively low doses which elicit these effects in fungi, most are of low toxicity to mammals – the oral LD_{50} of benomyl to rats is, for example, greater than 9000 mg/kg.

The fungitoxic action of all benzimidazoles is probably caused by one or both of two related but distinct effects, namely, an interference with DNA synthesis and an interference with the formation of spindle fibres in dividing nuclei. It has been found that low concentrations of benzimidazoles inhibit DNA synthesis within an hour or so of application to several species of fungi and, in the case of *Fusarium*, small quantities of certain purines reduce the toxicity of benzimidazole. Furthermore, Clemons and Sisler (1971) demonstrated that carbendazim does not kill fungi unless the test medium supports fungal growth. In the bacterium *Escherichia coli*, benzimidazole is incorporated into DNA and RNA in the place of guanine. All these observations are consistent with the possibility that these fungicides interfere with DNA synthesis in this bacterium because of their structural resemblance to adenine and guanine.

On the other hand, Kumari *et al.* (1977), working with *Fusarium oxysporum* and tritium-labelled carbendazim, could adduce no evidence for a direct interaction between carbendazim and *fungal* DNA. Labelled carbendazim was not incorporated into the DNA of *Fusarium* and benomyl did not alter the base composition of its DNA. Thiabendazole at 80 μM was found by Davidse and Flach (1978) completely to block mitosis in hyphae of *Aspergillus nidulans* but it did not stop the synthesis of DNA or RNA. Earlier evidence that mitosis in some organisms was more readily affected than DNA synthesis had led Davidse (1973) to postulate that the primary effect of these fungicides is on the microtubule assembly of the spindle fibres. Indeed, it has been proposed that tolerance in a resistant strain of *Penicillium expansum* is a consequence of an alteration in the structure of the tubulin protein of the spindle fibres.

Kumari *et al.* (1977) have pointed out that the duration of different stages in nuclear division (G1, G2, S) varies greatly in different species. Such differences could be in part responsible for the conflicting evidence just described. For *Fusarium oxysporum*, for example, the G1 and G2 stages are of long duration compared to the time it takes mitosis to occur so that most cells in unsynchronised cultures will be in the G stages at any one time. This enables the effects of fungicides on DNA synthesis and on mitosis to be separated in a way not possible if the G1 stage is very short. Kappas *et al.* (1976) have suggested that benzimidazoles may provoke mutations, not by misincorporation during replication but by misrepair of gaps in newly-synthesised DNA. The subject has been reviewed by Davidse (1977).

The propensity of benzimidazoles and other systemic fungicides to induce *resistance* in fungi has been referred to (section 10.1) and this tendency is further illustrated by the following examples. Hisada *et al.* (1979) reported that *Botrytis cinerea* on rose plants became highly resistant after the plants had been sprayed only five times. Chandler *et al.* (1978) isolated benomyl-tolerant strains of *Cladosporium carpophilum* from peach orchards in Georgia after benomyl had been applied between four and seven times a year for four years. They found that the strain retained its resistance for at least two years after isolation. Penrose *et al.* (1979) reported that tolerance of *Sclerotinia fructicola* to benzimidazoles was first recorded in New South Wales in 1976, since when it has been observed in eleven orchards of stone fruit trees. The incidence of infection was mapped and resistant strains were found to be present on scattered groups of trees, tolerant and susceptible strains coexisting on the same trees. Tolerance appeared to confer no competitive disadvantage and the authors concluded that resistance was in consequence likely to persist for a long time. Other examples include the development of resistance of *Venturia* (the cause of apple scab) and of *Cercospora* on sugar beet in Greece (the resistance of *Cercospora* has proved to be extremely persistent).

It was mentioned earlier that carbendazim is believed to be the primary toxicant irrespective of whether the applied fungicide is benomyl, thiophanate-methyl or carbendazim (figure 10.1). It is therefore not surprising that, with the occasional exception of thiabendazole, the development of resistance to one member of the group usually confers an automatic tolerance of most of the others. An illustration is provided by the work of Shabi and Katan (1979) who isolated nine cultures of carbendazim-resistant *Venturia pirina* from different pear orchards in which benomyl no longer controlled scab. Crosses between these resistant strains and sensitive wild types showed that carbendazim resistance is conferred by a mutation in a single gene and that the mutation is unaffected by modifying genes or by cytoplasmic components.

Although cross-resistance frequently occurs within the group there are nevertheless some exceptions, especially with regard to sensitivity to thiabendazole. Cross-resistance between benomyl and this substance is indeed common, but some mutants induced by ultraviolet light have been found to be resistant to thiabendazole while retaining high sensitivity to benomyl (van Tuyl *et al.*, 1974). On the assumption that mutation modifies microtubular protein it is reasonable to suppose that a modification that reduces affinity of thiabendazole for tubulin could also result in a greater than usual affinity for carbendazim.

It is, however, difficult to explain an observation by Nishijima and Smalley (1979) in a similar manner. These workers isolated three colonies of *Ceratocystis ulmi* from American elms, the growth of one of which (coded 99–T2) was hardly affected by a concentration of carbendazim ten thousand times that which inhibited typical sensitive wild strains. This carbendazim-tolerant strain was nevertheless sensitive to benomyl, thiophanate-methyl and thiabendazole, in potato-dextrose agar growth tests, at concentrations of about one hundredth of the concentration of carbendazim which was needed to inhibit its growth (figure 10.2). The tolerance appeared to be associated with a single gene mutation. The difference in concentrations of benomyl and carbendazim is so great that it seems incapable of explanation in terms of alteration in uptake, alteration in the rate of metabolism, or change in susceptibility of the site of action (if carbendazim is, indeed, the common ultimate toxicant for most benzimidazoles).

Despite the low acute oral toxicity of most benzimidazoles to animals, some fear has been expressed about possible toxic hazards associated with their use. This doubt has arisen because positive results have been reported in mutagenicity tests from time to time. Fairly recently Ficsor *et al.* (1978) have carried out a comprehensive series of mutagencity tests on benomyl, carbendazim and their commercial formulations. In these tests they used several different assay techniques with mutants of *Salmonella typhimurium*, captan and streptozotocin being the positive control compounds. In non-overlay spot tests, no mutagenicity was observed using benomyl over the range of 50–5000 μg per spot. In overlay spot tests weak mutagenic activity was observed only in one sample of carbendazim. In microsomal activation plate assays benomyl and carbendazim gave negative results over the range of 50–2000 μg per plate and in rodent host-mediated assays with benomyl all the tests gave uniformly negative results. The same authors give references to earlier work where the assay indications were positive.

Both benomyl and carbendazim are sorbed on to soil organic matter and clay particles. After root uptake, benomyl is hydrolysed to carbendazim and it is

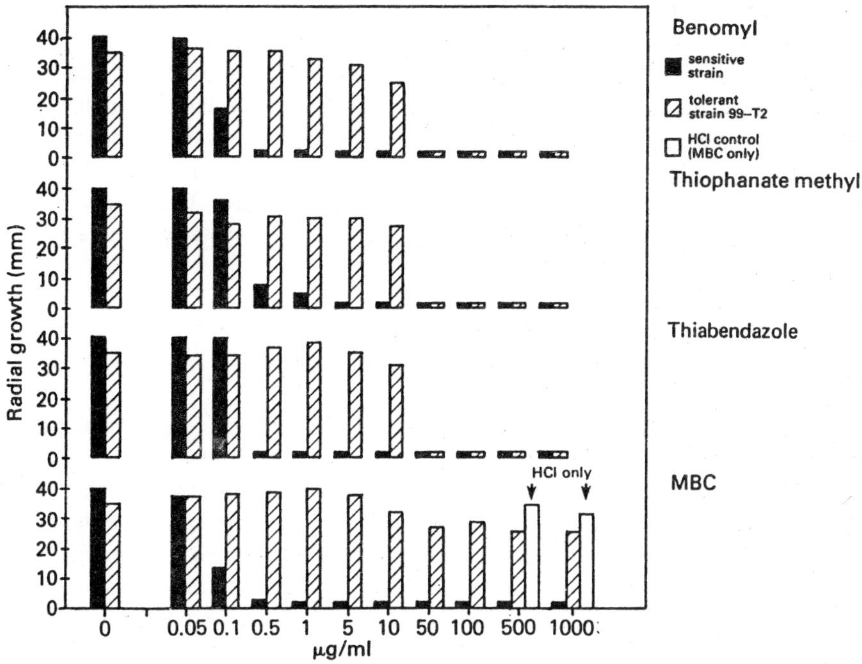

figure 10.2 MBC tolerance of a strain of *Ceratocystis ulmi* (after Nishijima and Smalley, 1979, with permission) It shows the radial growth of *C. ulmi* (sensitive and tolerant strains) on potato–dextrose agar containing the concentrations shown of one of the fungicides. At higher concentrations MBC (carbendazim) needed HCl to dissolve it so extra HCl controls are included at 500 and 1000 μg/ml MBC. These contained the same amount of HCl as the MBC treatments and show that the HCl alone was slightly inhibitory)

principally this substance which moves from roots to foliage. Indeed, Fuchs *et al.* (1972) have suggested that benomyl and thiophanate-methyl are bound to root constituents, forming a useful insoluble reservoir from which the active component is released more or less slowly into the transpiration stream. In contrast, thiabendazole is probably translocated unchanged. Since movement is in the xylem, the translocated substances pass predominantly to the leaves rather than to fruit (Siegel and Zabbia, 1972).

When uptake is via the roots, the main problem is the difficulty of getting compounds of low solubility near to the roots of crops. This difficulty increases as the root system becomes deeper, so that the root uptake method of application is, in practice, limited to shallow-rooted plants and to potted and greenhouse plants. The method of seed treatment ensures that the fungicide is readily available to the plant with minimal interference from soil constituents.

When applied to foliage, the main movement of these fungicides appears to be translaminar rather than from leaf to leaf (Solel and Edgington, 1973). In other words, systemic action is very limited following foliage application. Translaminar movement is facilitated by non-ionic detergents which not only act as spreaders to increase the area of contact, but also increase uptake and penetration perhaps by softening cuticular wax. Penetration is usually greater through the abaxial surface of the leaf, partly because the cuticular wax is thinner there and partly because there may be more stomata present. Penetration in the absence of detergent or oil is often inefficient; for example, Upham and Delp (1973) reported that only 5 per cent of a dose of benomyl penetrated the adaxial surface of cucumber leaves.

10.4 PYRIMIDINE AND PIPERAZINE DERIVATIVES

Three pyrimidine derivatives are commonly used as systemic fungicides for the control of *powdery mildews*; they are of little use against other types of fungi. **Dimethirimol** was introduced in 1968 and is very highly selective in its action, being used almost exclusively against *Sphaerotheca fulginea*, the powdery mildew of the cucumber family. It is particularly effective when uptake is through the roots and is applied to the soil around the plant. It is less used nowadays because of the tendency for resistant strains of the fungus to appear, although it has been reported that, when applied together with benomyl, resistance is much less likely to develop (Ebben and Spencer, 1973).

Ethirimol (figure 10.3a) is also highly specific in its action, being particularly

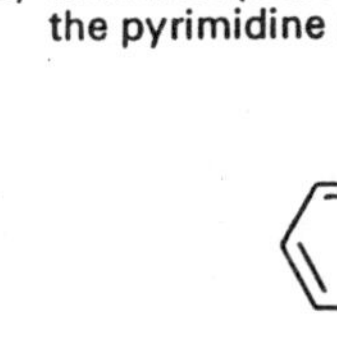

(a) Ethirimol (based on the pyrimidine ring)

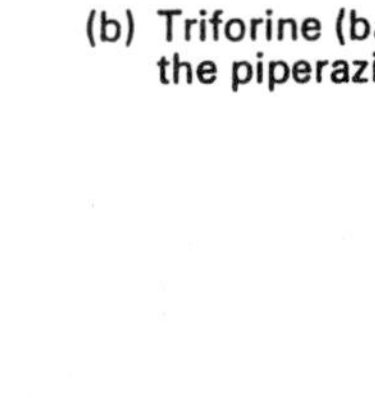

(b) Triforine (based on the piperazine ring)

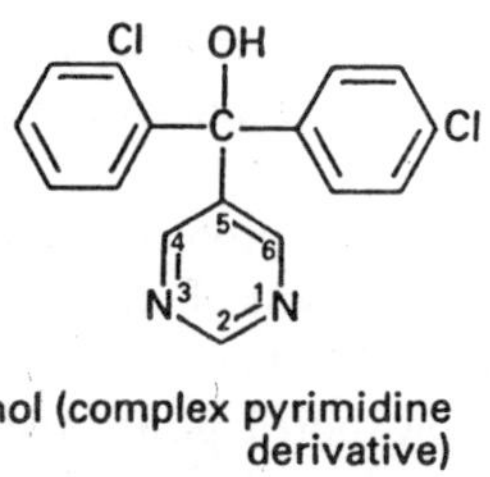

(c) Fenarimol (complex pyrimidine derivative)

figure 10.3 Examples of pyrimidine and piperazine fungicides. ((a) 5-butyl-2-ethylamino-6-methyl pyrimidin-4-ol; (b) 1,4-di(2′,2′,2′-trichloro-1′-formamidoethyl)-piperazine; (c) (2-chlorophenyl)-(4-chlorophenyl)-(pyrimidin-5-yl) methanol) effective against *Erysiphe graminis*, the powdery mildew of barley. In Britain it is normally formulated as a liquid seed dressing but it is also available as a liquid for foliage application and as granules for application in the furrow at the time the barley is drilled. Since relatively tolerant strains of the fungus are normally present in barley and are given a selective advantage when ethirimol is used, the incidence of tolerant strains can be reduced by restricting its use as a seed treatment to spring (rather than to winter) barley (Shephard *et al.*, 1975). It can also be applied as a foliage spray to wheat, barley and oats. Early control of mildew can encourage root growth and tillering, and this proliferation is probably responsible in part for the 10 per cent or more increase in yield which commonly accompanies its use (Brooks, 1971).

Bupirimate is a dimethylsulphamoyl $(CH_3)_2N.SO_2.O$- ester of ethirimol, this group being attached to C-4 of the pyrimidine ring instead of the hydroxyl group. It has a somewhat wider spectrum of action than the previous two compounds and is effective against the powdery mildews of apples, roses and some other horticultural crops. It is normally applied to foliage and stems for it is much less effective than ethirimol and dimethirimol when uptake is through the roots.

Ethirimol is readily absorbed by roots of plants and moves upwards in the apoplastic system. No symplastic movement out of sprayed leaves usually occurs so neither leaves formed after spraying nor lower parts of the plant are protected (i.e. it is not systemic when applied to foliage). Movement in leaves is towards their margins.

The primary metabolism of ethirimol is similar in plants and animals and takes routes that can be anticipated from the general reactions described in chapter 3. It can be N-de-ethylated to a primary amine of low but measurable biological activity (in the case of dimethirimol, where two methyl groups are attached to the C-2 amino group, mono-N-demethylation mildly activates and double demethylation greatly reduces activity). Alternatively, the butyl group can be hydroxylated in the position shown in figure 10.4, although this route does not appear to be as important in plants as it is in animals. Conjugation (which for many groups of compounds is a secondary stage in metabolism) can occur on hydroxyl C-4 *without* any preliminary change. In plants, a glucoside is usually formed, whereas in animals glucuronic acid is attached to C-4. A consequence of these changes is that ethirimol has a half-life of less than a week in barley plants when uptake is through the roots.

Ethirimol acts on *Erysiphe* when present at less than 5×10^{-7}M within plants and on *Sphaerotheca* at less than 10^{-8}M *in vitro*. Such low toxic concentrations imply a specific effect on some metabolic process. One possible candidate is some aspect of purine metabolism for Bent (1970) observed that the fungitoxic action of ethirimol and of dimethirimol was reduced in the presence of purines and by addition of folic acid. For the latter reason Sampson (1969) suggested that it might interfere with the one-carbon transfer process which is essential in purine biosynthesis, Hollomon (1979) also observed that adenine

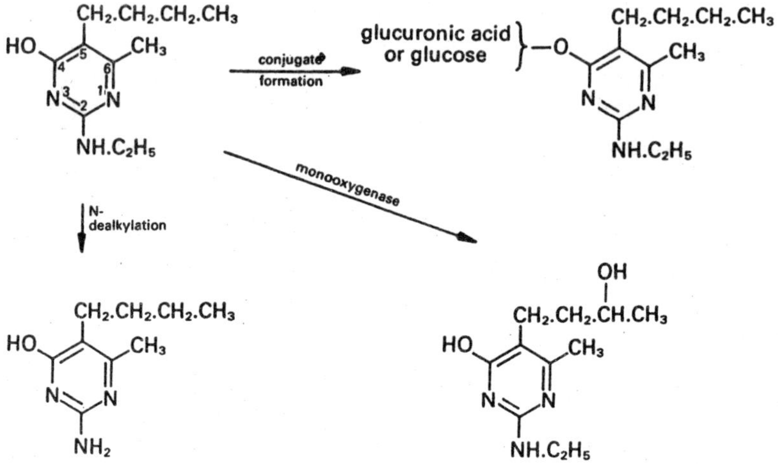

figure 10.4 Metabolism of ethirimol

reduces the fungitoxicity of ethirimol and concluded that it may interfere with adenine metabolism in the barley powdery mildew.

When ethirimol is absorbed through plant roots it inhibits the development of mildew spores prior to their penetration of the leaf and causes the vegetative hyphae to grow away from the leaf instead of being adpressed to it. Excellent electron micrographs which illustrate these morphological features have been published by Bent (1978).

In addition to the compounds described above there is a second group of pyrimidine compounds the members of which have systemic properties. The best-known fungicides of this family are **triarimol, fenarimol** and **nuarimol.** These are, in effect, methanol derivatives in which one methyl hydrogen atom has been replaced by pyrimidine and the other two have been replaced by phenyl derivatives. They are effective against powdery mildews on cereals, apples and cucumbers and also against apple scab (Green *et al.*, 1977). Typically, they do not prevent spore germination but they cause hyphae to swell and to branch abnormally. Sporidia develop to an abnormal size instead of dividing, although each enlarged cell only contains a single nucleus. Pearson *et al.* (1978) reported that fenarimol was the most effective of several substances tested for the suppression of pycnial formation by the cedar apple rust fungus. *Gymnosporangium juniperi-virginianae*.

Little biochemical information exists on the fenarimol group of pyrimidines and they will not be considered in detail. However, it is of interest that their mode of action appears to resemble that of the piperazine compound, triforine, which is considered below, rather than that of the pyrimidines of the ethirimol family.

Triforine is the most important piperazine derivative with systemic fungicidal properties. The (saturated) piperazine ring structure is related to that of pyrimidine but the two nitrogen atoms occupy different positions with respect to each other in the two ring systems (figure 10.3b). Triforine is recommended for use in Britain (MAFF, 1980) for the control of powdery mildew on spring barley. When used as a seed treatment it is often applied in a solvent which facilitates direct uptake through the testa (this technique is commonly used for benzimidazoles and other seed protectants; organic liquids are superior to water where testa penetration is desired but can reduce seed germination unless carefully selected). Triforine is also used against other powdery mildews as well as apple scab, brown rot of plums and rusts on cereals and roses.

Following soil uptake, triforine rapidly moves above ground and accumulates in leaves. There is some evidence that, when applied to leaves, there is a limited movement down to the roots (Green *et al.*, 1977, p. 135). In this respect it differs from many systemic fungicides, including the pyrimidines and benzimidazoles, the movement of which is little more than translaminar when applied to leaves. Triforine therefore presumably moves in the apoplast up the plant but can also move in the symplast with food materials in the phloem. Downward movement does not occur in the absence of active photosynthesis (Marsh, 1977). It only breaks down slowly in plants, piperazine being a major product.

It is believed that triforine (and also triarimol, fenarimol and triadimefon) are fungitoxic because they disrupt a late stage of steroid biosynthesis. An important reaction leading from steroid precursors to ergosterol (a characteristic fungal steroid) is the removal of the methyl group attached to C-14 of the steroid nucleus (figure 10.5). If this step is blocked, there is an accumulation of

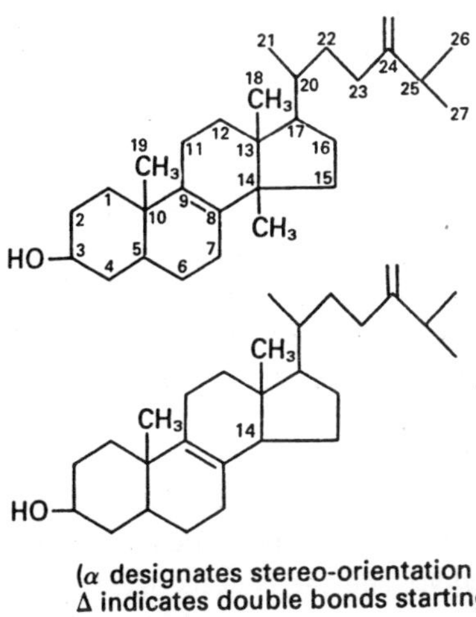

14-α-methyl-$\Delta^{8,24}$-ergostadienol

$\Delta^{8,24}$-ergostadienol

(α designates stereo-orientation at asymmetric C14;
Δ indicates double bonds starting at C8 and at C24)

figure 10.5 The step in sterol synthesis which may be inhibited by triarimol and by triforine some of the steroids that are intermediates before the blockage while at the same time there is a shortage of the finished steroids the fungus really needs. Thus, while total steroid content in poisoned fungi may actually rise, the amount of ergosterol declines (Ragsdale, 1977). Similar conclusions were reached by Schmidt and Benveniste (1979) who studied the effect of fenarimol not upon fungi but upon cell suspensions of the bramble (*Rubus fruticosus*). Such cultures normally contain sitosterol, campesterol and similar sterols. When, however, the cells were grown in a medium supplemented with fenarimol, steroids containing an α-methyl group in the C-14 position accumulated. Eight such sterols were recognised and it was concluded that fenarimol probably prevented C-14 α-demethylation.

It is of interest that C-14 demethylation is brought about by the action of the same microsomal mono-oxygenase system that metabolises pesticides (section 3.4). It has thus been suggested that one form of mono-oxygenase is specifically attacked by triforine and by triarimol. Alternatively, it has been suggested that steroids are moved to the active sites of different variants of mono-oxygenase by a steroid carrier protein (compare acyl carrier protein in fatty acid biosynthesis) and that one type of carrier is prevented from binding with its steroid when these fungicides are present (Ragsdale, 1977).

10.5 MORPHOLINE AND OXATHIIN DERIVATIVES

These two groups of compounds are related in that the morpholine ring contains atoms of carbon, oxygen and nitrogen whereas the oxathiin ring contains atoms of carbon, oxygen and sulphur (figure 10.6a,b). In both cases the non-carbon elements of the ring are *para* to one another but whereas the morpholine (oxazine) ring is fully saturated, the oxathiin ring is not.

Tridemorph and **dodemorph** are well-known systemic fungicides derived from morpholine. They are both used against *powdery mildews* but whereas tridemorph (figure 10.6c) is primarily employed against *Erysiphe graminis*, the powdery mildew of barley, dodemorph is used in Britain as a foliage spray to control powdery mildew on roses. Both are often mixed with benzimidazoles to widen their spectrum of activity. They are of low mammalian toxicity (LD_{50} > 1000 mg/kg) but dodemorph, in particular, can irritate the skin. It is of interest that the promising molluscicide, trifenmorph is also a morpholine derivative.

Published data upon the metabolism of the morpholines are limited but those on tridemorph (and trifenmorph) have been collated by Menzie (1978). In one study, after tridemorph had been used on cereals, residues fell to less than 0.05 ppm after 48 days, with a half-life of about 6 days. In another investigation, 80 per cent of the tridemorph added to Limburgerhof soil was degraded within a month. Metabolic products included an unstable amine oxide (compare 5b, table 3.1), one or more hydroxylated derivatives and the N-dealkylated product, 2,6-dimethylmorpholine. All these products resemble those which can be formed by NADPH-supported mammalian hepatic microsomes (section 3.4).

Although several different effects upon various organisms and organelles have been reported, the primary *mode of action* of tridemorph and dodemorph on fungi remains uncertain. At concentrations of about 2 ppm (7×10^{-6}M) it inhibits the oxidation of NADH in isolated mammalian mitochondria. Higher concentrations than this are needed to inhibit the oxidation of succinate. Concentrations exceeding 10^{-4}M are needed to inhibit oxidation *in vivo* (Lyr, 1977). Even recognising that it has a strong synergistic action when applied together with the electron transport inhibitor, antimycin A (figure 10.7) such concentrations seem very high in comparison to the ED_{50} of tridemorph for *Erysiphe*, which is of the order of 0.02–0.05 ppm ($\approx 10^{-7}$M).

More recently Kerkenaar *et al.* (1979) have provided evidence that it may interfere with lipid metabolism and with the synthesis of DNA. These workers found that it was not only active against members of most of the main groups of fungi but also against Gram-positive bacteria. At concentrations of about 5 ppm it inhibited the multiplication of sporidia of *Ustilago maydis*, the treated sporidia being swollen, multicellular and often branched. These are the same symptoms which follow the use of triarimol (section 10.4) and the authors concluded that tridemorph produces many of the symptoms characteristic of known inhibitors of sterol biosynthesis. Lipid synthesis was, in fact, first inhibited and then stimulated by tridemorph; unsaturated lipids, including α-tocopherol, alleviated the symptoms of tridemorph toxicity. The concentration having this effect is about a hundred times greater than that which affects powdery mildews but only one-hundredth of its LD_{50} when used against *Fusarium* (Lyr, 1977).

Carboxin (figure 10.6d) is an oxathiin derivative of importance because it is selectively toxic to many *Basidiomycetes*. It was introduced into Britain in 1966 to control rust, smuts and bunt on cereals and is used exclusively for seed treatment, usually in mixtures with other seed protectants. Its principal use is to control loose smut (*Ustilago*) and other seedborne diseases of barley and wheat (MAFF, 1980). For this purpose its systemic action is most valuable because

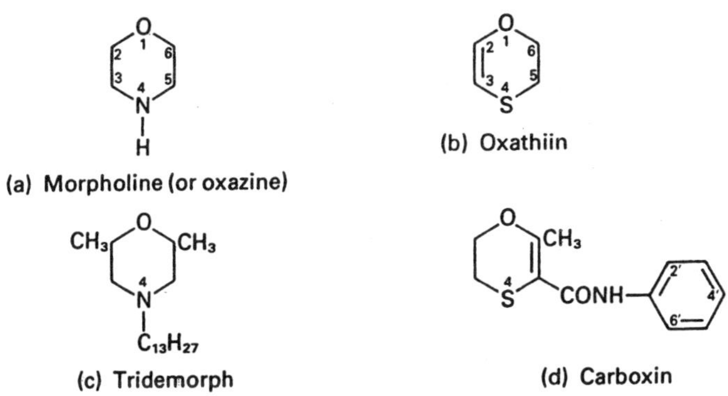

figure 10.6 Morpholine and oxathiin derivatives. ((c) 2,6-dimethyl-4-tridecylmorpholine; (d) 2,3-dihydro-6-methyl-5-phenylcarbamoyl-1,4-oxathiin)

loose smut mycelium penetrates deeply into seed and thus cannot be readily controlled by superficial protectants such as organomercury compounds used by themselves. Care is necessary to ensure that the moisture content of treated seed is not too high (above 16 per cent). Moreover, treated seed should not be sown immediately after treatment (Martin and Worthing, 1976). It is also active against the soil fungus *Rhizoctonia solani*, and Oyekan (1979) found that it controlled wet blight of cowpea, caused by this fungus, at one tenth of the dose of captafol needed to obtain similar results. The oral LD_{50} to mammals of carboxin is high but, in practice, the toxicity of the organomercury compounds with which it may be mixed will, in such cases, determine the safety precautions that are essential.

Metabolism of carboxin appears to proceed predictably (chapter 3). Hydroxylation products have been identified in both plants and animals, the hydroxyl group being in the 2′ or the 4′ position of the benzene ring of the phenylcarbamoyl moiety. These, and occasionally dihydroxy derivatives, are normally in large measure converted to conjugates. The sulphoxide has also been detected (compare section 4.7) but only small quantities normally oxidise further to the level of carboxin sulphone, i.e. to **oxycarboxin**. Instead, the sulphoxide is usually hydroxylated and then converted to conjugates. In addition, carboxin–lignin complexes are produced in some plants (e.g. barley), the total percentage of non-extractable residues increasing with time over a period of 35 days. Evidence for these conversions has been collated by Paulson (1977) and by Menzie (1974, 1978).

In relation to the mode of action of carboxin Mathre (1970) observed that fungi that are sensitive to carboxin are unable to metabolise acetate. On the other hand their ability to oxidise NADH is unimpaired. It will be recalled that oxidation of succinate in the tricarboxylic acid (TCA) cycle is not dependent upon NAD but is linked to the reduction of the flavin prosthetic group of the succinate dehydrogenase complex. This enzyme also contains a non-haem iron component and is present in electron transport particle II (figure 10.7). It is now

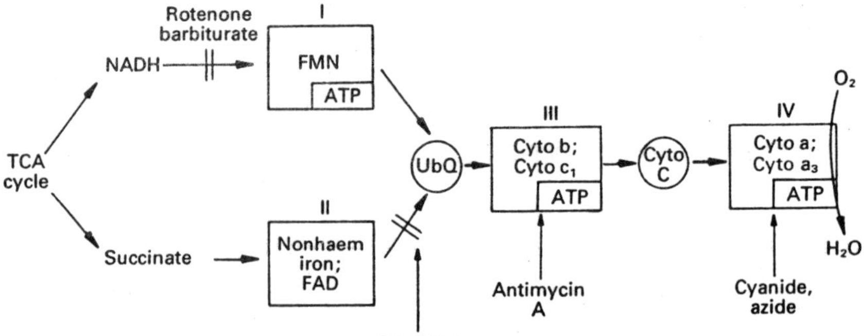

figure 10.7 Mitochondrial electron transport and proposed sites of action of some pesticides and other poisons

recognised that carboxin is a *potent inhibitor of succinate oxidation* in membrane preparations from animals, fungi and bacteria and probably acts somewhere near the interface between the non-haem iron component of succinate dehydrogenase and the oxidised form of ubiquinone (UbQ, figure 10.7).

It appears that there are both non-specific and specific binding sites for carboxin on the inner mitochondrial membrane, only the binding to specific binding sites being reversible (Coles *et al.*, 1978). Mitochondrial damage caused by carboxin has, in fact, been directly observed by electron microscopy, but so also has damage to other membranes. The probable importance of interaction with succinate dehydrogenase has been strikingly confirmed by White *et al.* (1978). These workers chose a number of strains of *Ustilago maydis* and of *Aspergillus nidulans* of varying resistance to carboxin and isolated succinate dehydrogenase complex from each. They found a marked parallelism between tolerance of the fungus to carboxin *in vivo* and the amount of carboxin needed to inhibit the succinate dehydrogenase from each of them *in vitro*.

The story is, however, still incomplete, for it does not explain the peculiar sensitivity to carboxin of many *Basidiomycetes*. Lyr (1977) has pointed out that fungi with low phosphofructokinase (PFK) activity are often those that are most sensitive. The *Basidiomycetes* are often fungi of this sort, for they possess a well-developed pentose phosphate pathway, whereas the *Peronosporales* and *Endomycetes* possess an active glycolytic system, for which PFK is a key enzyme. The glycolytic system is, in turn, linked to the machinery of the TCA cycle.

Lyr postulated that the non-haem iron site in electron transport particle II of PFK-deficient fungi might be more than usually sensitive to carboxin. The work on resistant strains mentioned above could be taken as a partial confirmation of this suggestion. Whether or not this is so, the relative importance of the glycolytic and pentose phosphate pathways appears to vary in different fungi.

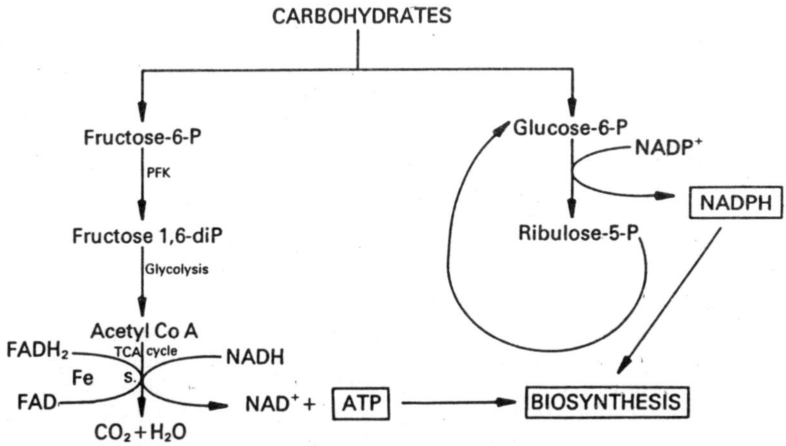

figure 10.8 Interrelation of some biochemical pathways. (s., succinate dehydrogenase complex; PFK, phosphofructokinase)

If a fungus well-endowed with the pentose phosphate system is simultaneously relatively deficient in the glycolytic pathway it might be expected to be especially vulnerable to poisons that interfere with succinate dehydrogenase activity. This follows from the fact that fungi, like other organisms, need a supply of both ATP and NADPH to drive their biosynthetic processes (figure 10.8).

10.6 OTHER SYSTEMIC FUNGICIDES

Limitation of space precludes a detailed treatment of each of the many other sorts of systemic fungicides that are known but several have special uses and must be briefly mentioned. In particular, the new fungicides based on acylalanine are worthy of note since they control pathogenic *Oomycetes*. The structures of several miscellaneous systemic fungicides are shown in figure 10.9.

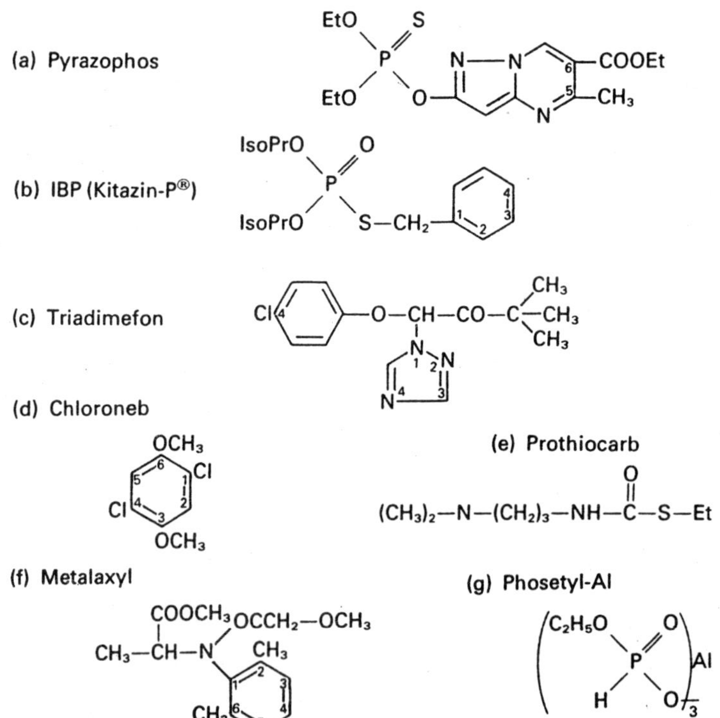

figure 10.9 Some miscellaneous systemic fungicides. ((a) 6-ethoxycarbonyl-5-methylpyrazolo-pyrimidin-2-yl-diethyl phosphorothionate; (b) di-*iso*propyl-S-benzyl phosphorothiolate; (c) (4-chlorophenoxy)-(*t*-butylcarbonyl)-(1,2,4-triazol-1-yl)methane; (d) 1,4-dichloro-3,6-dimethoxy benzene; (e) S-ethyl-N-(3-dimethylaminopropyl)-thiolcarbamate; (f) N-(2,6-dimethylphenyl)-N-(methoxyacetyl)alanine; (g) aluminium tris-(O-ethyl phosphonate)

Pyrazophos (figure 10.9a) is an organophosphorus compound used in Britain mainly to control powdery mildew on hops (*Sphaerotheca humuli*). Mixed with binapacryl, it is also used to improve the control of powdery mildew on apple trees (MAFF, 1980). According to Martin and Worthing (1976) its uptake by roots or from a seed dressing is insufficient for adequate systemic control. It is oxidised to the fungitoxic phosphate analogue by *Pyricularia*, a reaction that is reminiscent of the oxidative metabolism of insecticidal thionphosphates (section 4.2).

Another organophosphorus fungicide is used to control paddy blast of rice. It is di(isopropyl)-S-benzyl phosphorothiolate, **IBP**, and is often referred to in the scientific literature under its trade name of **Kitazin P**. It is marketed as a granular formulation which is put into the paddy water, from where, being soluble, it gains access to rice plants through their roots and leaf sheaths. Metabolic detoxication is principally brought about by cleavage of the P–S linkage. Resistant strains of the rice blast fungus, *Pyricularia oryzae*, and perhaps other strains also, are able to hydroxylate the benzene ring in the C-3 position (figure 10.9b). Sudo and Tamari (1969) found that an analogue of IBP lowered the carbohydrate content of treated *P. oryzae* cells and it is conjectured that it interferes with the biosynthesis of glycolipids.

4-n-butyl triazole is of interest in that it is effective against the wheat brown rust, *Puccinia recondita*, but not against other species of *Puccinia* or *Uromyces* causing rust diseases (Erwin, 1973). It has proved better for this purpose in the field than oxycarboxin, giving total control under bad epidemic conditions when used at a rate of 0.56 kg/ha.

Triadimefon (figure 10.9c) is one of a small group of complex triazole derivatives with systemic properties; other members of the group include **triadimenol**, **propiconazole** and **procloraz**. It is a moderately broad spectrum fungicide which is recommended in Britain for the control of powdery mildew, brown rust and leaf blotch (*Rhynchosporium*) on spring barley. It is also used against yellow rust and powdery mildew on winter wheat and against powdery mildew on rye. Powdery mildew on apples and grapes is also controlled by it. Mixtures with captafol (section 9.4) or with carbendazim (section 10.3) are widely used on cereals to broaden the spectrum of overall control. As a seed treatment for barley it can control powdery mildew for long periods but it may retard the emergence and growth of seedlings.

Buchenauer (1977) suggested that triadimefon may act by inhibiting the biosynthesis of ergosterol. A likely possibility is that it may prevent C-14 α-demethylation (compare figure 10.5); if so, the mode of action is similar to that of triforine, triarimol and fenarimol (section 10.4).

Chloroneb (figure 10.9d) has a rather limited systemic action but it is of interest in that it is used against a few *Oomycetes* that live in soil and attack through the roots of plants. Its fungitoxic spectrum of action is strictly limited, for while it is valuable for the control of *Rhizoctonia*, *Pythium*, *Sclerotium* and *Typhula*, disappointing results have been obtained in attempts to control *Phyto-*

phthora and *Fusarium*. It is normally applied as a seed treatment or as a soil drench, and has proved to be an effective method for controlling *Rhizoctonia solani* in cotton and soybeans and *Pythium ultimum* in peas. It enters plants readily but being poorly systemic, most of it remains in the roots, cotyledons and lower parts of the plant. This affords protection where it is most needed, since these are the regions which soil organisms normally attack (Rhodes *et al.*, 1971).

A principal route of *metabolism* in *Rhizoctonia* is O-demethylation of one of the methoxy groups (figure 10.9d). In plants, the β-D-glucoside of this metabolite is formed, as would be expected from the considerations in chapter 3. In both plants and fungi small amounts of the doubly O-demethylated hydroquinone and of its quinone oxidation product are also produced.

The mode of action of chloroneb is uncertain. Tillman and Sisler (1973) found that nucleic acid synthesis proceeded normally in *Ustilago* until the reproductive stage was reached, whereupon sporidial division was inhibited, as was protein and DNA production. On the other hand it appears to take part in what Lyr (1977) has called 'an unconventional attack on the respiratory system'. The mitochondria of *Mucor mucedo*, a sensitive fungus, undergo lysis and vacuolisation and at the same time a remarkable thickening of the cell walls occurs, with invagination of the plasmalemma. Lyr's article illustrates these events with electron micrographs produced by G. Casperson; the effects were obtained after exposure of *Mucor mucedo* to 30 ppm chloroneb.

In recent years attempts have been made to discover systemic compounds that are effective against pathogenic *Oomycetes* (compare section 8.1). Few detailed reports have yet appeared so compounds in this section cannot be considered at the length they will undoubtedly merit in a few year's time. However, a brief account follows of the compounds prothiocarb, phosetyl-Al and metalaxyl; several cyanoacetamides are also currently under test or have been marketed as systemic compounds active against certain *Oomycetes*.

Prothiocarb (figure 10.9e) is used as a soil drench or as an in-furrow spray near the roots of crops against *Pythium*, *Phytophthora*, *Pseudoperonospora* and some other *Oomycetes*. It enters plants readily and is in part translocated. In sandy soil it has a half-life of about 60 days and it persists longer in loams. Aqueous solutions give protection to cucumbers against the downy mildew, *Pseudoperonospora cubensis* (Cohen, 1979). It is toxic to sporangia and inhibits sporulation. In tomato plants it hydrolyses at the NH–CO linkage to give dimethylpropane-1,3-diamine as the major metabolic product. This has also been identified as a major metabolite in soil.

Work on phosphites which began in France in 1971 led to field trials at Villefranche-sur-Saône which, by the end of 1974, indicated that the substance, now known as **phosetyl-Al** (or fosetyl-Al) very effectively controlled *Plasmopara viticola*, the cause of downy mildew of vines. Since then it has been shown also to control some other Oomycetes; in particular, it has been used against *Phytophthora parasitica*, which causes heart rot of pineapples as well as gummosis of citrus, and against *Phytophthora cinnamoni*, the cause of root rot of avocadoes (Beach, 1979). It has now also been marketed for the control of wilt and die-back in such hardy nursery stock as conifers, Ericas and rhododendrons, and for the control of strawberry red core which is caused by *Phytophthora fragariae.*

Phosetyl-Al is a white water-soluble solid with a low vapour pressure and it possesses the structure shown in figure 10.9g. Its toxicology has been described by Williams and Beach (1977). Its acute oral toxicity for rabbits is 2680 mg/kg; for rats, quail and mice it is higher than this. Long-term trials have so far revealed no evidence of teratogenic or mutagenic effects. It breaks down in plants to innocuous phosphonic acid. It rapidly enters plants, where it usually persists, according to the crop, for four weeks to four months. In addition to its specification on some Oomycetes it is unusual in that it moves in plants both upwards and downwards; the downward movement enables foliar applications to be used to prevent infections in the lower parts of plants. For the control of *Plasmopara viticola* it is often formulated as a mixture with folpet (section 9.4); the two active ingredients appear to act synergistically.

Metalaxyl is an alanine derivative (figure 10.9f) which is selectively toxic to *Oomycetes*. In consequence, it can, for example, be used against late blight of potatoes caused by *Phytophthora infestans* but not against early blight that is caused by *Alternaria solani*. A formulation containing a mixture of metalaxyl with mancozeb (section 9.2) is now widely used on the potato crop in the United Kingdom, although resistant strains of *Phytophthora* have appeared in several countries including Eire and Holland. Indeed, one early formulation has already been withdrawn. Cohen *et al.* (1979) have reported that metalaxyl can achieve efficient control of *Phytophthora infestans* blight on tomato plants, so long as the plants are soil-drenched within two days of (experimental) inoculation. They observed that the fungicide did not prevent the fungus from penetrating into the plant and establishing itself in the leaves and fruit, but the lesions were small, restricted in number, and sterile.

Busch and Smith (1978) reported that metalaxyl (and also phosetyl-Al) effectively controlled a crown and root rot of the African violet, *Saintpaulia ionantha*, caused by *Phytophthora nicotianae*. Zentmyer (1978) referred to the possibility of using these same two substances by foliar application to control root rots caused by *Phytophthora* since, unlike most other systemic fungicides, some downward translocation appears to occur. On the other hand Aleksic *et al.* (1978) did not find metalaxyl any better than a standard captafol root dip for the control on pepper of *Phytophthora capsici*. Metalaxyl has also partly replaced Bordeaux mixture for the protection of grape-vines in France.

Metalaxyl has, in addition to foliar application, a possible role to play in the treatment of grain and other seeds. Exconde and Molina (1978) evaluated the use of metalaxyl as a seed dressing for the Philippine maize downy mildew, *Sclerospora philippinensis*. Applying the fungicide at rates of from 2 g active ingredient per kilogram of seed it was found that all rates gave total protection from emergence to harvest, whereas untreated seed had 36–100 per cent incidence of infection. Any initial depressive effect on plant growth had disappeared by the end of the second week of growth. Similarly, Lal *et al.* (1979) reported that seed treatment of maize at 4 g/kg of seed adequately controlled *Perono-sclerospora sacchari* on maize without affecting either seed germination or plant vigour.

In view of its narrow antifungal spectrum, it seems more likely that, in practice, metalaxyl will normally be used as one component of a seed treatment mixture. For example, Edgington *et al.* (1980), in a review of recent advances in the chemistry and use of systemic fungicides, report that it could prove valuable as a component of a triple protectant mixture for barley seeds, the metalaxyl being effective against *Pythium*, the other components acting against *Fusarium* and against smut fungus.

In conclusion, great progress has been made in recent years in the development of many kinds of systemic fungicides. Their value is undeniable as short-term remedies in potential emergencies and perhaps also for small scale use such as in the home garden. On the other hand, the development of resistance, often of a particularly refractory kind, has occurred with worrying frequency. In the author's personal opinion, the widespread and repeated use of systemic fungicides, if practised in non-emergency situations, could lead to long-term complications of a serious nature.

REFERENCES

Aleksic, D., Marinkovic, N., Aleksic, Z. and Sutic, D. (1978). *Zast. Bilja*, **29**, 317

Axness, M. E. and Fleeker, J. R. (1979). *Pestic. Biochem. Physiol.*, **11**, 1

Ayanaba, A., Verstraete, W. and Alexander, M. (1973). *Soil Sci. Soc. Amer. Proc.*, **37**, 565

Backman, P. A., Rodrigues-Kabana, R., Hammond, J. M. and Thurlow, D. L. (1979). *Phytopathology*, **69**, 562

Beach, B. G. W. (1979). *Proc. 1979 British Crop Protection Conference – Pests and Diseases*, p. 319

Bent, K. J. (1970). *Ann. appl. Biol.*, **66**, 103

Bent, K. J. (1978). *The Powdery Mildews* (ed. D. M. Spencer), Academic Press, London, Chapter 10

Bourgois, J.-J., Bronchart, R., Deltour, R. and de Barsy, Th. (1977). *Pestic. Biochem. Physiol.*, **7**, 97

Brooks, D. H. (1971). *Ann. appl. Biol.*, **70**, 149

Buchenauer, H. (1977). *Pestic. Biochem. Physiol.*, **7**, 309

Busch, L. V. and Smith, E. A. (1978). *Can. Pl. Dis. Surv.*, **58**, 73

Chandler, W. A., Daniell, J. W. and Littrell, R. H. (1978). *Pl. Dis. Rep.*, **62**, 783

Clemons, G. P. and Sisler, H. D. (1971). *Pestic. Biochem. Physiol.*, **1**, 32

Cohen, Y. (1979). *Phytopathology*, **69**, 433

Cohen, Y. Reuveni, M. and Eyal, H. (1979). *Phytopathology*, **69**, 645

Coles, C. J., Singer, T. P., White, G. A. and Thorn, G. D. (1978). *J. biol. Chem.*, **253**, 5573

Corbett, J. R. (1974). *Biochemical Mode of Action of Pesticides*, Academic Press, London

Cremlyn, R. J. (1977). *Herbicides and Fungicides: Factors Affecting their Activity* (ed. N. R. McFarlane), Chemical Society (London) Special Publication No. 29, Alden Press, Oxford

Cremlyn, R. J. (1978). *Pesticides: Preparation and Mode of Action*, John Wiley, Chichester

Davidse, L. C. (1973). *Pestic. Biochem. Physiol.*, **3**, 317

Davidse, L. C. (1977). *Neth. J. Pl. Path.*, **83**, (Suppl. 1), 135

Davidse, L. C. and Flach, W. (1978). *Biochim. Biophys. Acta*, **543**, 82

Dekker, J. (1977) *Systemic Fungicides*, 2nd edn (ed. R. W. Marsh), Longmans, London, p. 176

Ebben, M. H. and Spencer, D. M. (1973). *Proc. 7th Brit. Insectic. Fungic. Conf., Brighton*, Vol. 1, p. 211

Eckert, J. W. (1969). *Wld Rev. Pest Control*, **8**, 116

Edgington, L. V., Martin, R. A., Bruin, G. C. and Parsons, I. M. (1980). *Pl. Dis.*, **64**, 19

Edwards, N. A. and Hassall, K. A. (1980). *Biochemistry and Physiology of the Cell*, 2nd edn, McGraw-Hill, Maidenhead

Erwin, D. C. (1973). *Ann. Rev. Phytopath.*, **11**, 389

Exconde, O. R. and Molina, A. B., Jr. (1978). *Philipp. J. Crop Sci.*, **3**, 60

Ficsor, G., Bordes, S. and Stewart, S. J. (1978). *Mutat. Res.*, **51**, 151

Fuchs, A., van den Berg, G. A. and Davidse, L. C. (1972). *Pestic. Biochem. Physiol.*, **2**, 191

Fuchs, A. and de Vries, F. W. (1978). *Antonie van Leeuwenhoek J. Microb. Serol.*, **44**, 283; 293

Georgopoulos, S. G. (1977). *Antifungal Compounds, Vol. 2* (ed. M. R. Siegel and H. D. Sisler), Marcel Dekker, New York, Chapter 12

Green, M. B., Hartley, G. S. and West, T. F. (1977). *Chemicals for Crop Protection and Pest Control*, Pergamon Press, Oxford

Hisada, Y., Takaki, H., Kawase, Y. and Ozaki, T. (1979). *Ann. Phytopath. Soc. Jap.*, **45**, 283

Hollomon, D. W. (1979). *Pestic. Biochem. Physiol.*, **10**, 181

Kappas, A., Green, M. H. L., Bridges, B. A., Rogers, A. M. and Muriel, W. J. (1976). *Mutat. Res.*, **40**, 379

Kerkenaar, A., Barug, D. and Kaars Sijpesteijn, A. (1979). *Pestic. Biochem. Physiol.*, **12**, 195

Kumari, L., Decallonne, J. R., Meyer, J. A. and Talpaert, M. (1977). *Pestic. Biochem. Physiol.*, **7**, 273

Lal, S., Bhargava, S. K. and Upadhyay, R. N. (1979). *Pl. Dis. Rep.*, **63**, 986

Lyr, H. (1977). *Antifungal Compounds*, Vol. 2 (ed. M. R. Siegel and H. D. Sisler). Marcel Dekker, New York, Chapter 8

MAFF (1980). *List of Approved Products for Farmers and Growers*. For full reference see p. 334

Marsh, R. W. (1977). *Systemic Fungicides*, 2nd edn, Longmans, Edinburgh

Martin, H. and Worthing, C. R. (1976). *Insecticide and Fungicide Handbook*, 5th edn, Blackwell, Oxford

Mason, D. T. and Dennis, C. (1978). *Hortic. Res.*, **18**, 41

Mathre, D. E. (1970). *Phytopathology*, **60**, 671

Menzie, C. M. (1974, 1978). *Metabolism of Pesticides*; Updates. For full references, see p. 119

Nishijima, W. T. and Smalley, E. B. (1979). *Phytopathology*, **69**, 69

Ogunsusi, R. A. (1978). *Res. Vet. Sci.*, 25, 251
Oyekan, P. O. (1979). *Pl. Dis. Rep.*, **63**, 574
Paulson, G. D. (1977). *Antifungal Compounds*, Vol. 2 (ed. M. R. Siegel and H. D. Sisler), Marcel Dekker, New York, Chapter 4
Pearson, R. C., Szkolnik, M. and Meyer, F. W. (1978). *Phytopathology*, **68**, 1805
Penrose, L. J., Davis, K. C. and Koffman, W. (1979). *Austr. J. agric. Res.*, **30**, 307
Ragsdale, N. N. (1977). *Antifungal Compounds*, Vol. 2 (ed. M. R. Siegel and H. D. Sisler), Marcel Dekker, New York, Chapter 9
Rhodes, R. C., Pease, H. L. and Brantley, R. K. (1971). *J. agric. Fd Chem.*, **19**, 745
Ricci, P., Coleno, A. and Fevre, F. (1979). *Ann. Phytopath.*, **10**, 433
Roa, S. and Grollman, A. P. (1967). *Biochem. Biophys. Res. Commun.*, **29**, 696
Sampson, M. J. (1969). *Proc. 5th Br. Insectic. Fungic. Conf.*, **2**, 483
Schatz, G. and Mason, T. L. (1974). *Ann. Rev. Biochem.*, **43**, 51
Schmidt, P. and Benveniste, P. (1979). *Phytochemistry*, **18**, 1659
Shabi, E. and Katan, T. (1979). *Phytopathology*, **69**, 267
Shephard, M. C., Bent, K. J., Woolmer, M. and Cole, A. M. (1975). *Proc. 8th Br. Insectic. Fungic. Conf.*, **1**, 59
Siegel, M. R. and Zabbia, A. J., Jr (1972) *Phytopathology*, **62**, 630
Solel, Z. and Edgington, L. V. (1973). *Phytopathology*, **63**, 505
Sudo, T. and Tamari, K. (1969). *Proc. Meet. agric. Chem. Soc. Jap.*, p. 14
Tillman, R. W. and Sisler, H. D. (1973). *Phytopathology*, **63**, 219
van Tuyl, J. M., Davidse, L. C. and Dekker, J. (1974). *Neth. J. Pl. Path.*, **80**, 165
Upham, P. M. and Delp, C. J. (1973). *Phytopathology*, **63**, 814
White, G. A., Thorn, G. D. and Georgopoulos, S. G. (1978). *Pestic. Biochem. Physiol.*, **9**, 165
Williams, D. J. and Beach, B. G. W. (1979). *Proc. 1977 British Crop Protection Conference – Pests and Diseases*, p. 565
Woodcock, D. (1977). *Systemic Fungicides* (ed. R. W. Marsh), Longmans, London, p. 32
Zentmyer, G. A. (1978). *Proc. 3rd Int. Congr. Pl. Path.*, Munich, p. 362

11 Herbicides: General Considerations

11.1 INTRODUCTION

The global weed control problem is somewhat different from that relating to the control of insects and pathogens. Weeds only seriously reduce yield or quality of produce when they compete with the crop for available moisture, nutrients and light, and when soil quality does not, in itself, limit food production. In some areas, soil or climate may be so inhospitable that weeds may fare little better than crop plants. Elsewhere, especially in areas of subsistence farming, the hands and hoes of the family unit may be sufficient to provide adequate weed control; pathogens and insects cannot be controlled so readily. When cash crops are grown in simpler agricultural communities, the area under cultivation may be small and labour cheap, making chemical weedkillers at best a luxury.

The turning point probably occurs when groups of neighbouring farmers cooperate to cultivate larger areas (e.g. ujamaa villages in Tanzania or farming cooperatives in Libya) especially when local labour is scarce or more profitably employed on other activities. Moreover, even in some less developed countries, weedkillers are often necessities on tea, sugar or coffee plantations run by private, national or international organisations. In advanced farming communities, where densely packed crops are grown on good soil over vast areas, the financial loss caused by weeds can be enormous (Shaw, 1963). Moreover, when the area of monoculture is extensive, as in the wheat belt of America or some of the potato and sugar beet areas of Europe, the use of chemical weed control is usually imperative.

Crop rotation has been practised in many parts of the world for hundreds of years, e.g. the Norfolk four-course rotation (wheat, root crops, barley, seed crops). Although originally introduced to protect soil fertility, rotation often also acted as a means of weed and pest control. However, modern agricultural methods tend to make farms more profitable when they are larger and more specialised, because specialisation reduces the number of types of costly and more specialised equipment that are needed (Fryer and Makepeace, 1977). A related requirement for profitable agriculture is that farms should be less labour dependent, and labour can be reduced if row crops are not grown except on farms that specialise in such produce. In consequence, rotation is less often practised nowadays and herbicides provide an alternative mechanism for weed control. Furthermore, chemicals have in some circumstances found use in minimal cultivation or 'ploughless' farming, where shallow cultivators or even direct drilling replace the energy-consuming and tedious use of the plough. This can be particularly useful in areas where ploughing leads to excessive water loss.

Herbicide manufacture is big business – more so than the manufacture of insecticides (compare section 1.3). Of the £2600 m or so currently spent world-wide on toxic agricultural chemicals some 40 per cent is spent on weedkillers (Green *et al.*, 1977). The most used herbicides are members of the 2,4-D family, certain carbamates, propachlor and atrazine. From a farmer's point of view this huge expenditure is justifiable – often representing a four-fold return on outlay. It has been estimated that the use of 2,4-D for weed control in wheat saves hundreds of millions of man-hours of work (U.S. Dept. Agriculture, 1970). This in turn can mean that if an appropriate political climate exists, excess produce can be exported to less fortunate parts of the world.

In this chapter some of the fundamentals of chemical weed control are considered and then, in two subsequent chapters, the chemistry, mode of action and uses of specific foliage-applied and soil-applied herbicides are described. The toxicities of weedkillers to animals are listed in table 1.2 and general references are included in the Bibliography. A cross-reference list giving some proprietary names under which common active ingredients are marketed appears in Appendix B.

11.2 NON-SELECTIVE AND SELECTIVE HERBICIDES

Herbicides are used in agriculture to remove weeds that would otherwise compete with the crop. This can, however, be done in a number of ways, not all of which require the weedkiller to possess an intrinsic selectivity between weed species and crop plant. By 'intrinsic selectivity' it is meant that, if the chemical were to be applied to crop and weeds in such a way that equal amounts were uniformly deposited, the weed species would succumb but the crop would not. Some herbicides do indeed possess such selectivity. Many others are relatively non-selective, yet may nevertheless be used to control weeds by exploiting a variety of methods of application at various stages of crop development (sections 11.3 and 11.6). Whether a herbicide is best described as selective or non-selective is often a matter of convenience, dependent on what happens when the substance is used at economic dosage. If the dosage of a selective compound is increased, the selectivity is eventually lost. However, reduction of the dosage of non-selective compounds does not necessarily make them act selectively.

Sometimes, in a domestic or industrial context, total and **non-selective** weed control is wanted for as long a time as possible – for example, to keep garden paths or railway lines free from weeds. In such cases the poison does not need to – and usually should not – differentiate between one type of plant and another. Weed control in these circumstances is usually straightforward, except when weeds are well established, deep-rooted, or near to a cropped area. Non-selective soil-acting herbicides can be chosen, and their dosage so calculated, that the length of time that the substance remains in a particular type of soil can be predicted with reasonable precision (section 11.5). For example, of the compounds in figure 11.2, dichlobenil at about 60 g active ingredient per 100 m^2 of soil surface normally remains active for up to a year while diuron and simazine (200 g/100 m^2) may well persist for several years. In addition, sodium chlorate (4000 g/100 m^2) can give effective weed control for up to a year.

Compounds used for non-selective weed control differ greatly in their structure, in their physical properties and in the way that they are applied. Highly insoluble substances (e.g. diuron, simazine) can have a long-term residual action and, if misused, may incapacitate land for up to a decade. Others such as trichloroacetic acid (TCA) and sodium chlorate are water soluble and at appropriate concentrations are of relatively low persistence; these are capable of giving a 'short, sharp, shock' to weed-infested land. They are therefore used for certain land clearance programmes or for pre-planting treatments for certain crops. The same is true of some foliage-applied substances (e.g. paraquat) applied in a non-directed manner.

Most agricultural weed problems, however, require the destruction of weeds without simultaneous damage to the crop amongst which the weeds are growing. This is known as **selective** control. On some occasions the crop may comprise relatively large established plants, but on others it may consist of seeds or of seedlings that have not broken through the surface of the ground at the time that the spray is applied. In the former case there is a great biological similarity between the crop and the weeds growing amongst it and selective weed control may consequently be more difficult than, say, the control of insects living upon crops.

Biochemical differences can seldom be exploited to achieve selectivity because of the diversity of types of weeds usually found growing in one crop. In such cases the crop plant would need to possess a defence mechanism absent in most of the competing weed species. A few such cases are in fact known; for example, the absence of β-oxidase in some legumes enables them to withstand 2,4-DB (section 12.1) while a special defence system in maize protects the plants from the effects of atrazine (section 13.4). It is perhaps possible that genetic engineering may soon enable these comparatively rare examples to be multiplied. For the time being, however, less subtle attributes must usually be exploited. In practice, 'selectivity' often occurs as a result of an environmental or other situation such that the crop plant receives a lower effective dose than the weeds, or is so constructed that delicate tissue or vulnerable physiological systems are in some way protected from the main impact of the application.

Morphological differences may enable selectivity to be shown by a substance which, in other circumstances, might be almost equally toxic to all plants. Thus dicotyledonous plants often have meristematic tissue vulnerably exposed to a foliage spray. This is particularly evident when the leaves spread out and direct the toxicant to the growing point situated at the centre of a rosette (figure 11.1a). In contrast, monocotyledonous plants often have upright leaves which retain little of the spray and form a protective sheath around the meristem. This, in part, explains the selectivity of herbicides based on phenoxyacetic acid (section 12.1). Since grasses provide the staple diet for most of mankind and for many farm animals, this form of selectivity, characteristic of several translocated post-emergence herbicides (section 11.4) is of great importance to the human race.

Another form of selectivity exploits the fact that in certain crops (e.g. sugar beet) weeds are shallower-rooted and grow more rapidly than the crop plants (figure 11.1b). In consequence, many of the potentially more competitive weeds emerge before the crop and can be sprayed by a foliage spray such as diquat or glyphosate (sections 12.4 and 12.5). This is an example of **chronological selectivity**, in that the timing of the application is all-important. If it is done too early, many of the germinating weed seedlings will escape for they will not have broken through the surface; if it is done too late, non-selective foliage sprays such as those mentioned above will damage the crop. A modification of the method is the 'stale seed-bed' technique, in which the ground is prepared for sowing but the seeds are not put in immediately. Instead, the weeds are allowed to germinate before the crop is sown. Eventually, sowing is done with minimal soil disturbance, after which spraying is carried out in the time interval before crop emergence. Soil disturbance by hoeing would, of course, enable another batch of weed seeds to germinate.

A soil-acting herbicide, usually of low water solubility, may destroy weed seeds germinating in the top few millimetres of the soil whereas a suitable crop is protected from the herbicide because it is sown deeper in the soil. Positional selectivity (or 'depth protection') is evidently of particular value for pre-emergence application to weeds in large-seeded crops such as beet, swedes and turnips (figure 11.1c). Even very sparingly soluble chemicals slowly move down in the soil, but two factors combine to reduce the potential hazard to the crop. The first of these is the time factor; when used at economic concentrations bacteria and other micro-organisms attack and inactivate most herbicides. Secondly, there is the dispersion factor; under field conditions, a nearly insoluble substance is seldom washed down as a single concentrated layer (as it might be in a suitable chromatographic column) but instead becomes more and more diluted by admixture with soil as some of it penetrates more deeply into the soil while the rest is held back by adsorption on soil constituents.

A somewhat different example of positional selectivity is provided by the use of triallate to control wild oat (*Avena fatua* and *A. ludoviciana*) germinating in wheat (figure 11.1d). The triallate is incorporated carefully into the soil so that a couple of centimetres of untreated soil is left over the wheat seeds. When the contaminating wild oat germinates, its mesocotyl elongates rapidly so raising the nearly unprotected meristematic region into the triallate-treated zone. This does not happen in the case of the wheat, which has no extending mesocotyl, and by the time the meristematic region reaches the treated soil it is protected by

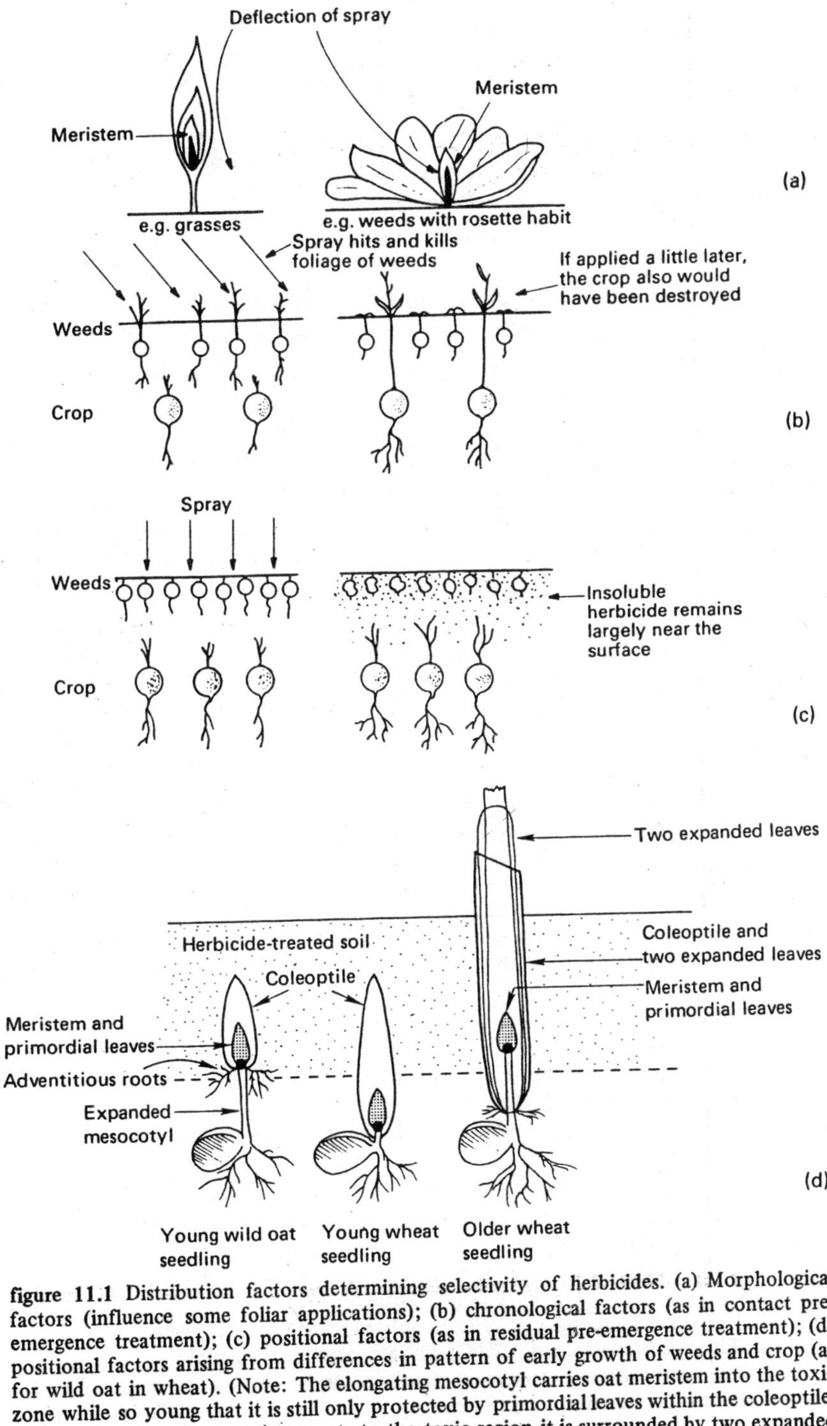

figure 11.1 Distribution factors determining selectivity of herbicides. (a) Morphological factors (influence some foliar applications); (b) chronological factors (as in contact pre-emergence treatment); (c) positional factors (as in residual pre-emergence treatment); (d) positional factors arising from differences in pattern of early growth of weeds and crop (as for wild oat in wheat). (Note: The elongating mesocotyl carries oat meristem into the toxic zone while so young that it is still only protected by primordial leaves within the coleoptile. By the time the wheat meristem gets to the toxic region it is surrounded by two expanded leaves, four primordial leaves and the coleoptile)

11.3 CLASSIFICATION OF HERBICIDES

Herbicides can be classified in several ways. In table 11.1 they are initially grouped according to whether they are normally applied to leaves or to soil. The herbicides mentioned under these two headings are considered in chapter 12 and chapter 13 respectively. The foliage-applied substances are subsequently subdivided into two groups, namely, those that are translocated and those that are not. Each of these groups is then further divided according to the chemical structure of the herbicides involved. A comprehensive list of herbicides and their formulae has been provided by Brian (1976).

Such a classification loses some of its usefulness if it is adopted too rigidly. Within one chemical family it is not unusual for individual substances to have uses that do not accord with the classification. Thus herbicides in the 2,4-D group are applied to leaves but the closely related 2,4-dichlorophenoxyethylsulphate (2,4-DES) is not only applied to soils but was specifically designed for this purpose. Similarly, the carbamates are best known for their action through the soil but some, such as barban (section 12.8) are foliage sprays. Again, there is no clear-cut distinction between **translocated** and **non-translocated** substances.

Table 11.1 Classification of herbicides

Foliar application (see chapter 12)

A. Systemic or translocated herbicides

1. **Phenoxyalkanoic acids**, e.g. 2,4-D, mecoprop, MCPB
2. **Growth regulators based on benzoic acid**, e.g. dicamba, 2,3,6-TBA
3. **Quaternary ammonium compounds**, e.g. paraquat, diquat
4. **Aminoacid derivatives**, e.g. glyphosate
5. **Triazoles**, e.g. aminotriazole
6. **Chlorinated aliphatic acids**, e.g. dalapon
7. **Translocated carbamates**, e.g. barban, asulam

B. Contact herbicides

1. **Petroleum oils**
2. **Hydroxyaryl nitriles**, e.g. bromoxynil, ioxynil
3. **Dinitrophenols**, e.g. dinoseb
4. **Miscellaneous**, e.g. phenmedipham, propanil, pentanochlor

Soil application (see chapter 13)

1. **Substituted ureas**, e.g. diuron, fluometuron, linuron
2. **Triazines**, e.g. simazine, atrazine, desmetryne
3. **Uracil** (pyrimidine) **derivatives**, e.g. bromacil, lenacil, terbacil
4. **Pyridazine derivatives**, e.g. pyrazone
5. **Phenylcarbamates**, e.g. chlorpropham, carbetamide, swep
6. **Thiolcarbamates**, e.g. EPTC, diallate, triallate, butylate
7. **Nitroanilines**, e.g. trifluralin, dinitramine
8. **Nitrophenyl ethers**, e.g. nitrofen
9. **Miscellaneous**, e.g. diphenamid, dichlobenil, chlorthiamid

Nothing is absolutely insoluble in water or totally adsorbed on plant constituents. Therefore, after root application of a herbicide of low aqueous solubility, radioisotope techniques often indicate a degree of translocation where other less sensitive analytical methods would suggest an entirely local distribution.

When a herbicide is used in the presence of a crop, there are two possibilities; it may be applied before or after the crop breaks through the surface. These two possibilities are termed **pre-emergence** and **post-emergence** situations respectively. It should be noted that the use of these terms gives no indication as to whether the weeds are being attacked by a foliar spray or by a soil-applied substance – in fact, either of these is possible for both a pre-emergence and a post-emergence situation.

A few general uses of the major groups of foliage and soil applied herbicides are briefly summarised at the end of the chapter in tables 11.2 and 11.3 respectively.

11.4 UPTAKE OF HERBICIDES BY LEAVES OR BY ROOTS

It is frequently found that the physicochemical properties of herbicides that are applied to foliage are very different from those of compounds normally applied through the soil. The factors influencing the uptake of herbicides are complex and interacting and the total process is, in consequence, poorly understood. However, some of the physical and chemical characteristics determining uptake by, and distribution within, plants can often be predicted in general terms by assuming that the upper part of the plant is hydrophobic by virtue of the waxy cuticle of the leaves whereas the lower part is essentially hydrophilic since the major function of roots is to take in water and various water-soluble substances.

When a toxic substance with a measure of oil solubility is applied to leaves, it will tend to penetrate through the waxy cuticle, especially if it is so formulated that it is soluble in a covalent carrier solvent (section 2.7). Similarly, weak acids or bases will tend to penetrate most easily if the formulation favours low polarity. This can be achieved by using covalent esters or by suppressing the ionisation of suitable salts by the addition of an activator which alters the pH of the total spray medium. If a leaf-applied herbicide is sufficiently soluble in both oil and water, it may enter the waxy cuticle, penetrate it, and then leave on its inside, so as to enter the aqueous phase of the plant.

The sieve tubes of the phloem constitute a special part of the inter-connected living parts of the plant, or symplast (Crafts and Crisp, 1971). Photosynthetic products (assimilates) move through the inter-communicating plasmadesmata of mesophyll cells and thence to the companion cells of the sieve tubes. It is assumed that leaf-applied herbicides move with such products. Very young leaves do not export photosynthetic products, but assimilates from leaves placed low on the stem often pass to storage areas in the roots and those from leaves higher up the plant pass to the apical meristems.

Movement of solutes in the phloem reflects the general metabolic activity of the plant and the majority of systemic foliage-applied herbicides are translocated most readily when the weeds are growing rapidly. Similarly, herbicides usually move out of the leaves rather slowly if poor light intensity restricts photosynthesis. However, some herbicides damage leaf tissue in such a way as to reverse the natural flow of water in the xylem. When this happens, xylem transportation of herbicides from foliage can occur, as it does, for example, at certain levels of application of quaternary ammonium compounds.

Naturally, many of these generalisations are, at best, rather facile over-simplifications. Corrosive or caustic substances may gain access in successive stages by first destroying cuticle of cells in their vicinity, then soaking further in, thus enabling the process to be repeated (e.g. sulphuric acid, which has been used for weed control in onions and kale). Another complication is that a water-insoluble herbicide may be oxidised after deposition, giving rise to a more water-soluble but still toxic substance capable of being translocated. This is reminiscent of what occurs to certain acaricides (section 7.10). Again, salts of weak acids or bases may be hydrolysed so that their apparently high ionisation may be misleading (e.g. amine salts of 2,4-D).

The situation as it relates to root uptake is somewhat different. Roots exist to absorb water and some solutes, but the relative amounts of solutes within plants is quite different from their relative amounts in soil water. This difference implies a selective uptake of natural soil constituents and it would not be surprising if herbicidal molecules were also absorbed in a discriminating manner. Some may move in passively, others may masquerade as natural substances; some may be actively pumped out as they enter, while others may destroy cells and so nullify the plant's normal protective system.

Once within the root, the substance may be strongly adsorbed on to cellular or extracellular components near to the point of entry. Such compounds may kill roots by contact action. Others may not be strongly bound, in which case they will tend to move upwards in the transpiration stream. Substances usually enter roots via the non-living, or **apoplastic** part of the plant, a system which not only includes the cell walls and the intercellular spaces, but also the xylem vessels of the transpiration stream. The water-impermeable Casparian strip of the root presents an apparent barrier to free entry. It is probably by-passed by aqueous solutions entering the symplast of the cells and then passing back into the apoplast via the parenchyma cells of the xylem.

The rate of upward movement of herbicide will be influenced by factors such as elevated temperature or reduced humidity which affect transpiration, as well as by any tendency of the herbicides to bind to macromolecules in root cells or to constituents in the walls of xylem vessels. For very slightly soluble substances, the slowest process in the uptake sequence may in practice be physicochemical, namely, the rate at which 'insoluble' herbicide particles in soil dissolve in the surrounding water and thus acquire the opportunity to reach plant roots. Movement in the xylem normally carries most of the herbicides to the leaves, a process that enables those that are photosynthetic inhibitors to reach their site of action. However, a proportion of the total dose absorbed usually escapes from the xylem to the phloem on the way up, a movement that allows growth-inhibiting herbicides such as trifluralin to reach meristematic tissue by conduction in the phloem.

It is perhaps useful to visualise the plant as a selective ion exchange column, for it has fixed negative sites on the cell walls and on some other cellular components. Such a column allows some ions and some polar molecules to move readily while others move with various ease but in a way determined by the extent of their attraction to the fixed charges of the stationary phase. Yet others remain firmly bound to cell constituents near to the point of entry. In normal circumstances, soil-applied herbicides move upwards in the xylem whereas foliage-applied substances are carried in the phloem out of the leaf and may then move either upwards or downwards. However, numerous complications are possible, including escape from xylem into phloem, damage to tissues in the leaf, and metabolic changes which render the herbicide more polar. Upward movement should depend on environmental factors such as humidity rather than on the level of metabolic activity of the weed, but stomatal damage can alter the rate of transpiration. In addition, biochemical events can lead to activation, deactivation or adsorption at any point in transit.

Many of the physicochemical factors which affect the penetration of active ingredients into higher plants have been examined from a mathematical viewpoint by Hartley and Graham-Bryce (1980, p. 544).

11.5 PERSISTENCE OF HERBICIDES IN SOIL

Stable substances with low vapour pressures and with low solubilities in water usually tend to persist in soil. On the other hand, if a substance is volatile, or chemically unstable, and often if it is soluble in water, it is unlikely to persist for long, although extent of persistence is considerably influenced by temperature, soil type and soil microbiology.

Volatile materials such as dichlobenil (section 13.9), certain thiolcarbamates (section 13.7) and nitroanilines (section 13.8) have relatively short half-lives in many soils. Trifluralin (section 13.8) illustrates a general point of some importance. At normal rates of application it may persist in soil for as little as 15 weeks or for as long as eight months. In common with many other volatile soil-acting herbicides, it is lost more rapidly from moist soil than it is from the same soil when it is dry. The explanation is that water competes with herbicide molecules for adsorption sites on soil constituents. Consequently, the molecules of herbicides that are strongly adsorbed on dry soils are often readily dislodged under damp conditions. When the substance is volatile, dislodgment allows diffusion to the surface where loss can occur by evaporation and sometimes also by photodecomposition.

The physicochemical characteristics of a substance *in vitro* may not accurately portray what occurs in 'living soil'. Some substances are strongly adsorbed on to soil organic matter (e.g. urea derivatives, triazines, carbamates and nitrophenyl ethers) while others (the quaternary ammonium compounds) are powerfully adsorbed on to certain types of soil particles. Yet others, although apparently stable, can be attacked by the oxidative, reductive or hydrolytic enzymes of soil micro-organisms. In such cases the substance may persist for a long time if used at such a high concentration that potentially destructive micro-organisms cannot recolonise the area (e.g. some very insoluble triazine* and urea derivatives).

Herbicides vary widely in their physical properties and biochemical stability and therefore the persistence in soil of different substances varies greatly. It is thus necessary to select with care a herbicide for a particular purpose, especially when using soil-acting compounds. If a substance of too low a persistence is used, there may be the additional cost and effort of multiple application. However, to employ a substance which is too persistent for, say, pre-planting clearance, could render the land unusable for one or many growing seasons. Furthermore, some of the family names of groups of herbicides are somewhat similar (soil-acting triazines and foliage-acting triazoles: soil-acting amides and foliage-acting anilides). It is therefore essential that users should double-check the names and the instructions on containers before applying the product.

Figure 11.2 illustrates on a logarithmic time-scale the probable range of persistence for each of several herbicides. It is important to remember that rate of application, rainfall, soil type, temperature and microbial population all affect persistence, so such schemes can only be indicative of probable persistence. In addition, some crops are relatively tolerant of some of these compounds so can be sown or planted a shorter time after application than would be possible for other crops.

A further complication is that **soil type** not only influences the time a particular substance persists but it also affects the toxicity of a given dose of that chemical to weeds. For instance, carbamate and urea herbicides are considerably less toxic to weeds in a heavy organic soil than to the same weeds in a light soil. Similarly, 2,4-D is only about one-eighth as effective in a black acid clay as it is in a light soil. Corresponding concentration ratios of 43 and 80 (the concentration needed for unit effect in heavy soil compared to that in light soil) have been observed for monuron and pentachlorophenol (Hartley, 1964). As would be expected, persistence of such herbicides as ureides and carbamates is usually greater in organic soils than in others, but such soils may also contain more bacteria capable of degrading them.

*** The author recalls upsetting half a teaspoonful of a solid simazine formulation on the much-prized lawn of the University of Reading in 1964. Twice since, the lawn has been returfed over several square metres, yet the area still shows. Several members of Senior Common Room have been heard condemning the Rag-day vandalism of irresponsible undergraduates.**

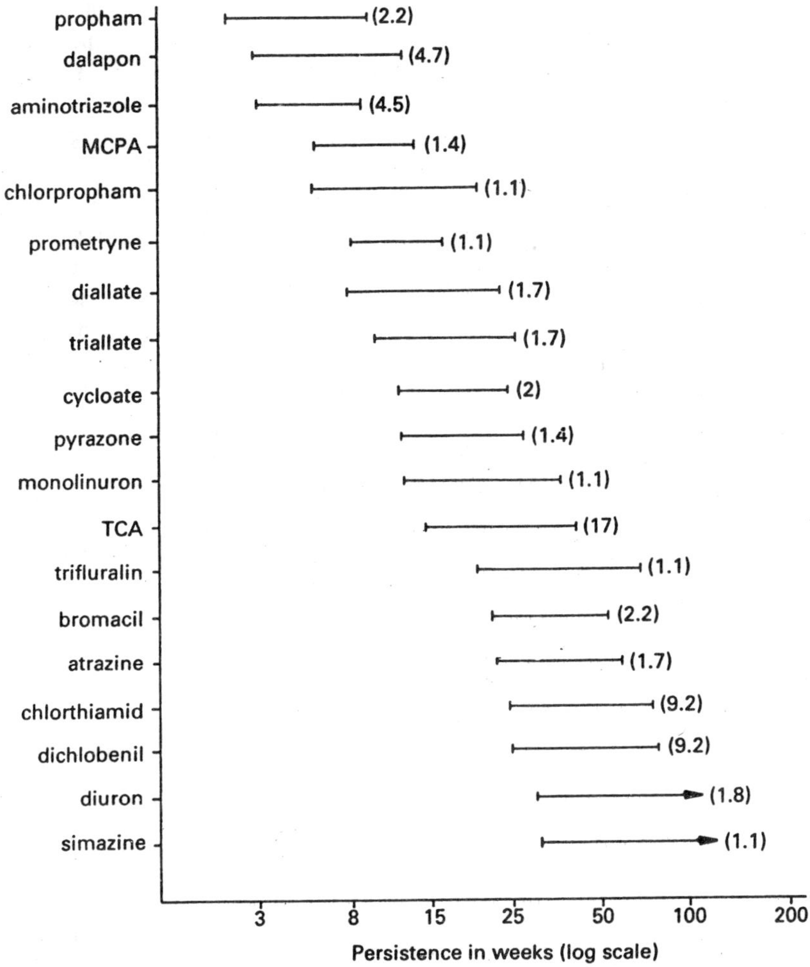

figure 11.2 Persistence of herbicides in soil: approximate maximal and minimal persistence at levels toxic to susceptible plants when used in average soils at normal dosage. (Notes: (a) Figures in parentheses indicate the dose (kg/ha) often used. (b) Specific crops may possess a level of tolerance which allows incorporation into soil much nearer to sowing or planting than the shortest times shown. For trifluralin, for example, brassicas and carrots can be drilled 1–14 days after incorporation. Similarly diallate can be used shortly before sowing brassicas and some types of beet)

The effectiveness of many herbicides with solubilities in the range of 1–100 ppm is greatly dependent on the level of rainfall within the few weeks following soil application. Unlike soil type, this factor cannot be allowed for in advance of application. Difficulties can therefore arise in seasons where the rainfall in the period shortly after spraying is atypical. If rainfall is light, wash-down may be insufficient to bring the chemical into contact with even shallow-rooted weeds. On the other hand, if rainfall is excessive, even nearly insoluble substances may be washed down to the level of the germinating crop and so become phytotoxic. For a limited but practically important range of rainfalls Hassall and Jewell (1956) observed that, in different seasons, as total rainfall in the 6-week period after spraying doubled, so the amount of a urea herbicide needed to kill a given percentage of weeds was roughly halved. Residual pre-emergence treatment is evidently somewhat hazardous in regions where climatic conditions are unpredictable.

The **mobility** of a given herbicide in a particular type of soil is a function of the intensity with which it is adsorbed by one or more kinds of soil constituents and the level of rainfall after application. Attempts have been made to quantify the relative mobility of various herbicides either by measurement of adsorption constants or by measuring R_f values in thin layer chromatography systems for which the thin layer is a standard soil.

The first approach is illustrated by the work of Rhodes *et al.* (1970) who measured the relationship between the amount of a herbicide adsorbed and the amount in equilibrium in the circumambient solution. On theoretical grounds the Freundlich plot should apply in simple cases:

$$\frac{x}{m} = KC^{1/n} \tag{11.1}$$

where x is the amount (μg) adsorbed by m grams of soil, C is the circumambient concentration and K, n are constants. By plotting the logarithm of this relationship

$$\log \frac{x}{m} = \frac{1}{n} \log C + \log K \tag{11.2}$$

the intercept is evidently the logarithm of the adsorption constant K. Using this equation Rhodes *et al.* (1970) determined the following values for Keyport silt loam: bromacil, 1.5; terbacil, 1.7; monuron, 2.6; chloroneb, 20. The larger the constant, the stronger the adsorption and the lower the mobility in soil.

The second approach is illustrated by the work of Helling (1971). He determined the R_f values of some 40 pesticides on Hagerstown silty clay loam, using thin layer plates. The pesticides were then divided into five groups according to the (increasing) magnitudes of the observed R_f values. As examples, chloroxuron and trifluralin are in group 1 (highly immobile); diuron, dichlobenil and chlorpropham are in group 2; atrazine and monuron are in group 3; bromacil, aminotriazole and 2,4-D are in group 4; and TCA, dicamba and dalapon are in group 5 (highly mobile).

The importance of the effective **depth of soil penetration** of a herbicide is apparent when it is remembered that there may well be, per hectare of soil, tens of millions of viable weed seeds at any one time. Most of those which germinate are probably in the top 3 cm of soil although seeds of some troublesome weeds can develop from a depth of up to 12 cm. It is clearly desirable to know the distribution of the more important weed species in the soil before calculating what dose of a soil-acting poison to apply, or before deciding to what depth a soil-mixed pre-planting herbicide should be incorporated into the soil.

Persistence in soil is often of some importance in relation to the use of foliage-applied herbicides, since a considerable proportion of the applied substance may fall on to the soil. Moreover, plants killed by foliage weedkillers may inadvisedly be collected for compost, and any remaining active substance may thus be concentrated into a smaller area of land. For soil treatment, persistence is evidently important; indeed, the field usefulness of such substances often depends on a predictable duration of effective action. Thus, for pre-planting treatment, disaster would occur should the time of effective herbicidal action not be matched with the time which will elapse between applying the herbicide and sowing a suitable crop. Various aspects of soil treatment are considered in the next section.

11.6 SITUATIONS INVOLVING THE APPLICATION OF HERBICIDES TO SOIL

The preceding two sections illustrate that soil application procedures may be hazardous unless the properties of a soil-acting herbicide are appropriate for the purpose for which it is to be used. This section will describe the uses of soil-applied chemicals and summarise some of the applications of pre-emergence and post-emergence treatments (section 11.3).

The following situations in which soil-applied herbicides can be employed should be carefully distinguished:

(a) Non-selective weed control (e.g. on paths). Here long-term persistence is often desirable.

(b) Non-selective clearance some time prior to sowing (i.e. a non-crop situation, but only temporarily so).

(c) Pre-planting and pre-sowing treatments (i.e. as (b) but nearer to the deadline, so rate of disappearance even more critical and the herbicide must be matched to a tolerant crop).

(d) Soil application after sowing the crop but before the crop emerges (residual pre-emergence treatment).

(e) Soil application to kill seedling weeds in an established crop (residual post-emergence treatment).

(a) **Non-selective** soil-acting herbicides are used in non-crop situations such as on paths and railways embankments. Persistent substances are normally employed to avoid the necessity of frequent re-application. At appropriate application rates, inorganic borates, sodium chlorate, various ureides and certain triazines all remain active for upwards of a year. Since such chemicals are mostly ineffective by leaf uptake it is customary to kill foliage by a leaf-acting herbicide (e.g. diquat) before applying the soil-acting chemical. Prior to the use of soil-acting substances in this mode it is obviously necessary to ascertain that the land will not be needed for crops for several seasons.

(b) For non-selective **clearance of land** before sowing a crop usually in the next growing season, it is essential to use a soil-acting herbicide of such a nature and at such a concentration that it has largely disappeared before the crop is planted. As in the case above, a non-selective foliage-acting herbicide is usually applied before the soil-acting substance. Dichlobenil and chlorthiamid (section 13.9) are often applied in autumn to clear ground that is to be sown in spring.

(c) **Pre-sowing and pre-planting** applications are made between about three weeks and three months before sowing or planting the crop. Exact timing and correct application rate may be essential to success and nearly always a herbicide must be chosen to which the would-be crop shows a degree of tolerance. The advantage of this type of treatment is that it can reduce the density of troublesome perennial weeds which are difficult to deal with once the crop is present. An example of a herbicide often used in this manner is trifluralin (section 13.8) which is used to control weeds in land where brassicas are to be grown within a few weeks. In addition to soil-acting substances, downward-translocated foliar herbicides such as aminotriazole and glyphosate are used in pre-sowing and pre-planting situations (sections 12.5 and 12.6).

(d) **Residual pre-emergence** treatment involves killing weeds by uptake through their roots or subterranean shoots at a time when the crop is still beneath the surface. It contrasts with *contact* pre-emergence treatment, where a foliar application kills emerged seedling weeds before the crop appears. Some of the problems associated with the residual pre-emergence method are considered in section 11.5. Unless the crop is tolerant (e.g. as maize is tolerant of atrazine), it must be protected against wash-down, and this is usually done by deep planting. In practice, this usually means that the method is primarily applicable to large-seeded crops such as beans as well as to sets such as bulbs, corms and tubers.

(e) **Residual post-emergence** sprays are applied to soil to control germinating weeds when a crop is visible. The chemicals employed are often the same as those used for residual pre-emergence weed control although the application procedure or concentration of herbicide may be different. The method is often used for weed control in orchards and in established fruit bushes. Chloroxuron (section 13.1) can be used in established celery and strawberries, and ametryne (section 13.3) is used after planting pineapples. In some crops, greater safety can be achieved by the use of placement spraying. Granular formulations are sometimes an advantage for residual post-emergence treatment because the granules bounce off foliage of the crop plants and come to rest on the ground beneath them.

11.7 THE USE OF MIXTURES OF HERBICIDES

It is evident from the preceding section and from the classification of herbicides in table 11.1 that considerable specialisation exists in respect to herbicidal function. Some are applied to foliage of well established plants while others attack germinating weed seeds. Others preferentially attack dicotyledonous weeds in the presence of foliage of a monocotyledonous crop while a few selectively destroy seedling monocotyledons. Yet others destroy existing weeds but have no long term effect, and so on. It follows that if certain biological and chemical compatibilities exist, the application of appropriate mixtures containing two or more herbicides can achieve a versatility not possessed by either active ingredient used singly, thereby saving time and effort in weed control and diminishing the overall cost of weed control measures.

Problems of a biological and of a chemical nature can in practice restrict the types of substances that can be usefully mixed together. One biological requirement is that active ingredient must be applied at a time and in a manner that allows its herbicidal potential to be realised. It is, for example, not economically sensible to apply a foliage-acting compound so early that little foliage is present, in order that the soil-acting compounds with which it is mixed can fulfil its function. Similarly, the mode of action of the active ingredients should be complementary rather than antagonistic. On the other hand, chemical compatibility must exist not only between all the active ingredients but also between all the supplementary substances that may be present. Usually, of course, these problems will have been resolved by the manufacturer of a purchasable mixture, but they become important if the farmer makes his own mixtures (a practice which, for safety reasons, is often undesirable).

In order to illustrate the usefulness and versatility of herbicidal mixtures, one or two examples will now be given to illustrate each of the four principal purposes for which such mixtures are used. Section references are given to indicate where the individual active ingredients are discussed in more detail.

First, a number of refractory weeds are very difficult to control at economic rates of application of individual chemicals, but they often succumb to 'cocktails' of herbicides. The following examples are amongst the Recommendations of Fryer and Makepeace (1978). These Recommendations provide a valuable source of information on special weed control problems. Ground elder (*Aegopodium podagraria*) is a particularly annoying weed in many parts of the British Isles; it has been found that a mixture of aminotriazole (section 12.6) and 2,4,5-T (section 12.1) applied in late summer controls it moderately well. Again, the docks *Rumex obtusifolius* and *Rumex crispus* are usually quite readily controlled by a mixture of dicamba (section 12.1) and MCPA (section 12.1) applied around midsummer.

A second common requirement is for a mixture of two foliage-acting herbicides, one of which attacks grasses. For example, total but temporary weed control can be obtained with mixtures of dalapon (section 12.7) and 2,4-D (section 12.1) for clearing rather neglected ground that is required for growing crops in the next growing season. The dalapon acts on the grasses and the 2,4-D kills the dicotyledonous weeds.

Thirdly, mixtures containing a soil-acting and a foliar herbicide are employed for several different purposes (Fryer and Makepeace, 1978). Some of these mixtures are designed so that planting can be done fairly soon afterwards, while others contain active ingredients that keep the ground weed-free throughout a complete season. A specific example is the use of a mixture of ioxynil (section 12.10) and linuron (section 13.1) for post-emergence treatment in onions; the ioxynil controls annual dicotyledonous weeds by foliar action and the linuron provides a degree of permanence by killing newly-germinating weeds by soil action. A dinoseb (section 12.11) and monolinuron (section 13.1) mixture is used rather similarly for pre-emergence weed control in dwarf beans and potatoes. An aminotriazole (section 12.6) and simazine (section 13.3) mixture is used for weed control in apple orchards, the former removing mature weeds around trees while the simazine provides longer term control.

Finally, it is sometimes useful to apply a mixture of two soil-acting herbicides, either because they tend to control a different spectrum of weeds, or because one has a quick clearing action while the other persists to prevent new weeds from recolonising the area. An example of the first of these is a mixture of diallate (section 13.7) and pyrazone (section 13.5) as a pre-sowing treatment in ground later to be sown with fodder beet or sugar beet. The diallate is particularly efficacious against wild oat (*Avena fatua* and *A. ludoviciana*) while the short-acting pyrazone controls many other annual weeds. An example of the second is a mixture of bromacil (section 13.5) and borate. In this case, the bromacil rapidly removes annual and shallow-rooted perennial weeds, while the borate helps to prevent re-infestation.

11.8 THE MODE OF ACTION OF HERBICIDES

Many herbicides, but not all of them, are specifically toxic to plants, as is evident from the sizes of LD_{50}s for oral administration to rats (table 1.2). Important exceptions are dinoseb and paraquat. One reason for this selectivity of action is that many herbicides attack processes of unique significance to plants, outstanding examples of which are those which occur within the chloroplasts and those that are dependent on the action of plant growth regulators. Any chemical which acts upon such processes may be expected to be relatively safe to animals unless, of course, it can at low concentration attack animals in some other way (and, in practice, very few toxicants, whether medical, social or agricultural, are totally devoid of side effects at a high enough concentration or after exposure for a long enough period of time).

Other herbicides attack more general processes, such as some aspect of nuclear division, or interfere with lipid metabolism. In such cases it is not always evident why selective phytotoxicity exists but, if the site of action is identical in plants and animals, selectivity must be related to some factor determining its selective availability there. In some cases, such as that of the dinitrophenol herbicide, dinoseb, the lack of selective action (as shown by its low LD_{50} to rats) is to be expected precisely because the mode of action is such that the substance attacks a process of universal biological importance.

The modes of action of herbicides, insofar as they are known, are discussed in relation to the chemistry of specific herbicides in chapters 12 and 13. Some of the more likely phytotoxic effects are listed in tables 11.2 and 11.3, although it must be stressed that the evidence in favour of particular modes of action often, in the author's opinion, leaves much to be desired. In addition, other actions than those shown have been proposed for some groups of herbicides.

Table 11.2 Foliar herbicides – some uses and possible modes of action

Group/examples	Uses	Symptoms/possible mode of action
Phenoxyacetic acids e.g. 2,4-D	Broadleaf weeds in monocot. crops	Contorted growth. Persistent auxins Aberrant nucleic acid metabolism; crushing of vascular system
Benzoic acids e.g. dicamba	Aggressive spp. in monocots. e.g. *Achillea*, *Rumex*, *Urtica*	Growth regulation; possibly similar to 2,4-D
Phenoxybutyrics e.g. 2,4-DB	Broadleaf weeds in some legumes	Only toxic if a β-oxidase in weeds converts them to 2,4-D, etc.
Quaternary ammonium compounds e.g. paraquat	Powerful non-selective action on foliage. Deactivated by clay. Used pre-em. or by placement	Replace ferredoxin as electron acceptor from PS I. Auto-oxidise. Membrane damage by lipid peroxidation
Glyphosate	Powerful foliage poison. Deactivated by soil iron? Grasses very susceptible	Breaks apical dominance – tillering. Decreases availability of phenylalanine
Aminotriazole	Aggressive weeds (e.g. *Agropyron*, *Equisetum*, in uncultivated land	Chlorosis. Via 70S ribosomes, blocks development of new chloroplastic membranes
Carbamates e.g. barban	Wild oat in wheat; *Alopecurus*, *Phalaris*	Shoot apices swell. May affect microtubule protein in nucleus
Benzonitriles e.g. ioxynil	Contact foliage action. No soil action. For new lawns. Some 2,4-D-resistant spp.	Quick effect needs light – may affect PQ in e.t.s. Chronic effect – shortage of ATP
Dinitrophenols e.g. dinoseb	Contact foliage action. Broadleaf weeds in some tolerant legumes	Raised resp. Uncoupling of oxidation; and phosphorylation in mitochondria

(All have systemic action unless otherwise stated; for abbreviations, see table 11.3)

Table 11.3 Soil-acting herbicides – some uses and possible modes of action

Group/examples	Uses (s.w. = seedling weeds)	Symptoms/possible mode of action
Ureas, *Uracils*, *Triazines*, e.g. diuron, bromacil, ametryne, simazine	s.w. pre-em. of large-seeded crops. In orchards, established crops. Soil adsorbed	Inhibit Hill reaction in isolated chloroplasts. Affect PS II near entry of electrons into e.t.s. Do not stop photosyn. in bacteria
Phenylcarbamates e.g. chlorpropham	s.w., esp. grasses, pre-em. of tolerant crops. Orchards. Also estd. *Stellaria media*	Bulbous shoot apex and stunted growth. Affects microtubule protein. Nuclear components in cytoplasm
Thiolcarbamates e.g. triallate, diallate	Volatile so soil-mixed. Fast action. s.w. pre-drilling. Some estd. perennials	Dark shiny leaves with thin cuticles. Affect lipid metabolism – inhib^{n} of biosynth. unsat^{d} fatty acids
Nitroanilines e.g. trifluralin	Volatile so soil-mixed. s.w. prior to brassicas. Pre-em. beet. Grass in cotton	Seedlings stunted, lateral roots fail. Arrested mitosis, multi-nucleate cells, cf. colchicine
Nitrophenol ethers e.g. nitrofen	Put on soil surface; s.w. esp. grasses, pre-em. Attack emerging shoot	Affects form. ATP in chloroplasts – action light-stimulated (the e.t.s. is not affected – so uncoupled?)
Amides e.g. diphenamid	s.w., esp. grasses, pre-planting or pre-em. of crop	Inhibit root dev. Seedlings die before they emerge. Alters uptake and distribution of Ca, Mg ions
Nitriles e.g. dichlobenil (also chlorthiamid)	Volatile so soil-mixed. s.w. 3m prior to crop. Controls est^{d}. perennials, e.g. *Aegopodium* (by inhibiting underground buds)	Stunted growth, melanin-blackened tissues. Dark green leaves. Action unknown but cell division restricted and chromosomes blotchy appearance

(estd., established; e.t.s., electron transport system; PS I, PS II, photosynthesis I and II; PQ, plastoquinone)

REFERENCES

Brian, R. C. (1976). *Herbicides: Physiology, Biochemistry, Ecology*, 2nd edn, Vol. 1 (ed. L. J. Audus), Academic Press, London, Chapter 1

Crafts, A. S. and Crisp, C. E. (1971). *Phloem Transport in Plants*, W. H. Freeman, San Francisco

Fryer, J. D. and Makepeace, R. J. (1977). *Weed Control Handbook, Vol. 1: Principles*, 6th edn, Blackwell, Oxford

Fryer, J. D. and Makepeace, R. J. (1978). *Weed Control Handbook, Vol. 2: Recommendations*, 8th edn, Blackwell, Oxford

Green, M. B., Hartley, G. S. and West, T. F. (1977). *Chemicals for Crop Protection and Pest Control*, Pergamon Press, Oxford

Hartley, G. S. (1964). *Physiology and Biochemistry of Herbicides* (ed. L. J. Audus), Academic Press, London, Chapter 4

Hartley, G. S. and Graham-Bryce, I. J. (1980). *Physical Principles of Pesticide Behaviour*, 2 vols, Academic Press, London

Hassall, K. A. and Jewell, C. A. (1956). *Rpts. 3rd Brit. Weed Control Conf.*, p. 663

Helling, C. S. (1971). *Soil Sci. Soc. Amer. Proc.*, **35**, 737

Rhodes, R. C., Belasco, I. J. and Pease, H. L. (1970). *J. agric. Fd Chem.*, **18**, 524

Shaw, W. (1963). *Natn agric. Chem. Ass. News*, **21** (2), 14

U.S. Dept. Agriculture (1970). *Economic Report No. 194*, Agriculture Research Service

12 Herbicides Applied to Foliage

Section 1: Translocated Compounds

Some foliage-applied herbicides are translocated within the plant and others kill by an action at or near the region where they make contact with the leaves. Translocated herbicides are considered in sections 12.1 to 12.8 and contact weedkillers in sections 12.10 to 12.12. A general classification of foliage- and soil-applied herbicides appears in the previous chapter (table 11.1) and some of their modes of action and general uses are summarized in tables 11.2 and 11.3.

12.1 PHENOXYALKANOIC ACIDS: STRUCTURE, FORMULATION AND USES

These translocated compounds are the most extensively used of all weedkillers (table 1.1), for they are used to control broadleaved weeds in cereal crops and grassland – and these crops play a crucial role in the diet of man and farm animals. Moreover, grass and cereals are often grown in large areas of monoculture, a method of farming which makes manual control of weeds impracticable but favours the use of aircraft for spraying. Naturally, aerial spraying is not possible except in extensive agriculture because of the likelihood of damage to dicotyledonous crops in nearby fields. For this reason, phenoxyalkanoic acids are not approved for use in this way in Britain.

Phenoxyacetic herbicides were introduced as a result of work carried out both in the USA and in Britain around 1944, although the substance now known as 2,4-D was used earlier as a plant growth regulator by Zimmerman and Hitchcock (1942). Formulae of principal members of the group are shown in figure 12.1. All members of the group have a chlorine atom attached to C-4 of the benzene ring and either a chlorine atom or a methyl group on C-2. Sometimes an additional chlorine atom is present on C-5. In various homologues of phenoxyacetic herbicides the alkanoic side-chain may contain three or four carbon atoms but, if three are present, one is always so placed that the substance is a methyl acetic acid rather than a *n*-propionic acid derivative.

Formulation is usually either as esters emulsified in oil or as water-soluble amine or other salts. Amine salts in particular are, with one or two exceptions, very soluble in water and such formulations are convenient for low volume application. Ester formulations are insoluble in water and more toxic to weeds than are the ionised forms. They are, however, much less selective. Originally the esters were made using alcohols of low molecular weight but problems sometimes arose from *vapour drift*, resulting in damage to such sensitive crops as tomatoes growing nearby. Nowadays, esters containing long chain alcohols such as iso-octanol are normally employed since their volatility is much lower (a different problem, *spray drift*, can of course still occur if weather conditions at the time of application are unsuitable). Sodium and alkylamine salts are more ionised and more selective than oil-emulsified esters and are safer to use on less tolerant crops. A typical alkylamine is $(HO.CH_2.CH_2)_3NH^+$ but other examples have been tabulated by Loos (1975).

Molecular polarity largely determines the ease of entry of phenoxyalkanoic acid derivatives through cuticular lipids. Covalent forms, and forms soluble in carrier oil, usually enter more readily than ionic or polar forms. In contrast, ionised forms are in equilibrium (at constant pH) with a fixed concentration of free, un-ionised acid and they enter more slowly. This slower or less complete entry of ionic forms is in part responsible for their selectivity, for in such circumstances the plant's enzymic defence system is more likely to be able to cope with the compound by detoxifying it as fast as it enters (section 11.4). In addition, morphological characteristics contribute significantly to the lower susceptibility of most grasses compared to broadleaved plants (section 11.2).

Physicochemical factors similarly determine the relative ease of entry of various salts of phenoxyalkanoic acids. Salts of strong bases (e.g. MCPA-Na) are less hydrolysed at a given pH than are salts of weaker bases (e.g. amine salts of MCPA). Consequently, amine salts are often intermediate in phytotoxicity and selectivity between covalent esters and highly ionised sodium salts. A well-known relationship, the Henderson–Hasselbalch equation, states the connection between the pH of the medium, the pK value of a weak acid (or base) and the ratio of ionised to un-ionised forms:

$$pK - pH = \log \frac{[\text{un-ionised molecules}]}{[\text{ionised particles}]} \qquad (12.1)$$

In a spray mixture, the uptake is further complicated by the fact that surfactants can increase availability and penetration of toxicant by reducing run-off and by softening surface lipids.

MCPA (figure 12.1a) is extensively used in Europe but **2,4-D** is often preferred in the United States. The difference probably reflects the relative costs of raw materials at a regional level. The sodium salt of MCPA is more soluble in water than is that of 2,4-D (27 per cent as against 4.5 per cent) and may cause less jet blockage when the volume of application is low. MCPA-Na is more selective than 2,4-D-Na but it is also less effective against many (but not all) weed species. When comparisons are made between various phenoxyalkanoic acid

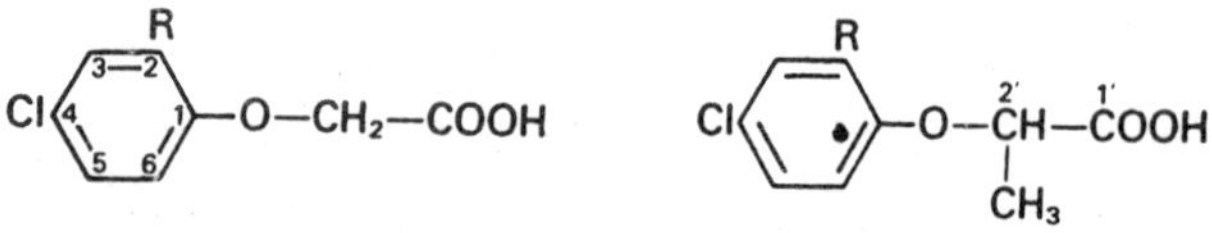

(a) R=CH_3; MCPA
(b) R=Cl; 2,4-D

(c) R=CH_3; Mecoprop
(d) R=Cl; Dichlorprop

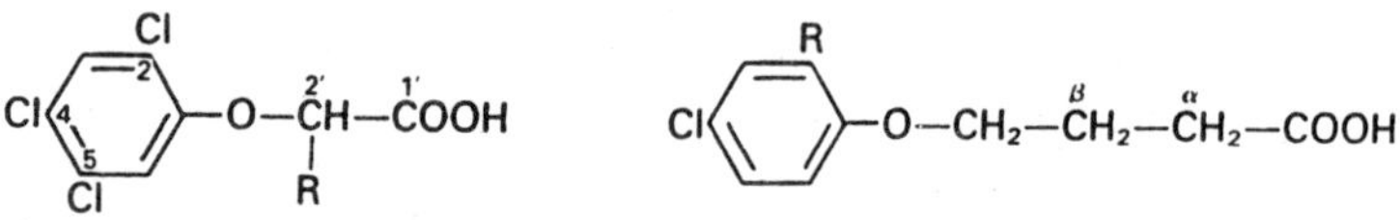

(e) R=H; 2,4,5-T
(f) R=CH_3; Fenoprop

(g) R=CH_3; MCPB
(h) R=Cl; 2,4-DB

figure 12.1 Phenoxyalkanoic acid herbicides. ((a) 2-methyl-4-chlorophenoxyacetic acid; (b) 2,4-dichlorophenoxyacetic acid; (c) 2′(2-methyl-4-chlorophenoxy) propionic acid; (d) 2′(2,4-dichlorophenoxy)propionic acid; (e) 2,4,5-trichlorophenoxyacetic acid; (f) 2′(2,4,5-trichlorophenoxy) propionic acid; (g) 2-methyl-4-chlorophenoxybutyric acid; (h) 2,4-dichlorophenoxybutyric acid)

herbicides, or between different formulations of one such compound, all concentrations are expressed in terms of *acid equivalents* of active ingredients per gallon of applied liquid.

The alkali metal and amine salts of MCPA are extensively used in European countries for the control of dicotyledonous weeds in cereal crops. The same is true of the amine salts of 2,4-D in the USA. In various countries these compounds are used on wheat, barley, oats, rye, rice, maize and sorghum. Both substances are frequently used in mixtures with one or more of the following herbicides: aminotriazole, 2,4-DB, dichlorprop, dicamba and dalapon. At lower concentrations, salts of MCPA can be used on flax and those of 2,4-D are used as growth regulators to reduce fruit drop.

Ester formulations of MCPA and 2,4-D are often used on non-cropped land as well as on grassland and turf; they are more persistent and control a wider range of weeds than do the sodium or amine salts. Ester-in-oil formulations are used to control some refractory weeds, especially in circumstances where selectivity may not be essential (e.g. for the control of woody plants). Where selectivity is a requirement, the application rate for ester formulations is usually about two-thirds of that for the corresponding salt form.

Various formulations of 2,4-D and MCPA have been used for broadleaved weed control on perennial grass pastures. An unusual hazard to farm animals is that they render toxic plants such as common ragwort (*Senecio jacobea*) palatable to cattle, whereas normally the animals avoid them. Another aspect of weed control in pastures is the quality of the forage after use of 2,4-D, dicamba and other growth regulators. The evidence on this subject is somewhat conflicting, possibly because of differences in the nature of the sward or the rate of application. For example, on the basis of protein content and the *in vitro* dry matter disappearance, McCarty (1979) concluded that smooth brome (*Bromus inermis*) and intermediate wheatgrass (*Agropyron intermedium*) suffered no deleterious effects from the application of 2,4-D and other substances to control musk thistles (*Carduus nutans*). On the other hand Mislevy (1978), using 2,4-D and dicamba on bahiagrass (*Paspalum notatum*), concluded that these herbicides, separately and together, significantly reduced dry matter, digestible forage and yield of crude protein.

Mecoprop can be applied to cereals somewhat earlier than can MCPA (i.e. before the five-leaf stage). Its weed control spectrum is similar to that of MCPA but it is more effective against chickweed (*Stellaria media*) and against cleavers (*Galium aparine*).

The relative safety of mecoprop, and indeed of other foliar applied substances, often depends on the fact that the soil acts as a barrier which reduces or prevents root uptake. This is illustrated by the work of Tottman and Davies (1978) who showed that herbicidal concentrations of mecoprop and of MCPA had adverse effects on lateral branches of roots of wheat plants grown in sand culture; the swollen lateral roots grew out to form tassels. In the case of mecoprop, tassels on roots were even visible after foliar application followed by downward translocation.

Dichlorprop, often in the form of mixtures with 2,4-D, is used to control woody plants and *Polygonum* spp. (e.g. knotgrass and redshank). **2,4,5-T** and **fenoprop** are also often used as mixtures with 2,4-D and are well-known as 'brushwood killers'. Used alone and diluted with water rather than dissolved in oil 2,4,5-T esters are also available for selective use in conifers (MAFF, 1980). They are also used to control nettles (*Urtica dioica* and *U. Urens*) in grassland and woody weeds in non-crop situations.

MCPB and **2,4-DB** (figure 12.1h) have very different uses from the foregoing substances, for they are used to control weeds in certain legumes. There is a biochemical explanation for this, for whereas the phenoxyacetic and phenoxypropionic acid compounds are the actual toxicants (except that esters probably undergo hydrolysis) the butyric acid homologues are not. Instead, they are of low toxicity until they are converted by the plant to the acetic acid member of the series. The enzyme responsible appears to be a specific β-oxidase, although the degradation involves the formation of acyl coenzyme A derivatives analogous to those formed in plants and animals during fatty acid oxidation. The oxidation results in the removal of two carbon atoms in the form of the acetyl group of acetyl coenzyme A (figure 12.2). The activity of this specific β-oxidase in some legumes is low and in consequence they are not damaged by concentrations of MCPB or 2,4-DB that kill weeds possessing a more active enzyme system. This is an example of lethal metabolism, the weeds being destroyed as a result of the activity of their own enzyme (other aspects of lethal metabolism are mentioned in sections 3.4 and 4.6).

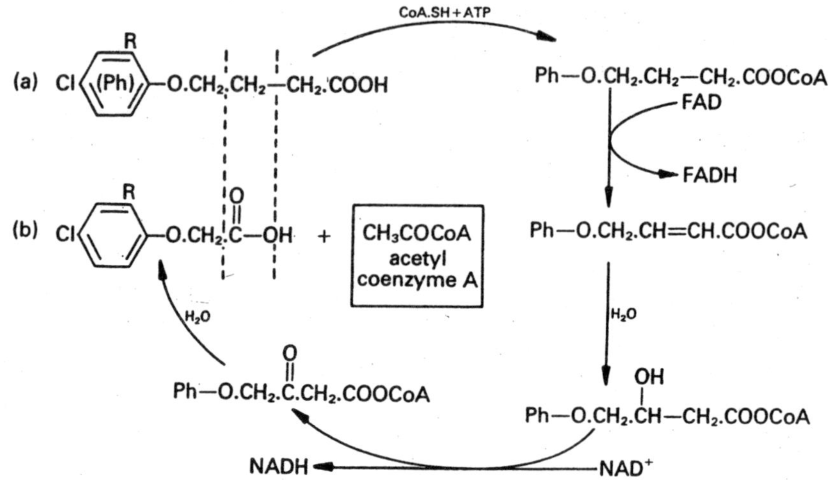

figure 12.2 The β-oxidation of phenoxybutyric acid herbicides. (a) MCPB or 2,4-DB (see figure 12.1); (b) MCPA or 2,4-D (see figure 12.1).) Compare the coenzyme A derivatives with those formed in fatty acid catabolism (compare Edwards and Hassall, 1980, chapter 10)

12.2 PHENOXYALKANOIC ACIDS: TOXICOLOGY, BIOLOGY, MODE OF ACTION AND METABOLISM

Toxicological aspects of the phenoxyalkanoic acids have been discussed by Barnes (1976) and by Garraway and Wain (1976). At very high dosage, animals lose the ability to maintain their body temperature when moved into either a hot or a cold environment. The reason for this loss of homeostasis is unknown. At all dosages relevant to accidental exposure to spray or to spray residues, evidence for the occurrence of side effects in man is tenuous. A mild neural disorder (peripheral neuropathy) has been reported in some individuals but it is by no means certain that exposure to 2,4-D by most members of the population would result in similar symptoms. There is, in any case, no reason to believe that residues in food could lead to such symptoms; such residues are often undetectable and any substance absorbed by the body does not accumulate but is rapidly excreted.

By far the most important hazard relates to possible **teratogenic** effects of these chemicals or of impurities in them. In 1970, it was shown that the use of some samples of 2,4,5-T could lead to teratogenic effects; that is to say, when fed to pregnant animals the offspring were born with often serious deformities. The problem aroused great adverse publicity as a result of the use of 'agent orange' as a defoliant during the Vietnam war; this was, in fact, impure 2,4,5-T containing some 25–30 ppm of the highly teratogenic 'dioxin' (figure 12.3). This impurity is formed in a side-reaction in the chemical preparation of 2,4,5-

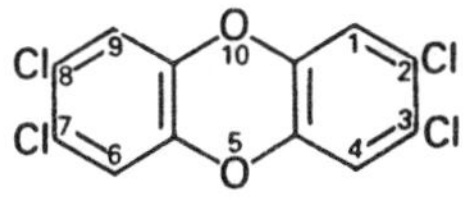

TCDD; 2,3,7,8-tetrachlorodibenzo-p-dioxin
(closely related analogues are probably present also)

figure 12.3 Dioxin – a trace impurity sometimes present in 2,4,5-T

trichlorophenol, itself a precursor of 2,4,5-T. Nowadays, the level of dioxin in 2,4,5-T seldom reaches 0.1 ppm, but the fear has been aroused in the public mind, and the use of 2,4,5-T is likely to be increasingly restricted in many countries.* There have, additionally, been reports that at extremely high dosage (up to half the LD_{50} in some cases), embryo-deforming effects have been observed with apparently pure samples of the free acids of 2,4-D, MCPA and 2,4,5-T when fed to rats and mice (Garraway and Wain, 1976).

Typical members of the phenoxyalkanoic acid family are foliar poisons. At high dosage they can, however, penetrate through some types of soil to plant roots, which they readily enter (compare mecoprop above). Translocation from the leaves is normally in the *symplast* (section 11.4). Factors which favour photosynthesis, such as warmth, good lighting and high humidity, also favour translocation of both food materials and phenoxyalkanoic acid herbicides. This is illustrated by the work of Schultz and Burnside (1980) who demonstrated that 2,4-D was absorbed by, and translocated out of, leaves of hemp dogbane (*Apocynum cannabinum*) for 12 days after application, that higher light intensity resulted in more 2,4-D reaching the roots, and that autoradiographs showed a typical symplastic movement. These considerations are particularly important if the object is to kill subterranean parts of perennial weeds – the best time to do this is evidently when food reserves are being deposited in late spring and in early autumn. Sometimes, however, the herbicide damages phloem or other tissues with consequent reversal of the direction of movement of both assimilates and herbicide. Since high concentrations of herbicide may cause premature collapse of aerial tissues, two or three applications at lower rates may be more advantageous than one application at a high rate.

If applied to soil, salts of MCPA and 2,4-D are readily washed away or decomposed by micro-organisms. Degradation is promoted by environmental factors

* Although the author does not consider a rearguard action in support of 2,4,5-T is worth the effort, it should be added that several detailed evaluations in Britain – the most recent by the Advisory Committee on Pesticides (MAFF) at the end of 1980 – have all concluded that the herbicide is safe to use. However, a Swedish survey, just published in the *British Journal of Industrial Medicine*, concluded that workers who spray 2,4,5-T have a seventeen-fold greater risk than usual of developing cancer of the soft tissues. More disturbingly, a sevenfold increase in risk apparently attended those who sprayed 2,4-D, and this risk did not seem to arise from the presence of dioxin as contaminant (– footnote added after a press report appearing in March, 1981).

that favour microbial growth, including suitable temperature, moisture and air content of the soil. Consequently, at low rates of application the herbicides usually disappear in 1–4 weeks but they can persist much longer when the soil is dry or cold. Low doses or 2,4-D affect root nodule formation in legumes, but this probably reflects toxicity to the host plant rather than to *Rhizobium*.

When soil organisms are exposed to 2,4-D a phenomenon termed 'enrichment' occurs. On single exposure of a culture of soil bacteria to a herbicide a lag phase occurs during which little herbicide is decomposed. On subsequent exposure no lag phase is evident and the herbicide disappears much more rapidly. During the initial exposure the population of micro-organisms alters in such a way that the surviving organisms are able to tolerate and to degrade larger amounts of MCPA and 2,4-D than could the organisms comprising the earlier population. 2,4,5-T does not readily cause 'enrichment' and the greater persistence of 2,4,5-T is in part attributable to this fact.

Typical symptoms of the phytotoxic effects of MCPA and similar compounds include contorted growth, malformed leaves and severe epinasty of stems and petioles. Such symptoms probably reflect the fact that these herbicides are chemically related to the auxin group of plant growth regulators. Unfortunately, little is known about how natural auxins function (their effects cannot be demonstrated in cell-free systems). However, they promote cell enlargement in shoots but not in roots, promote the growth of lateral roots and of fruit, and delay senescence. It would appear that 2,4-D and MCPA act as **persistent auxins**, the concentration of which does not fluctuate in response to stimuli in the way characteristic of natural auxins. High concentrations of indolylacetic acid, a natural auxin, produce symptoms similar to those elicited by 2,4-D.

The primary effect of phenoxyacetic acid herbicides on plants is to cause aberrant growth of young, rapidly growing tissues near the meristem. The problem arises in trying to explain why such abnormal growth occurs, for much work over many years has shown a plethora of physiological and biochemical systems to undergo change. There is considerable evidence that auxins stimulate cell wall development (van der Woude *et al*., 1972) and, if so, phenoxyalkanoic acid herbicides may interfere with the metabolism of pectin methyl esters or of some other component of young cell walls. There is also little doubt that these herbicides can upset nitrogen metabolism. A recent example is the work of Samoilov and Ladonin (1979) who showed that 2,4-D and dicamba, separately or together, altered the distribution of nitrogen so that nitrogenous substances accumulated in the stems while at the same time they inhibited the incorporation into proteins of inorganic nitrogen.

A theory which is currently popular is that abnormal growth following application of 2,4-D is a consequence of aberrant nucleic acid metabolism (Cherry, 1976). When hypocotyl cells are exposed to these herbicides, ribonucleic acid production increases several-fold. It seems probable that the extra RNA, and protein translated from it, leads in susceptible plants to cell proliferation which crushes the vascular system and causes death.

It has been postulated that phenoxyacetic acid herbicides have the effect just described because they interact with the plasma membrane so that an unknown factor is released. Natural auxins probably behave in this way also, releasing a factor which makes its way to the nucleus where it modifies the rate of transcription of DNA, perhaps by modifying RNA polymerase. Presumably phenoxyacetic acid herbicides modify the (natural) unknown factor or cause it to stimulate transcription for too long a time.

The data of Hardin *et al.* (1972), in part summarised in table 12.1, seem to support this hypothesis. Using a system containing calf thymus DNA, they added supernatant fluid prepared by incubating soybean plasma membranes with 0.1 μM indolylacetic acid (IAA), with 2,4-D or with the inactive substance 3,5-D. Using tritium-labelled uridine monophosphate, UMP, they studied the rate of its incorporation into RNA in the presence of the DNA template. The results in the table suggest that something liberated from cell walls that had been exposed to natural or to synthetic auxin did indeed increase RNA synthesis. Interestingly, cell walls exposed to 3,5-D did not release anything which significantly increased the incorporation of UMP into RNA. However, if a suspension of walls treated with 3,5-D was centrifuged and then treated with 2,4-D, a stimulatory factor was still released, indicating that it had survived the 3,5-D treatment. In view of the interest attached to these observations it is somewhat surprising that little independent developmental work appears to have been reported in the last decade.

Metabolism of phenoxyacetic acid derivatives in plants conforms to the principles outlined in chapter 3. **Side-chain degradation** (figure 12.4) is probably a minor pathway in most plants, although it may be relatively more important

Table 12.1 Evidence for the liberation of a transcription-stimulating factor from soybean cell membranes upon treatment with natural or synthetic auxins (after Hardin *et al.*, 1972)

Calf thymus DNA, buffer, plus:	Picomoles UMP incorporated (two expts)
RNA polymerase (no cell wall exposure)	88
+ s/n from (PM in IAA)	146
+ s/n from (PM in 2,4-D)	138
+ s/n from (PM in 3,5-D)	96
RNA polymerase (no cell wall exposure)	160
+ s/n made from (PM in 3,5-D)	151
+ s/n made by (a) incubating PM in 3,5-D; centrifuging; resusp. in 2,4-D; recentrifuging	249

Method. Isolated plasma membrane (PM) was incubated with 0.1 μM IAA, 2,4-D or 3,5-D and the solution centrifuged. The supernatant (s/n) was added to a system containing buffer, DNA, ^{3}[H]-UMP and plant RNA polymerase. Auxins added directly to the system did not increase incorporation of UMP.

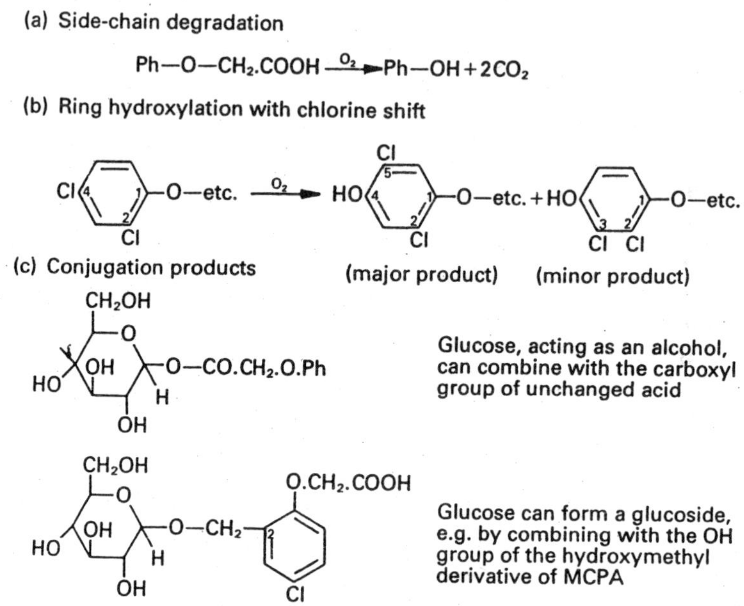

figure 12.4 Detoxication of phenoxyacetic acid herbicides

in redcurrants, bedstraws (*Galium* spp) and some varieties of apples. A substituted phenol is the final product. **Ring hydroxylation** is usually the major metabolic route. Aromatic compounds chlorinated on C-4 are often hydroxylated in this position and, in such cases, the chlorine atom is not substituted but 'migrates' to the C-3 or to the C-5 position (figure 12.4). **Conjugation** can involve the direct union of glucose with the carboxyl group of the phenoxyacetic acid. Alternatively, after a hydroxyl group has been inserted by primary metabolism, glucose can form glucosides with the newly-inserted hydroxyl group. When aspartic acid is the conjugant, union occurs between the carboxyl group of the phenoxyacetate and the amino group of the aspartic acid. Literature on the metabolism of phenoxyalkanoic acid herbicides has been reviewed by Loos (1975).

12.3 PLANT GROWTH REGULATORS DERIVED FROM BENZOIC ACID

Several derivatives of benzoic acid and of benzonitrile are used for various forms of weed control. The most important translocated foliage-applied herbicides of this group are dicamba and 2,3,6-TBA and these are briefly considered in this section. Bromoxynil and ioxynil are contact foliage poisons, not regulators, and are considered in section 12.10, while dichlobenil and chlorthiamid are soil-acting herbicides and are described in section 13.9.

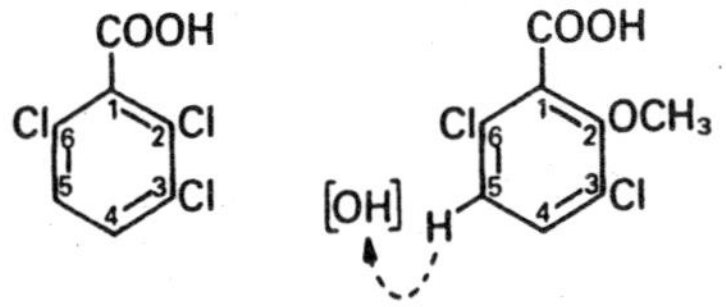

(a) 2,3,6-TBA (b) dicamba and a hydroxymetabolite

figure 12.5 Growth regulators based on benzoic acid. ((a) 2,3,6-trichlorobenzoic acid; (b) 3,6-dichloro-2-methoxybenzoic acid)

Dicamba and **2,3,6-TBA** (figure 12.5) resemble the phenoxyalkanoic acid herbicides in that they act as growth regulators and elicit the same sorts of growth responses in broad-leaved weeds (section 12.2). Although primarily foliar herbicides they are much more effective than are most phenoxyalkanoic compounds when applied to the soil. A closely-related compound, **chloramben**, is a minor herbicide used for pre-emergence weed control in legumes, and this is very active through both the soil and foliage.

In Britain, dicamba and 2,3,6-TBA (both in the form of potassium or sodium salts) are only available in mixtures with various phenoxyalkanoic acids, so their purpose appears to be to add range and versatility to the herbicidal spectrum of these compounds. According to the Recommendations of Fryer and Makepeace (1978) they are used in circumstances where one or more of the following aggressive species is present:

black bindweed (*Polygonum convolvulus*)
camomile (*Anthemis arvensis* and *A. cotula*)
chickweed (*Stellaria media*)
cleavers (*Galium aparine*)
corn spurrey (*Spergula arvensis*)
docks (*Rumex spp*)
knotgrass (*Polygonum aviculare*)
mouse-ear (*Cerastium holosteoides*)
pineapple weed (*Matricaria matricarioides*)
redshank (*Polygonum persicaria*)
scented mayweed (*Matricaria recutita*)
scentless mayweed (*Tripleurospermum maritimum*)
yarrow (*Achillea millefolium*)

Both substances are used for the control of broadleaved weeds in cereals and grass seed crops. Dicamba is also used for similar weed control in established apple orchards (MAFF, 1980). Holroyd and Thornton (1978) reported dicamba to be the most active of the herbicides they tested against bracken (*Pteridium aquilinum*) in the absence of live bracken fronds although it was less effective than lower doses of **asulam** (section 12.8) for bracken with live foliage.

Auch and Arnold (1979) reported that 31 per cent of soybean farmers they surveyed in South Dakota used dicamba on this crop. The yield of crop was adversely affected if dicamba was used at the flowering stage and germination was reduced if it was used at the pod-fill stage. Residues could be detected on foliage seven days but not 18 days after application. Behrens and Lueschen (1979) investigated the volatility of dicamba formulations and found that soybean exposed to vapours from foliar-treated maize developed symptoms of dicamba toxicity. The use of less volatile formulations did not eliminate the vapour problem for the volatile component was not the formulated salt but free dicamba acid.

The major route of **dicamba** metabolism in several plant species studied involves the hydroxylation of position C-5 of the benzene ring. Little of this hydroxylated metabolite accumulates in most species because it is rapidly conjugated with glucose. In soil, dicamba is poorly-adsorbed on to soil constituents so it washes down nearly as rapidly as does the wash water. Adsorption is strongest in soils of low pH value. It is of intermediate persistence in most soils, its disappearance being most rapid under conditions which favour bacterial growth.

In contrast to dicamba, **2,3,6-TBA** does not appear to be readily metabolised by plants nor does it usually disappear rapidly from soil. Dowler *et al.* (1963) reported that active residues were still present in a North Carolina soil 18 months after applying 9 kg/ha. On the other hand, adsorption onto most soil constituents is very limited, so it tends to disappear from damp soils by leaching. Loss by volatilisation probably also limits soil persistence.

12.4 QUATERNARY AMMONIUM COMPOUNDS

Paraquat and **diquat** are bipyridyl herbicides and were discovered around 1955 at the Jealott's Hill Research Station of I.C.I.; more recently, the mono-pyridinium compound, **cyperquat**, has been introduced (figure 12.6). In view of what is now known about their mode of action it is of interest that paraquat was used as early as 1933 as the redox indicator methyl viologen because of its ability to form a rather stable free radical.

Salts of paraquat and diquat are nearly colourless in the dry state. They are very soluble in water and the solutions are corrosive to some metals. A solution of paraquat absorbs light at 260 mμ and that of diquat absorbs at 310 mμ. On reduction, paraquat is reduced to a purple coloured free radical and diquat to one that is coloured green. These free radicals *auto-oxidise* in the presence of oxygen. The standard redox potentials, E'_0, are respectively −0.45 V and −0.35 V, values that are, perhaps significantly, similar to the E'_0 value for the biological electron carrier, ferredoxin.

A free radical is a substance that contains an unpaired electron – in the language of classical valency theory, it contains an 'un-satisfied valency bond'.

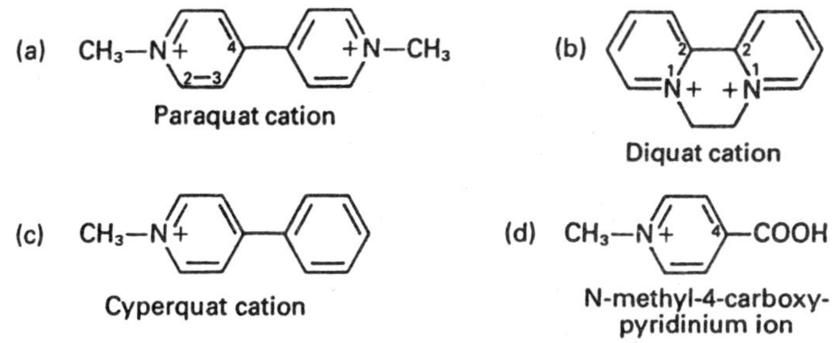

figure 12.6 Quaternary ammonium herbicides. ((a) 1,1′-dimethyl-4,4′-bipyridinium ion; (b) 1,1′-ethylene-2,2′-bipyridinium ion; (c) 1-methyl-4-phenylpyridinium ion

Such substances, although common, are usually so unstable that their half-lives are measured in micro- or milliseconds. When, however, the molecule contains a system of alternating double bonds (π-electrons), the energy can be delocalised to give molecules of much greater stability; the stability is said to be provided by 'resonance energy'.

The unique attribute of the quaternary ammonium group of herbicides is that they kill exclusively by **foliage contact** so long as the plants are growing in soil.* They cannot enter plant roots in such circumstances because they are inactivated by total adsorption on clay minerals. Montmorillonite and similar minerals adsorb these substances most strongly because they hold the doubly-charged planar cations firmly within the negatively charged crystal lattice. Except in highly sandy soils, the cation exchange capacity of the soil is little diminished by such uptake (Knight and Tomlinson, 1967). The affinity of clay minerals for bipyridinium cations is so high that neither fertilisers nor lime normally displace them.

Unlike the urea herbicides (section 13.2) they are not strongly bound to soil organic matter. This is demonstrated by the fact that soil that has been treated with peroxide loses little of its paraquat-adsorbing capacity. Paraquat and diquat slowly disappear from soil, although it is not clear whether physical or biochemical processes contribute most to this loss.

Some of the uses of quaternary ammonium herbicides are listed in table 12.2 and the following selected examples will illustrate the versatility of these materials. Firstly, paraquat can be used as a **between-crop application**. For example, it can be used to clean up land before drilling cereals or kale (Halliday, 1973). Another example is its use to destroy an old sward prior to renovating grassland by re-seeding. Mat grass, ***Nardus stricta***, one of the least valuable British

*** That is, plants growing in culture solutions are rapidly killed by root uptake. Similarly, seedling crops can be killed before emergence above the soil when that soil is very light and lacks clay particles.**

Table 12.2 Some general uses of quaternary ammonium herbicides

Herbicide	Uses (all are used after emergence of weeds; they may be used pre-emergence of a crop or applied to an established crop by placement techniques)
Diquat	Often preferred where broadleaved weeds predominate. Non-crop situations, e.g. clearing heavily infested land. For pre-harvest desiccation of foliage. To control submerged aquatic weeds
Paraquat	Often preferred where grassy weeds predominate. Non-crop situations. e.g. clearing heavily infested land. Applied after harvesting one crop and before planting the next. To destroy an old sward before reseeding. Control of aerial parts of perennials. For ploughless (minimal cultivation) farming
Cyperquat	To control annual weeds in cereals. Control of nutsedges (*Cyperus*) in many crops

grasses, is killed by paraquat, although, in the absence of cultivation, such treatment may encourage the growth of *Agropyron repens* (couch grass; also called quack grass in the USA).

A second range of uses requires that the herbicide be applied to the foliage of fast-growing weeds **before the crop emerges.** Thus paraquat can be sprayed over dormant bulbs, maize, rhubarb, soybeans and some fruit bushes. Thirdly, paraquat is often applied by **directed spraying** around established fruit trees, forestry trees and sugar canes. Rather similarly, it can be used for inter-row spraying to reduce weed competition so long as protective shields are used to prevent the spray touching the crop.

Diquat is also used as a **pre-harvest desiccant** to dry up foliage and so to make harvesting easier and cheaper. It is particularly useful for potato haulm destruction but it is also employed as a desiccant for seed clovers, oil seed rape and peas (MAFF, 1980). In the USA it is similarly used for alfalfa, sorghum and soybeans (Klingman and Ashton, 1975).

Quaternary ammonium compounds are rainfast soon after application because they are rapidly absorbed by foliage. An interesting interaction has been observed between these compounds and herbicides such as the triazines which are inhibitors of photosynthesis; in the presence of triazines the herbicidal effectiveness of quaternary ammonium compounds is enhanced. At high light intensities they tend to be contact poisons whereas at low light intensities they are often translocated. This difference arises from the fact that, in strong light, the leaf tissue is killed before appreciable translocation can occur. Symptoms of their action include a rapid discoloration and scorch of target leaves, followed by desiccation. Chlorosis of unsprayed leaves follows somewhat later.

Electron microscopy has shown that the tonoplast of cells in paraquat-treated leaves is damaged in the vicinity of chloroplasts within six hours of application (Harris, 1970). It is probable that peroxidation of membrane phospholipids causes chloroplastic and other membranes to rupture, with loss of potassium ions from the leaf. Such damage probably accounts for the effect of high light intensity referred to above, as well as for the atypical apoplastic movement of quaternary ammonium compounds.

Baldwin (1963) observed that ^{14}C-labelled diquat stayed localised in diquat-spotted leaves so long as the plant was kept in the dark, but that it moved into all aerial parts upon illumination. It is now known that death occurs after some days even in the dark, but the more dramatic toxic effects mentioned earlier require the presence of light, oxygen and a photosynthetic tissue. The importance of light makes the light reaction of photosynthesis (figure 12.7 and section 13.2) an obvious candidate as the primary target of bipyridinium herbicides, yet paraquat does not inhibit the **Hill reaction** in isolated chloroplasts.

Around 1973 Hill found that isolated chloroplasts upon illumination were unable to reduce carbon dioxide. They could, however, remove electrons from water and pass them to a suitable artificial electron acceptor. Consequently, not only was the artificial electron acceptor reduced but oxygen was simultaneously evolved. Hill himself originally used ferric potassium oxalate as the terminal acceptor, but it is now known that numerous electron acceptors can replace carbon dioxide in this reaction, amongst them being indigo carmine (E'_0 = -0.12 V), phenazine methosulphate (E'_0 = -0.032 V) and methyl viologen (which, as was mentioned earlier, is now known as paraquat).

Paraquat (E'_0 = -0.45 V) inhibits the reduction of $NADP^+$ and thereby prevents the reduction of carbon dioxide in the photosynthetic carbon cycle. It does this by replacing ferredoxin (E'_0 = -0.43 V) as electron acceptor in photosystem I (figure 12.7). Thus light energy is re-routed, paraquat being reduced by a one electron transfer and is automatically reoxidised by oxygen (air) to form superoxides – Mees (1960) showed that oxygen was essential for the action of quaternary ammonium compounds. The superoxide free radicals attack unsaturated fatty acids in membrane lipids and lead to the deleterious consequences described earlier. Since it is the auto-oxidation which does the damage, quaternary ammonium herbicides would be expected to act *catalytically* rather than stoichiometrically. Structural and physicochemical requirements for herbicidal activity of quaternary compounds have been discussed by Moreland and Hilton (1976).

Quaternary ammonium compounds have rather low LD_{50}s (table 1.2) but their correct field use probably presents no serious hazard. Small amounts can irritate the skin or, if inhaled, can cause nose bleeding, but such amounts are rapidly and completely eliminated from the body. However, when larger

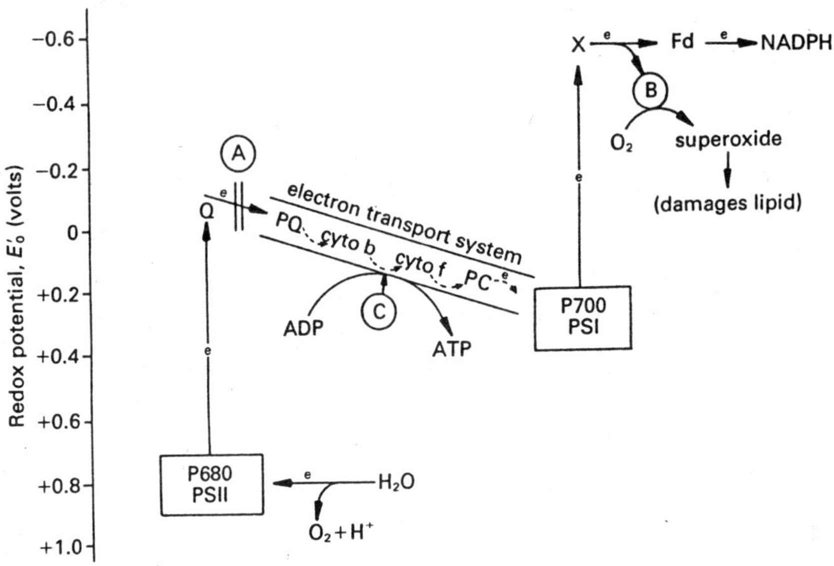

figure 12.7 Light 'reaction' of photosynthesis and the probable sites of action of some herbicides. PQ, plastoquinone; cyto f, cytochrome f; cyto b, cytochrome b559; PC, plastocyanin. Q and X are hypothetical redox intermediates. A, primary site of inhibition by ureides, triazines, uracils; B, site of electron deflection by quaternary ammonium compounds; C, approx. site of action of uncouplers such as anilides, nitrophenols and benzonitriles

amounts are ingested there is an irreversible and traumatic delayed action effect – symptoms are minimal for several days but later a non-cancerous proliferation of lung cells occurs. This proliferation continues long after all traces of the herbicides have disappeared; it leads to respiratory failure and death. There is no known antidote, although medical treatment is possible if administered sufficiently soon after ingestion. Clearly, the concentrate must be handled with the greatest care and never sub-sampled into unauthorised containers. In particular, the removal of samples into beer or soft drink bottles has been responsible for numerous tragedies. For home gardeners a granular formulation is available, which is to be preferred on the grounds of safety; when put into water the granules disintegrate and the herbicide dissolves.

In contrast to what happens to most other pesticides, the bipyridinium herbicides undergo little metabolism in plants or in animals. When paraquat is applied to plants kept in the dark, very little loss occurs, whereas in the light it rapidly disappears. Similarly, *in vitro*, diquat and paraquat keep almost indefinitely in the dark, but 85 per cent of the radioactivity is lost from ^{14}C-diquat within seven days upon exposure to ultraviolet light (Coats *et al.*, 1965). In the case of paraquat, it is believed that in both of these cases ultraviolet light causes one of the pyridine rings to split open to leave eventually the N-methyl-4-carboxy-pyridinium ion (figure 12.6d).

12.5 GLYPHOSATE

This organophosphorus herbicide was introduced in 1971 for foliar application to many annual and perennial weeds. It is employed in non-crop situations as well as for pre-emergence treatment. It can also be used in appropriate established crops where shielding or placement is possible. It resembles the quaternary ammonium compounds in that it is **inactivated by soil** and consequently the material that runs off weed leaves cannot damage crop plants by being taken up by their roots (in the absence of soil, root uptake is rapid and lethal). In the case of glyphosate, inactivation may result from interaction with iron in the soil.

Glyphosate is an acidic substance containing phosphonate and derived from glycine (figure 12.8a). It is usually formulated as the water-soluble isopropyl-amine salt. There have been several reports of apparent incompatibility with formulations containing other pesticides so considerable care is necessary to ensure spray apparatus is clean before glyphosate is used. For example, Rieck *et al.* (1974) reported that the presence of one per cent montmorillonite or organic matter could reduce glyphosate activity by 80–90 per cent, presumably as a consequence of physical binding analogous to that which takes place in soils. Stahlman and Phillips (1979) confirmed this effect, finding that

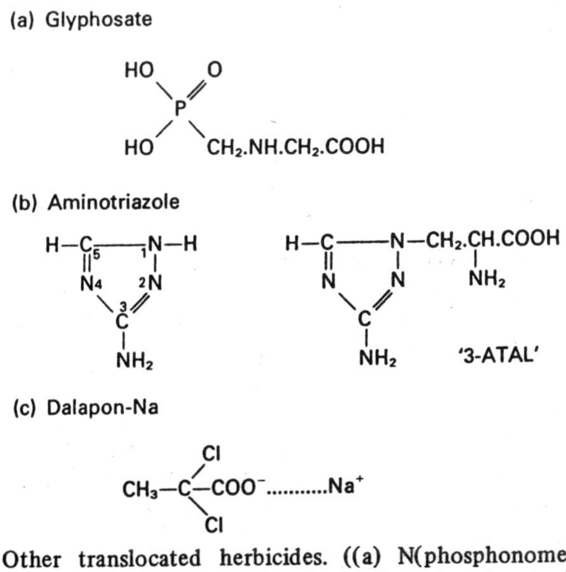

figure 12.8 Other translocated herbicides. ((a) N(phosphonomethyl) glycine (acid form); (b) 3-amino-1,2,4-triazole and conjugate; (c) 2,2-dichloropropionate (sodium salt))

various formulations (not all based on wettable powders or containing clays) of all seven herbicides they tested decreased the activity of glyphosate. These workers also observed that the inhibitory effect could be considerably reduced by using distilled water to dilute in the spray tank, i.e. by decreasing the cation content of the spray liquid.

Much experimental work using plant species has demonstrated that glyphosate is readily absorbed and translocated and considerable agreement exists about the conditions most likely to achieve maximal benefit from its action. Richard and Slife (1979) showed that uptake by detached leaves of hemp dogbane (*Apocynum cannabinum*) was rapid, the quantity absorbed being proportional to the external glyphosate concentration over a limited concentration range. For isolated cells, the uptake was almost complete within 15 min and the adsorbed glyphosate was not tightly bound to cell constituents. Also working with *A. cannabinum*, Schultz and Burnside (1980) found that uptake by whole plants and translocation out of leaves continued for only three days (in contrast to 12 days for 2,4-D under identical conditions). After 12 days, over 90 per cent of the ^{14}C recovered was still in the form of ^{14}C-labelled glyphosate. Translocation was greater at a higher temperature (30°C *v.* 25°C), the amount leaving the leaves more than doubling (39 per cent *v.* 18 per cent).

Moosavi-Nia and Dore (1979) found that plants growing in shaded surroundings were usually more damaged than similar plants growing in open ground. They explained their results by morphological changes which occur in the leaves of plants under shaded conditions. The results could, however, also relate to differences in humidity, for Chase and Appleby (1979b) working with the purple nutsedge, *Cyperus rotundus*, found that glyphosate was more effective at 90 per cent relative humidity than it was at 50 per cent. Similarly, it was more effective at −2 bars plant water potential that it was at −11 bars. Furthermore, these differences could be ascribed to consequential differences in translocation, for three times as much was translocated to underground parts at 90 per cent humidity *v.* 50 per cent, and twice as much was translocated at −2 bars water potential *v.* −11 bars. Coupland and Caseley (1979) similarly observed rapid absorption by *Agropyron repens* and symplastic movement to roots and rhizomes. Eight days after leaf treatment, ^{14}C from ^{14}C-glyphosate accumulated at rhizome nodes and apical buds and some was exuded from intact roots.

Best results are usually obtained when weeds are mature enough to present a large target for foliar uptake and when rain does not fall for several hours after spraying. For some types of weeds a combination of chemical treatment and tillage is very effective. For example, Chase and Appleby (1979a) found that the optimal time interval between applying glyphosate to *Cyperus rotundus* and tillage was three days. It is relatively non-selective in its action but grasses are often more susceptible than broadleaved plants.

The efficient downward translocation of glyphosate enables it to break the dormancy of buds on rhizomes of grasses. This may well be a factor contributing to its effective control of *Agropyron repens* and of similar rhizomatous species.

For pasture renovation it is therefore used to kill perennial, stoloniferous or rhizomatous grasses prior to re-seeding. It is relatively slow-acting and serious symptoms may take ten days to appear. Sowing can, however, proceed once the foliage has started to turn yellow. At lethal dosage plants yellow before dying, while at sub-lethal doses it affects apical dominance and leads to marked extra tillering at the base of treated shoots.

Another use of glyphosate is to clear stubble from ground where grass is to follow cereals. In this case its use allows direct drilling some 14 days after spraying. In various countries it is also used to control *Sorghum halepense* (Johnson grass), *Cynodon dactylon* (Bermuda grass) as well as to eliminate such troublesome weeds as *Aegopodium podagraria* (ground elder) water lilies, various reeds, *Convolvulus* spp and *Polygonum* spp. It can also be used in an established and susceptible crop by directed inter-row spraying. For example, Frank and King (1979) used it for weed control in strawberries; good results were obtained but the effective concentration was higher than that at which paraquat did the same task.

The mechanism whereby glyphosate exerts its phytotoxic action is still uncertain but unlike the quaternary ammonium compounds (which have similar practical uses) it does not appear to affect photosynthesis. Richard *et al.* (1979) found that glyphosate has no effect on the electron transport system between coupled photosystem II and photosystem I in stroma-free thylakoids from *Pisum sativum* (figure 12.7). Nor does it affect cyclic or non-cyclic photophosphorylation. On the other hand there is increasing evidence that glyphosate may have an effect, direct or indirect, on the metabolism of **aromatic amino acids.** Jaworski in 1972 was the first to suggest that it might inhibit aromatic amino acid biosynthesis but not all later work supported this contention. However, Hoagland *et al.* (1979) have shown that if soybean seedlings grown in the dark until three days old are then transferred to glyphosate in water culture, the amount of extractable phenylalanine–ammonia lyase activity increases. This enzyme is important in the breakdown of phenylalanine and tyrosine. The higher the glyphosate concentration, the greater is the activity of this enzyme, the lower the phenylalanine pool and the higher the concentration of ammonia. In effect, glyphosate appears to decrease the available phenylalanine (and therefore tyrosine) by stimulating this enzyme which destroys it. Shaner and Lyon (1980) also observed an interaction between the phytotoxicity of glyphosate and the availability of aromatic amino acids. They showed that glyphosate at 15 mmol/g fresh weight inhibited transpiration of excised shoots of beans (*Phaseolus vulgaris*). However, if 5×10^{-4} M tyrosine and 5×10^{-4}M phenylalanine were jointly present in the transpiration liquid, the inhibition by glyphosate of the transpiration process did not occur during the 8 hour period of the experiment. They also observed that, within six hours of treatment, the levels of endogenous aromatic amino acids were 50 per cent lower in leaves of shoots transpiring water containing 60 μmoles of glyphosate than they were in the controls (table 12.3).

Table 12.3 Effect of 60 μmoles/litre glyphosate on the aromatic amino acid content of transpiring cuttings of *Phaseolus vulgaris*. (After Shaner and Lyon, *Weed Sci.*, 28, 31 (1980))

Amino acid	n moles/mg leaf dry weight		Percentage of control value
	Control	Glyphosate	
Phenylalanine	1.19	0.56	47
Tyrosine	0.73	0.42	58

12.6 AMINOTRIAZOLE (AMITROLE)

The practical uses of aminotriazole in some ways resemble those of the compounds in the previous two sections, for it is predominantly a foliar herbicide which rapidly disappears after it has destroyed weeds by systemic action. It is also particularly efficacious against stoloniferous grasses and certain refractory weeds. However, it is not inactivated by soil and it is therefore able to reach roots of crops under conditions when quaternary ammonium compounds and glyphosate would be harmless. It was introduced in 1954 as a cotton defoliant. Although non-selective, its relatively low persistence in soil can be exploited for pre-planting and pre-sowing clean-up. In Britain, it is often mixed with simazine and MCPA and applied to foliage for the control of weeds in orchards or mixed with atrazine, diuron or dichlorprop for weed control on land not intended for cropping (MAFF, 1980).

Aminotriazole is a weak base which exists in tautomeric forms, one of which is shown in figure 12.8b. It is a solid which is soluble in water but of low solubility in hydrocarbon oils. It is usually formulated with detergent to increase spreading and with ammonium thiocyanate, $NH_4.SCN$, to act as a synergist (see below).

A major use of aminotriazole (in the absence of soil-acting herbicides as adjuncts) is to control weeds in land which is uncultivated but in which a crop is to be planted some weeks after application. It works more effectively in early summer, when weed foliage is extensive but when some growth is still occurring. For example, it is used for the control of such troublesome weeds as couch grass, horsetails (*Equisetum* spp), rushes (*Juncus* spp) and reeds (e.g. *Phragmites communis*).

For the control of grasses with stoloniferous rootstocks such as couch grass (*Agropyron repens*) and bermudagrass (*Cynodon dactylon*) it is more effective in the presence of ammonium thiocyanate. Rhizomatous grasses cannot be selectively controlled in the presence of young seedling grasses, so old swards must be destroyed in pastures before reseeding (compare glyphosate). For couch grass control, it is desirable to cultivate the ground after application to increase contact and to encourage dormant buds on rhizomes to develop. In general, about six weeks should elapse before reseeding.

Aminotriazole can also be used before sowing various cereal and other crops, although a compromise may then be necessary with regard to the optimal time of application. It can be applied in autumn prior to the sowing of winter wheat or oats, or of any spring crop; it can also be applied in the spring at a suitable time before sowing wheat, maize and potatoes (MAFF, 1980). For weed control before kale, it can also be applied in spring or summer when the foliage of weeds is about three or four inches high, the ground being ploughed some three to six weeks later. If this ploughing completely inverts the soil it is possible to plant kale almost immediately (Fryer and Makepeace, 1978).

Since aminotriazole is able to penetrate soil it is able to reach plant roots. It is therefore *seldom employed in the presence of a crop* unless, like some trees, the root system is deep. It is somewhat sorbed on to soil constituents and its phytotoxicity is roughly proportional to the amount which is readily recoverable from a particular type of soil. Its persistence in soil varies but is usually between one and three months; sowing can sometimes, but not always, be done about one month after application.

Chemical and microbial processes, as well as leaching, play a part in the disappearance of aminotriazole from soil. When soil is autoclaved, the rate of release of carbon from the aminotriazole ring is decreased fifty-fold. However, the reinoculation of autoclaved soil does not always restore its metabolic activity (Kaufman *et al.*, 1968). In some soils, up to half the aminotriazole may be destroyed by microbial action but, in others, non-biological degradation by a free radical mechanism possibly predominates.

In plants, detoxication involves conjugation reactions that result in metabolites that are of larger molecular weight and are less readily translocated than aminotriazole itself (Ashton and Crafts, 1973). The best authenticated of these is aminotriazolyl aminopropionic acid ('3-ATAL') which may be formed from aminotriazole and serine (figure 12.8b). The abbreviation presumably reflects the fact that the conjugate can be regarded as the 3-aminotriazole (3-AT) derivative of alanine (AL).

Any substance able to decrease the rate of conversion of aminotriazole into 3-ATAL would be expected to make a given dose of aminotriazole more effective. **Ammonium thiocyanate** appears to act in this way, for Carter (1975) reported that only 6.7 per cent of a dose of aminotriazole was converted to 3-ATAL in the presence of thiocyanate whereas 63 per cent was converted to it in the absence of this supplement. Thus ammonium thiocyanate is an example of a substance with a synergistic action.

Typical symptoms of the herbicidal action of aminotriazole are chlorosis of leaves and retardation of growth. It moves in both symplast and apoplast and accumulates in active meristems. The chlorosis or 'albinism' is caused by an interference with chloroplast development – like the bipyridinium compounds, it does not affect the Hill reaction in isolated (i.e. fully grown) chloroplasts. Instead, it affects chloroplasts during the stages of their development. This is shown by the fact that if germinating plants are treated with aminotriazole in the dark, the leaves remain white on exposure to light. On the other hand, if plants are kept in darkness until mature leaves have formed and are then treated with aminotriazole, these leaves do turn green when the plant is illuminated (Hilton, 1969).

It therefore seems probable that aminotriazole hinders the development of the membranes of the chloroplast rather than damaging existing ones. This is probably brought about by a blocking of the development of certain chloroplastic ribosomes known as 70 S ribosomes. This appears to lead to a dearth of membrane proteins (e.g. catalase activity in aminotriazole-treated plants is only 5 per cent of that in the controls). How aminotriazole blocks this development is, however, uncertain. Chelation of trace elements, interference with purine or histidine biosynthesis, and shortage of enzymes for carotenoid biosynthesis, have all been implicated as primary effects by various workers. An apparent difficulty with the carotenoid theory is that a constant turnover is presumably occurring yet only developing leaves are severely affected by aminotriazole.

12.7 DALAPON

This substance is one of two chlorinated aliphatic acids commonly employed as herbicides, the other being trichloroacetic acid, TCA. Both compounds are used in the form of their sodium salts, but whereas TCA is usually applied preplanting to soil to control couch and other grasses, dalapon is internally translocated after application to foliage. Dalapon has a high LD_{50} but is irritant to the skin. It is soluble in water and hydrolyses slowly in aqueous solution with the loss of chlorine from the molecule.

Dalapon (figure 12.8c) is formulated at its sodium salt, dalapon-Na, and is particularly **effective against grasses**. In this respect it resembles aminotriazole, compared with which it is often somewhat cheaper for an equi-effective dose rate. It has to be applied at a rather high application rate (up to 16 kg/ha) and six months may need to elapse between the final spraying and the sowing of a new crop. It thus persists for considerably longer than aminotriazole.

Grasses, and especially perennial and stoloniferous grasses, have become increasingly troublesome in recent years as a consequence of the removal of broad-leaved competitors by the action of the very effective hormone-type weedkillers (section 12.1). In Britain, couch grass (*Agropyron repens*), wild oat (*Avena fatua* and *A. ludoviciana*) and blackgrass (*Alopecurus myosuroides*) are controlled by dalapon while in the United States it is frequently used against bermudagrass (*Cynodon dactylon*) johnsongrass (*Sorghum halepense*) and couch or quackgrass (*Agropyron repens*). For many of these, better control often follows the application of two smaller doses rather than one larger one, for this prevents too rapid a kill and hence allows greater time for subterranean translocation). Ploughing before dalapon application often improves the effectiveness of the treatment provided sufficient time is allowed for couch shoots to grow (a precaution dictated by the fact that uptake of dalapon from soil is poor). The improvement results from the fact that ploughing breaks up the rhizomes and encourages dormant buds to develop.

In addition to its use for pasture renewal, it is used pre-planting of several crops, primarily to control grassy weeds. It is also employed pre-emergence of potatoes and post-emergence of sugar beet and carrots. It can be used for dormant season application to fruit bushes (e.g. gooseberries), to rhubarb and to forestry trees, and for treating the ground in the neighbourhood of established apple and pear trees (MAFF, 1980). It is also used in non-crop situations against rushes and sedges, and bamboo is susceptible to a 1 per cent solution applied to the foliage (Kasasian, 1971).

Dalapon-Na can enter plants through their roots but it is usually uneconomic to use it in this manner. In the presence of adequate surfactant it enters leaves rapidly and thus, like aminotriazole, it is not very vulnerable to early rain. Once inside the plant it is not firmly bound to tissues, for it can be readily extracted by water. It is **readily translocated**, moving in both the apoplast and the symplast, and it accumulates in meristematic tissues. It not only appears to move in unchanged form but few well-defined breakdown products have been recognised in plants. Some soil bacteria can, however, decompose it (Foy, 1975). On the other hand, nitrifying bacteria (but not nitrogen-fixing bacteria) are somewhat sensitive to it.

Dalapon induces **chlorosis** of green tissues and is most effective when the plant is actively growing. Little is known about how this occurs, for, although its acidic solution is a protein coagulant, at cellular pH the alkali salts of this rather strong acid are very largely dissociated. Foy (1975) distinguished between an acute toxic action which may involve changes in cell membrane permeability and a chronic effect on growth. In respect to the latter, it appears to interfere with the production of cuticular wax in plants and with the synthesis of pantothenate (an essential part of coenzyme A) in some micro-organisms. Nothing is known about why it is selectively toxic to grasses, yet this is probably the most intriguing aspect of its herbicidal action.

12.8 TRANSLOCATED CARBAMATES

Most herbicidal carbamates act through the soil, exceptions being barban, asulam and phenmedipham. Of these barban and asulam are significantly translocated and are considered in the present section. Phenmedipham is primarily a contact poison (section 12.12).

Barban (figure 12.9a) is of low aqueous solubility (11 ppm) and it is formulated as an emulsifiable concentrate. It attacks such troublesome grasses as blackgrass (*Alopecurus myosuroides*) and reed canary-grass (*Phalaris arundinacea*) but it is marketed primarily for the control of wild oat (*Avena fatua* and *A. ludoviciana*) growing amongst wheat or amongst some varieties of barley. Such a use is pressing selectivity to its limit – in space-age parlance, there is a very narrow selectivity 'window' indeed, and precise timing is essential. The wild oat plants must usually be in the 1.5 to 2 leaf stage; if they are younger or older, growth may be retarded but recovery is likely. In Britain, spraying of winter cereals is undesirable in late January and in February, and the volume of the application is also important (Fryer and Makepeace, 1978). Oats, buckwheat and some ryes are not tolerant, but an increase in the yield of wheat of as much as 50 per cent has been reported as a result of the suppression of wild oats growing in it. Where both wild oats and dicotyledonous weeds are present, wheat and some other small grains can be sprayed with a mixture of barban and MCPA (section 12.1). Barban is also employed to control grass weeds in broad beans, flax, peas, safflower, sunflower, soybeans and sugar beet.

Following foliar application, barban readily enters leaves but its translocation is somewhat restricted and occurs principally in the **apoplastic** system. Translocation is much more marked when contact is with the axils of the leaves. It is readily degraded within the plant, probably to a chloroaniline derivative, the peak of formation of which occurs around the third day after application. Few of the water-soluble metabolites have been unequivocally identified, but it seems probable that aryl hydroxylation is not a major metabolic route in plants. In this respect, it is unusual for most other carbamates are hydroxylated (section 3.4). Instead, the 4′-chloro-2′-butynyl alcohol moiety (figure 12.9a) seems to be most vulnerable, although Lamoureaux *et al.* (1971) have reported that a glutathione conjugate can sometimes be formed. Whatever the mechanism of degradation, differential responses by different plants to barban are more likely to be related to metabolism than to differential uptake. Soil micro-organisms appear to destroy barban by hydrolysis to chloroaniline; an amidase appears to be responsible.

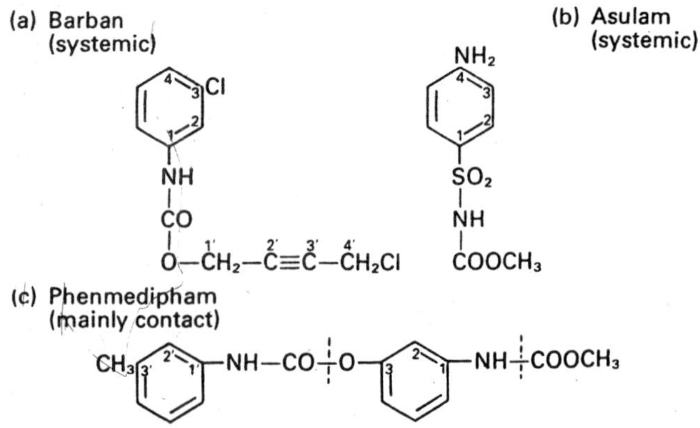

figure 12.9 Foliar-applied carbamates. (a) 4′-chloro-2′-butynyl-3-chlorocarbanilate; (b) methyl ester of N-(4-aminobenzenesulphonyl)carbamate; (c) methyl ester of 3-hydroxycarbanilate-3′-methylcarbanilate. (Dotted lines indicate sites of probable metabolic attack)

The cause of selective attack upon wild oats is uncertain but is a function of the precise molecular structure of barban. If the chlorine atom is removed to any other position in the ring the selective action (as well, very often, as the absolute toxicity) diminishes. It has also been shown that replacement of the triple bond by a double or single bond reduces the selective action.

A characteristic effect of barban is to cause the shoot apex of wild oats to swell. Development then ceases and death occurs some five weeks later. The affected apices contain abnormal cells with groups of chromosomes devoid of nuclear membrane. This may be caused by an alteration of the globular protein which forms the *microtubules* and which are responsible for alignment of the chromosomes during mitosis. Additionally, the foliage of poisoned plants often turns dark green and it is possible that this is an indication that barban interferes with protein synthesis.

Asulam (figure 12.9b) is a carbamate-like substance which is also a sulphonamide. It provides a means of controlling bracken (*Pteridium aquilinum*) on hill and grazing lands, and in non-cropped land, at a price which is economically acceptable. It is also used to control docks (*Rumex* spp) in various types of established orchards and in permanent grassland (MAFF, 1980). It has been recommended for the control of horsetails (*Equisetum* spp) in established grass that is later to be ploughed. For this purpose, the best results are obtained by spraying in late spring just before the shoots have reached maximal growth, and then to leave for a week before ploughing. Crops should not be sown for at least six weeks after the treatment (Fryer and Makepeace, 1978).

Section 2: Contact Herbicides

12.9 PETROLEUM OILS

These oils are used in large quantities as herbicides and for other crop protection purposes (table 1.2). Evidently, different mixtures of hydrocarbons are required when oil is being used as a foliar-applied insecticide from those required when it is applied to weed leaves in order to kill them. Similarly, when oils are used for non-selective weed control, the use of low grade oils may be acceptable, but highly refined oils are essential if taint is to be avoided when hydrocarbons are used as selective weedkillers on root crops.

Petroleum oils vary in composition according to their origin, but comprise a mixture of saturated hydrocarbons (mostly aliphatic), of unsaturated aliphatic hydrocarbons and of aromatic compounds related to benzene and naphthalene. Of these, the aromatic compounds are the most phytotoxic, the aliphatic hydrocarbons the least so. Oils are refined by a variety of processes, including distillation, acid treatment and catalytic reduction, to provide mixtures appropriate for their numerous industrial applications as fuels, lubricants and solvents. Amongst these applications are their uses in crop protection, both directly as toxicants and solvents and indirectly as fuels.

Petroleum spirits are refined petroleum distillates of high flammability. They are used for selective weed control in *Umbelliferae*. On the other hand, **diesel oil** has a high boiling point and a high viscosity and, together with **tractor vaporising oil** (TVO), has been used for non-selective weed control. Not infrequently, diesel oil is 'fortified' for this purpose by dissolving in it a powerful non-selective herbicide such as pentachlorophenol (PCP). **Kerosene ('paraffin')** varies greatly in composition; after refinement to remove excess phytotoxic aromatic and unsaturated hydrocarbons, it is frequently used as a solvent for pesticides formulated as emulsions or as emulsifiable concentrates. **Gasoline** ('petrol') should never be used in crop protection preparations because of the hazard associated with its low flash point.

Many umbelliferous crops show tolerance to viscous oils that are damaging to most broadleaved plants. For this reason – and if application is appropriately timed – highly refined oils can be used for weed control in parsnips and parsley; they are also occasionally used in carrots. If the oil is impure or contains unsaturated components there is a possibility that tainting will occur in plants that are actively depositing storage materials in tap roots.

Petroleum oils are of low chemical reactivity and thermodynamic aspects of their 'physical' action are discussed in section 7.8. It was there suggested that 'physical' poisons often interfere with various processes which have in common the characteristic that they are integrated at the surface of a membrane or are dependent on membrane integrity.

In relation to herbicidal action, the 'physical' effect can be at the crude macroscopic level of blocking the stomata and damaging cuticular lipids. In less extreme cases a more gradual action probably involves damage to the cell membrane or to a lipoprotein membrane within the cell (i.e. to the tonoplast or to membranes of the chloroplast, nucleus or mitochondrion), with consequent loss of cell compartmentation. Plants sprayed with oil do, for example, often lose cell sap into the intercellular spaces, with a consequent change in the overall appearance of the leaf.

12.10 BROMOXYNIL AND IOXYNIL

These contact herbicides have *little or no action through the soil.* Their practical importance rests in the fact that, while most grasses tolerate them, they attack the seedlings of several troublesome broadleaved weeds that are not readily controlled by phenoxyalkanoic acid herbicides. The range of species they attack when used at economic concentrations is limited, a disadvantage that is usually offset by using them in mixtures with other herbicides.

As well as being nitriles (figure 12.10) these sparingly soluble solids are also halogenated phenols and are, in consequence, acidic substances ($pK_a \approx 4.0$)

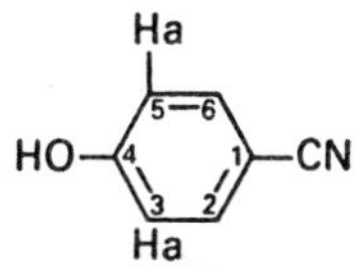

(a) Ha=Br; Bromoxynil
(b) Ha=I; Ioxynil

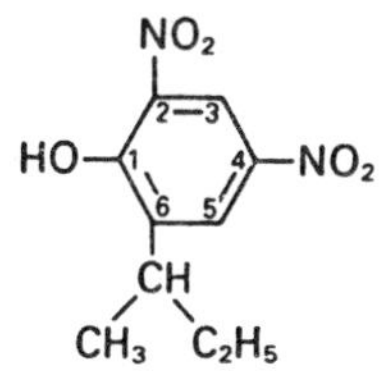

(c) Dinoseb

figure 12.10 Bromoxynil, ioxynil and dinoseb. ((a) 3,5-dibromo-4-hydroxybenzonitrile; (b) 3,5-diiodo-4-hydroxybenzonitrile; (c) 2,4-dinitro-*sec*-butylphenol. (a) and (b) are often in the form of octanoate ester)

capable of forming both salts and esters. As would be expected, the salts are more soluble in water and less persistent as foliar sprays than are the esters. Many formulations contain octanoic acid esters, together with adequate spreading agent to achieve the good coverage that is essential for successful contact action.

Bromoxynil and ioxynil enter leaves through the cuticle, killing by a rapid contact action. Nevertheless, studies with radioactive bromoxynil have shown that very small quantities are translocated. Most of the radioactivity of ^{14}C-bromoxynil does, however, remain in the treated leaves of both resistant and susceptible species (Schafer and Chilcote, 1970).

Herbicidal action occurs in two stages. There is a rapid contact action which causes local blisters or necrotic spots within 36 hours of spraying. Thereafter, the damage spreads, with deterioration of leaf tissue in target leaves and the development of chlorosis in untreated leaves. The latter occurs despite the fact that very little herbicide appears to move into untreated leaves, and it eventually leads to the death of the plant.

Bromoxynil and ioxynil are often used in spring cereals to control *weeds resistant to MCPA and 2,4-D*. Some of these weeds are listed in table 12.4. Timing of the application is critical for good results – when weed seedlings have passed beyond the fourleaf stage they often respond poorly. Established weeds are usually resistant.

These herbicides are amongst the very few substances that can be used in **newly-sown lawns**; they can be applied once the grass has reached the two-leaf stage. In cereals, the stage of development of the crop is usually of small importance. This has a valuable consequence, for it makes possible the use of bromoxynil as an emergency spray on winter cereals which have passed the stage where phenoxyalkanoic acids can be safely employed. Ioxynil has been used for weed control in onions and leeks.

With the exception of the use of ioxynil for weed control in newly-sown lawns, both bromoxynil and ioxynil are usually applied in mixtures with a (translocated) hormone-type herbicide or with a soil-acting substance. The former combination increases the range of broadleaved weeds that can be controlled but puts limits on the timing of the application. On the other hand,

Table 12.4 Weeds resistant to some phenoxyalkanoic acids but the seedlings of which can be controlled by bromoxynil or ioxynil

(a) *Weeds of frequent importance*	
Mayweeds, several kinds (*Matricaria* spp)	(1) Hempnettle (*Galeopsis tetrahit*)
Black bindweed (*Polygonum convolvulus*)	(1) Corn marigold (*Chrysanthemum segetum*)
Chickweed (*Stellaria media*)	(2) Fiddleneck (*Amsinckia intermedia*)
Knot grass (*Polygonum aviculare*)	(2) Mustards (resistant)
(3) Speedwells, several kinds (*Veronica* spp)	(2) Wild buckwheat
(b) (4) *Weeds of occasional importance*	
Fools parsley (*Aethusa cynapium*)	Bugloss (*Lycopsis arvensis*)
Scarlet pimpernel (*Anagallis arvensis*)	Forget-me-not (*Myosotis arvensis*)
Chamomile (*Anthemis* spp)	Corn poppy (*Papaver rhoeas*)
(5) Mouse ear (*Cerastium holosteroides*)	Redshank (*Polygonum persicaria*)
Fumitory (*Fumaria officinalis*)	Groundsel (*Senecio vulgaris*)

(1) Better control when mixed with linuron, (MAFF, 1980).
(2) In the USA (Klingman and Ashton, 1975).
(3) A good way to control slender speedwell (*Veronica filiformis*) in established lawns (Fryer and Makepeace, 1977).
(4) Selected from Table, pp. 15–20, Fryer and Makepeace (1978).
(5) Best control is at the young plant stage.

inclusion of a soil-acting compound gives longer-lasting activity. Thus, bromoxynil is often mixed with MCPA or dichlorprop, while ioxynil is formulated with linuron. A novel situation is the use of a mixture of bromoxynil with ioxynil to control corn marigolds and for use in undersown cereals (MAFF, 1980).

Both bromoxynil and ioxynil interfere with a wide range of biochemical processes in plants or plant organelles. They destabilise lipoprotein membranes, inhibit the Hill reaction in isolated chloroplasts, affect mitochondrial electron transport and inhibit the synthesis of proteins. In addition, they prevent the incorporation of carbon dioxide into acetyl coenzyme A, and thus prevent the formation of the malonyl coenzyme A needed for fatty acid biosynthesis (Edwards and Hassall (1980), p. 203). They also inhibit the reduction of $NADP^+$, and therefore interfere with the photosynthetic reduction of carbon dioxide.

For bromoxynil-treated plants to respond quickly, the treated plants must be illuminated. This fact, together with other effects just mentioned, favours the hypothesis that these herbicidal nitriles affect the light reaction of photosynthesis. This they appear to do at concentrations lower than 10^{-6}M. A likely target is the **electron transport system** that connects electrons from photosystem II with the positive moiety formed in photosystem I (figure 12.7). Some workers have suggested that a probable locus is the quinonoid carrier, plastoquinone. It is therefore of some interest that ioxynil, which interferes with mitochondrial oxidative phosphorylation, appears to exert its effect at or near the quinone carrier, ubiquinone (Kirkwood, 1976).

It thus seems that the reduced levels of ATP evident in bromoxynil-poisoned plants could be caused by interference with *both photosynthetic and oxidative* phosphorylation. Since almost all ATP is ultimately used to synthesise macromolecules, a lack of ATP could have the longer-term effect of slowing down the synthesis of nucleic acids and proteins, an effect likely to lead to the eventual death of the plant. This could account for the chronic second stage in the action of these herbicides.

Information on the metabolism of bromoxynil is limited and inadequate. Buckland *et al.* (1973) found that the octanoic acid ester was first hydrolysed by wheat seedlings and the nitrile group was then hydrolysed to give substituted benzamide and substituted benzoic acid. The latter then appears to undergo rapid decarboxylation. In addition, there is evidence that bromine is replaced by hydroxyl or by hydrogen in some metabolites. Micro-organisms present in some soils appear to be able to degrade bromoxynil all the way to carbon dioxide (Hsu and Camper, 1975). On the other hand a concentration of less than 50 ppm of bromoxynil strongly inhibits nitrification in soils (Debona and Audus, 1970).

Although a large number of morphological and biochemical factors has been implicated in the selective action of these nitriles, it is probably true to say that the main cause – if there is one – remains unidentified. In one study (Schafer and Chilcote, 1970), differential spray retention was excluded as the reason for the greater susceptibility of *Amsinckia intermedia* (fiddleneck or yellow borage) vis-à-vis wheat. Different rates of metabolism probably play a part, because the benzoic acid hydrolysis product has only one-twentieth of the activity of the parent nitrile. Another possibility is that an innate difference in sensitivity to these compounds exists at the site(s) of action.

12.11 DINOSEB

Various derivatives of 2,4-dinitro-6-alkyl phenols have been used as winter ovicides, as selective fungicides and as weed killers. The fungicides dinocap and binapacryl are discussed in section 9.5. Dinoseb (figure 12.10c) is a herbicidal member of the same family of compounds. It is marketed as a foliage contact poison to control weeds in some tolerant legumes and in certain other specified crops. It has very limited action through most soils.

Dinoseb is a yellow flammable solid. It is sparingly soluble in water but, being a weak acid ($pK_a \approx 4.4$) it forms much more soluble salts. It can be **formulated** in three ways, namely as an oily solution of the free acid, as an emulsifiable liquid containing the covalent acetate and as emulsifiable liquids containing partly ionised amine or ammonium salts.

For most herbicides that are weak acids or bases, maximal selectivity is usually obtained when ionisation is high. This is presumably because the rate of entry of covalent molecules through lipophilic barriers is greater than the rate of entry of anions. A consequence of such an assumption is that minor structural differences (e.g. in thickness or in composition) between such barriers in weeds and crop plants can best be exploited when some constraint (e.g. lipid insolubility or ionic charge) exists which impedes penetration. For dinoseb, as for the phenoxyalkanoic acids (section 12.1), covalent or oil-based formulations are more toxic to weeds, but also less selective, than the partly ionised amine or ammonium salts. Some uses of different formulations of dinoseb will illustrate these principles.

Oily solutions of dinoseb acid are used as foliage desiccants and also for pre-emergence weed control. In neither of these circumstances is selectivity a requirement. Oil-based formulations are used in Britain to facilitate harvesting by defoliating or by desiccating hops, potato haulms, clover and lucerne. They are also used before the crop emerges to control annual broadleaved weeds in beans, cucumbers, maize, peas and potatoes.

The acetate ester is often used in pre-emergence situations as well as post-emergence in certain tolerant crops such as forage legumes and onions. The ammonium and amine salts can be used pre-emergence in peas and beans as well as post-emergence in cereals undersown with clover or lucerne. When sprays are used in situations involving emerged legume crops it is necessary that two or three trifoliate leaves are present (in practice, although dinoseb has been an important selective herbicide for weed control in peas, its use for this purpose is declining). Similarly, dinoseb amine can be applied to control seedling weeds in new grass-legume mixtures, although some temporary damage is common (Fryer and Makepeace, 1978).

Dinoseb is usually applied using a middle to high volume of water. It is not so effective when the weather is cold and dry as when it is warm and moist, and the amount of active ingredient may need adjustment by up to 20 per cent to allow for this difference. An interesting side effect is that it sometimes stimulates crop growth or crop yield in a weed-free environment. This was first reported by Bruinsma (1962) in relation to the yield of winter wheat, but Kasasian (1971) has referred to the reduction of *Sclerotium* root rot of groundnuts.

Dinoseb and other 2,4-dinitrophenols are highly toxic to mammals, birds and insects, as well as to plants. The universality of this toxicity implies impairment of some fundamental biological process. The effect is, in fact, to uncouple oxidation from phosphorylation in the mitochondrial electron transport system. Respiration not only continues in poisoned organisms but actually speeds up as the organism produces more and more NADH in a vain attempt to make ATP (figure 12.11). A commonplace analogy is that of trying to drive a car at constant speed while depressing the clutch more and more (depression equals higher dinoseb concentration). Uncoupling occurs in isolated animal mitochondria at a dinoseb concentration of about 1–10 μM. This is probably also the main effect in plants, for plants kept in the dark are more susceptible than those that are illuminated, a result which Mellor and Salisbury (1965) attributed to the contribution to the ATP pool made by photosynthetic phosphorylation in the presence of light. Numerous other effects in plants have been reported (e.g. interference with phosphorus uptake and inhibition of RNA and protein synthesis) but these are probably secondary effects consequent upon a shortage of ATP.

In mammals, the symptoms of poisoning reflect the 'slipping clutch' effect and the squandering of energy. The patient breathes rapidly and starts to flush and sweat. In an emergency, it is essential to keep the patient lying down in a cool place while awaiting a doctor. It is appropriate to recall that the numerous deaths which occurred in the early years of the use of dinitrophenol pesticides were in large measure the catalyst for the enactment of pesticide safety regulations in many countries.

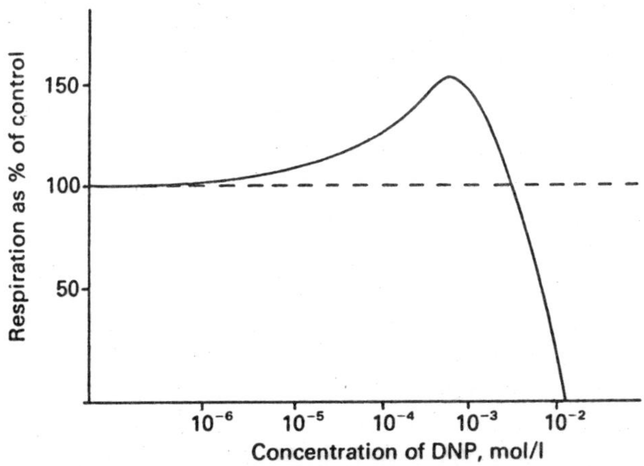

figure 12.11 Effect of dinitrophenol on the respiration of *Chlorella vulgaris*

12.12 MISCELLANEOUS HERBICIDES WITH CONTACT FOLIAGE USES

Several compounds with a contact foliage action belong to chemical families the majority of whose members are better known for their action through the soil.

Phenmedipham is a carbamate herbicide (figure 12.9c) which is marketed principally to control seedlings of broadleaved weeds in fodder beet, sugar beet and red beet (MAFF, 1980). It is also one of very few foliar herbicides which can be used, in appropriate circumstances, for seedling weed control in strawberries (Fryer and Makepeace, 1978). A single application seldom controls weeds that have developed much beyond the cotyledon stage but some larger weeds respond if the treatment is repeated after an interval of three days. Mixtures with barban give improved control of *Polygonum* spp in beet and, if wild oats are present, these can be controlled by increasing the proportion of barban. When used on beet and mangolds some crop damage may ensue if the temperature is above 21°C or if the light intensity is high; such damage is often minimised by evening spraying at a reduced application rate. The crop may also be damaged if it is suffering from manganese deficiency, frost or insect damage, or some other stress factor.

By the use of radiocarbon-labelled phenmedipham it has been shown that foliar uptake is rapid for about two hours, but thereafter the rate of uptake greatly decreases. Both susceptible and tolerant species of plants absorb it readily. Phenmedipham is sometimes classified as a translocated herbicide because movement within the plant can be detected some eight hours after foliar application, but movement is limited and is predominantly upwards in the transpiration stream (in the absence of soil it, like many other carbamates, can enter roots). The rate of absorption increases with increasing temperature (10–40°C) and increasing light intensity (2–90 lux), a fact which probably explains the phytotoxic damage mentioned above (Koch *et al.*, 1969).

The mode of action of phenmedipham, in common with that of other herbicidal carbamates, is probably on the light reaction of photosynthesis, where it inhibits electron transport and non-cyclic photophosphorylation (figure 12.7 and section 13.2). Arndt and Kötter (1968) showed that the Hill reaction carried out by isolated chloroplasts from tolerant and susceptible species was inhibited to an equal extent. Differences in uptake, distribution or metabolism must be responsible for its selective action, for the same workers found that there was a lower inhibition of photosynthesis in tolerant plants than in sensitive ones, following application of equal amounts to leaves of whole plants.

Little is known about the metabolism of phenmedipham in plants and soil, except that 3-hydroxyaniline is probably one of the products. In rats, Sonawane and Knowles (1972) found that the equivalent of 80 per cent of an oral dose was excreted in the urine within 24 hours. Hydrolysates of the urine contained the methyl ester of 3-hydroxyphenyl carbamic acid. Other products suggest that the latter is cleaved *in vivo* to give 3-hydroxyaniline (figure 12.9c) which is then acetylated. All three of these substances are probably present in the urine as various conjugates.

Propanil (figure 12.12a) is an anilide (a phenyl substituted amide) which is applied after the emergence of both weeds and crop. It is not much used in Britain but about 85 per cent of the rice crop in the United States is treated with propanil when the rice plants are six inches high (Anderson, 1977). It is primarily used to control barnyard grass, sedges and some broadleaved weeds. It is of

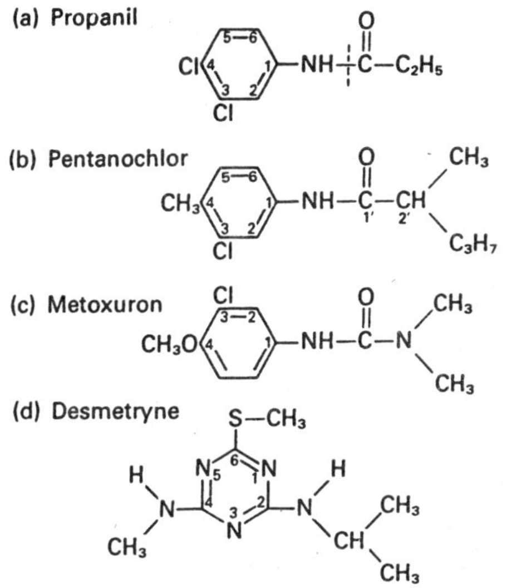

figure 12.12 Some miscellaneous foliar-acting herbicides. ((a) 3,4-dichloropropionanilide; (b) 3-chloro-2′,4-dimethyl valeranilide; (c) 3′-(3-chloro-4-methoxyphenyl)-1′,1′-dimethyl urea; (d) 2-isopropylamino-4-methylamino-6-methylthio-1,3,5-triazine)

interest and of possible importance that carbamate and organophosphorus *insecticides* inhibit an amidase in rice that in normal circumstances inactivates propanil. As a result, crop injury can occur if such insecticides are sprayed too soon after applying propanil. When amidase acts on propanil, 3,4-dichloroaniline is one of the products (Menzie, 1974).

Pentanochlor (figure 12.12b) controls annual weeds in carrots, parsnips and celery. Amongst weeds that are resistant to several other herbicides at the young plant stage but which respond to pentanochlor are fat hen (*Chenopodium album*), fumitory (*Fumaria officinalis*), cleavers (*Galium alparine*), various *Polygonum* spp., wild radish (*Raphanus raphanistrum*) and smooth sow thistle (*Sonchus oleraceus*). These weeds can also be controlled in established gooseberry, blackcurrant and some other fruit bushes and trees if contact with leaves of the crop is avoided by directed spraying. For this last purpose a mixture with chlorpropham is occasionally employed in Britain (MAFF, 1980).

Metoxuron (figure 12.12c), like most urea herbicides, can act through the soil (section 13.2) but, appropriately formulated, it can be applied after the emergence of certain crops. It is particularly useful for the control of blackgrass (*Alopecurus myosuroides*) and mayweed (*Matricaria* spp.) It can be applied after the emergence of carrots, of winter barley and of some varieties of winter wheat, in each case when the crop has reached the two-leaf stage.

Triazine herbicides are normally soil-acting compounds that are applied before the crop emerges (section 13.4), but several of them, when formulated with surfactant to promote good contact action, have also found uses as foliage sprays. Selectivity is limited and directed spraying is usually essential. **Desmetryne** (figure 12.12d) controls fat hen and many other weed seedlings in brassicas. Thus it can be used on cabbages and on brussel sprouts when the plants are about 13 cm high. **Prometryne** (figure 13.3) is used for weed control in carrots, parsley and celery after the crop plants have reached specified stages of leaf development. It is also used for weed control in early potatoes, a technique that is particularly useful when the potatoes are to be followed by another crop (Fryer and Makepeace, 1978). In the United States it is also used post-emergence in maize. **Cyprazine** (figure 13.3) is used to control seedling grasses and broad-leaved weeds in maize. The weeds should not be more than four to six cm high. **Simazine, atrazine** and **ametryne** (figure 13.3) are other triazines that are usually employed for pre-emergence treatment but which, when suitably formulated, have specialist uses as foliage contact poisons. The mode of action of triazines is discussed in section 13.4.

Endothal is a highly toxic substance sometimes used as a post-emergence spray to control emerged weed seedlings in sugar beet. It is also used as a pre-harvest desiccant (e.g. for soybeans) and defoliant (e.g. for cotton). It is the disodium salt of a hexahydrophthalic acid analogue which possesses an oxygen bridge ortho to each of the two carboxyl groups. It is not recommended for use in Britain.

REFERENCES

Anderson, W. P. (1977). *Weed Science: Principles*, West Publishing Co., New York, p. 291

Arndt, F. and Kötter, C. (1968). *Weed Res.*, **8**, 259

Ashton, F. M. and Crafts, A. S. (1973). *Mode of Action of Herbicides*, John Wiley, New York

Auch, D. E. and Arnold, W. E. (1979). *Weed Sci.*, **26**, 471

Baldwin, B. C. (1963). *Nature, Lond.*, **198**, 872

Barnes, J. M. (1976). *Herbicides: Physiology, Biochemistry, Ecology*, Vol. 2, 2nd edn (ed. L. J. Audus), Academic Press, London, chapter 13

Behrens, R. and Lueschen, W. E. (1979). *Weed Sci.*, **27**, 486

Bruinsma, J. (1962). *Weed Res.*, **2**, 73

Buckland, J. L., Collins, R. F. and Pullin, E. M. (1973). *Pestic. Sci.*, **4**, 149

Carter, M. C. (1975). *Herbicides: Chemistry, Degradation and Mode of Action*, Vol. 1, (ed. P. C. Kearney and D. D. Kaufman), Marcel Dekker, New York, p. 383

Chase, R. L. and Appleby, A. P. (1979a). *Weed Res.*, **19**, 207

Chase, R. L. and Appleby, A. P. (1979b). *Weed Res.*, **19**, 241

Cherry, J. H. (1976). *Herbicides: Physiology, Biochemistry, Ecology*, Vol. 1, 2nd edn (ed. L. J. Audus), Academic Press, London, chapter 17

Coats, G. E., Funderburk, H. H., Lawrence, J. M. and Davis, D. E. (1965). *Proc. 18th Ann. Meeting, S. Weed Conference*, p. 614

Coupland, D. and Caseley, J. C. (1979). *New Phytol.*, **83**, 17

Debona, A. C. and Audus, L. J. (1970). *Weed Res.*, **10**, 250
Dowler, C. C., Sand, P. F. and Robinson, E. L. (1963). *Weeds*, **11**, 276
Edwards, N. A. and Hassall, K. A. (1980). *Biochemistry and Physiology of the Cell*, 2nd edn, McGraw-Hill, Maidenhead, England
Foy, C. L. (1975). *Herbicides: Chemistry, Degradation and Mode of Action*, Vol. 1 (ed. P. C. Kearney and D. D. Kaufman), Marcel Dekker, New York, chapter 8
Frank, J. R. and King, J. A. (1979). *Weed Sci.*, **27**, 385
Fryer, J. D. and Makepeace, R. J. (1977). *Weed Control Handbook, Vol. 1: Principles*, 7th edn, Blackwell, Oxford
Fryer, J. D. and Makepeace, R. J. (1978). *Weed Control Handbook, Vol. 2: Recommendations*, 8th edn, Blackwell, Oxford
Garraway, J. L. and Wain, R. L. (1976). *Drug Design*, Vol. 7 (ed. E. J. Ariens), Academic Press, New York, chapter 3
Halliday, D. J. (1973). *Outl. Agric.*, **7**, 142
Hardin, J. W., Cherry, J. H., Morre, D. J. and Lembi, C. A. (1972). *Proc. natn Acad. Sci., U.S.A.*, **69**, 3146
Harris, N. (1970). *Studies on the Mode of Action of Paraquat and Diquat*, Ph.D. thesis, University of Bath
Hilton, J. L. (1969). *J. agric. Fd Chem.*, **17**, 182
Hoagland, R. E., Duke, S. O. and Elmore, C. D. (1979). *Physiol. Plant.*, **46**, 357
Holroyd, J. and Thornton, M. E. (1978). *Weed Res.*, **18**, 181
Hsu, J. C. and Camper, N. D. (1975). *Pestic. Biochem. Physiol.*, **5**, 47
Kasasian, L. (1971). *Weed Control in the Tropics*, Leonard Hill, London
Kaufman, D. D., Plimmer, J. R., Kearney, P. C., Blake, J. and Guardia, F. S. (1968). *Weed Sci.*, **16**, 226
Kirkwood, R. C. (1976). *Herbicides: Physiology, Biochemistry, Ecology*, Vol. 1, 2nd edn (ed. L. J. Audus), Academic Press, London
Klingman, G. C. and Ashton, F. M. (1975). *Weed Science: Principles and Practices*, John Wiley, New York
Knight, B. A. G. and Tomlinson, T. E. (1967). *Soil Sci.*, **18**, 233
Koch, W., Majumdar, J., Fykse, H. and Rademacher, B. (1969). Schering A. G., Proceedings of a Conference held in Berlin, p. 12
Lamoureaux, G. L., Stafford, L. E. and Tanaka, F. S. (1971). *J. agric. Fd Chem.*, **19**, 346
Loos, M. A. (1975). *Herbicides: Chemistry, Degradation and Mode of Action*, Vol. 1, 2nd edn (ed. P. C. Kearney and D. D. Kaufman), Marcel Dekker, New York, chapter 1
MAFF (1980). *List of Approved Products for Farmers and Growers* (for full ref. see p. 334)
McCarty, M. K. (1979). *Weed Sci.*, **27**, 415
Mees, G. C. (1960) *Ann. appl. Biol.*, **48**, 601
Mellor, R. S. and Salisbury, F. B. (1965). *Plant Physiol.*, **40**, 506
Menzie, C. M. (1974). *Metabolism of Pesticides: an Update* (see p. 119)
Mislevy, P. (1978). *Soil Crop Sci. Soc. Fla. Proc.*, **37**, 50
Moosavi-Nia, H. and Dore, J. (1979). *Weed Res.*, **19**, 321
Moreland, P. E. and Hilton, J. L. (1976). *Herbicides: Physiology, Biochemistry, Ecology*, Vol. 1, 2nd edn (ed. L. J. Audus), Academic Press, London, chapter 16
Richard, E. P. Jr., Goss, J. R. and Arntzen, C. J. (1979). *Weed Sci.*, **27**, 684
Richard, E. P. Jr. and Slife, F. W. (1979). *Weed Sci.*, **27**, 426
Rieck, C. E., Wright, T. H. and Harper, T. R. (1974). *Abstr. Weed Sci. Soc. Amer.*, p. 119

Samoilov, L. N. and Ladonin, V. F. (1979). *Fiziol. Biokhim. Kul't Rast.*, **11**, 35 (in Russian; quoted from abstract)
Schafer, D. E. and Chilcote, D. O. (1970). *Weed Sci.*, **18**, 729
Schultz, M. E. and Burnside, O. C. (1980). *Weed Sci.*, **28**, 13
Shaner, D. L. and Lyon, J. L. (1980). *Weed Sci.*, **28**, 31
Sonawane, B. R. and Knowles, C. O. (1972). *Pestic. Biochem. Physiol.*, **1**, 472
Stahlman, P. W. and Phillips, W. M. (1979). *Weed Sci.*, **27**, 575
Tottman, D. R. and Davies, E. L. P. (1978). *Ann. appl. Biol.*, **90**, 93
van der Woude, W. J., Lembi, C. A. and Morré, D. J. (1972). *Biochem. Biophys. Res. Comm.*, **46**, 245
Zimmerman, P. W. and Hitchcock, A. E. (1942). *Contr. Boyce Thompson Inst.*, **12**, 321

13 Herbicides that Act Through the Soil

Herbicides applied directly to the soil are frequently used to attack germinating weed seeds or very young plants although there are some exceptions in non-crop situations. Established weeds are usually resistant at economic application rates. Problems arising from soil application are considered in general terms in section 11.5 and five different circumstances in which soil application is appropriate are outlined in section 11.6. Briefly, these five include three non-crop situations and two in which a crop is present. When a crop is present, the soil-applied herbicide may be applied pre-emergence of deeply sown (or set) crops or it may be applied to certain established crops (e.g. orchard or forestry trees). In pre-crop situations the length of time the herbicide persists after a fixed rate of application is crucial to success and since some herbicide is usually still present at the time of sowing or planting, the following crop must usually be tolerant of the particular herbicide employed.

Soil-acting herbicides are for the most part structurally different from those applied to foliage, although in a few cases some overlap in function does exist. Where, however, the same substance can be used in either mode, it is normally formulated differently for the two purposes, much spreading agent being present in formulations intended for leaf application.

13.1 UREA HERBICIDES: CHEMISTRY, PROPERTIES AND USES

Urea is the amide of carbonic acid. In most members of the ureide (i.e. substituted urea) family of herbicides, urea is tri-substituted in the way indicated in figure 13.1. One of the amino groups carries either two methyl groups or one methyl and one methoxy group. The other amino group is substituted with a benzene ring which, in most cases, contains halogen atoms. Ureides are solids with a low vapour pressure (10^{-6}–10^{-8} mm Hg) at room temperature and possess aqueous solubilities ranging from 4 to 700 ppm.

Acute toxicity of ureides to mammals is usually low (table 1.2). On the other hand low concentrations of some ureides have been known to induce back mutations in *Salmonella typhimurium* in the Ames test and also to inhibit testicular DNA synthesis in mice (Seiler, 1979). Since both the Ames test and a test based on the inhibition of DNA synthesis can reveal potential carcinogenicity, there is a possibility that diuron (and perhaps other ureides) may be metabolised to an active carcinogen or mutagen. Seiler presented evidence for believing that the reaction sequence leading from herbicide to the induction of mutagenic effects might involve N-hydroxylation, followed by acetylation of the hydroxyl group. A carbonium ion in the N-acetoxy derivative may then upset cell division by reacting with nucleophilic centres in cell constituents.

Diuron, fluometuron and **linuron** are quantitatively the most important urea herbicides manufactured in the United States (Green *et al.*, 1977), although at least nine are used in Britain. Metoxuron resembles diuron (figure 13.1) except that it has a methoxy group in ring position 4. Monolinuron resembles linuron, and monuron resembles diuron, but, as the names partly imply, each lacks a chlorine atom from ring position 3. Metobromuron resembles chlorbromuron (figure 13.1) except that it lacks the chlorine atom and fenuron resembles diuron but is unchlorinated. For the structures of others, reference should be made to Table 5.059 in the Handbook edited by Fryer and Makepeace (1977). It is noteworthy that the names of all the ureide herbicides terminate in -uron.

Some of the non-selective, pre-emergence and post-emergence uses of urea herbicides are listed in table 13.1. Needless to say, all the compounds become less selective as the application rate is increased. Some control with equal effectiveness seedling grasses and a range of seedling broadleaved weeds whereas others are more selective in their action. Of the substances in the Table, diuron and monolinuron are amongst the most persistent when used at higher rates of application. Fenuron is exceptional in that it is rather soluble in water and is not strongly adsorbed on to soil constituents. For these reasons it can wash down to plane roots, a property that is exploited for the control of woody weeds.

Ureides are **adsorbed** on soil organic matter and, to a lesser extent, on clay particles, with the consequences described elsewhere (section 11.5). Adsorption

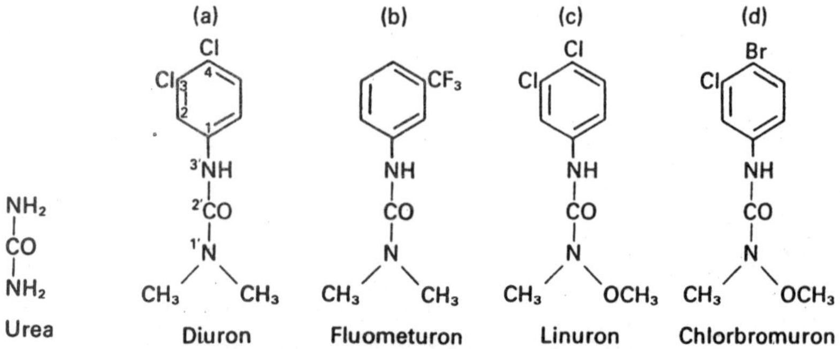

figure 13.1 Urea herbicides (ureides). ((a) 3′-(3,4-dichlorophenyl)-1′,1′-dimethyl urea; (b) 3′-(3-trifluoromethylphenyl)-1′,1′-dimethyl urea; (c) 3′-(3,4-dichlorophenyl)-1′-methoxy-1′-methyl urea; (d) 3′-(4-bromo-3-chlorophenyl)-1′-methoxy-1′-methyl urea)

Table 13.1 Uses of some urea herbicides

Herbicide[1]		Uses
Diuron[2]		Total weed control on paths and other non-crop situations.
		Annual weeds in alfalfa, maize, cotton, pineapple, sugar cane, sorghum.
		Annual weeds around some fruit trees and bushes
Fluometuron[2]		Annual weeds in cotton and sugar cane
Linuron[2]		Annual weeds pre-em. of potatoes, carrots, sorghum, soybeans.
		Kills corn marigold, pre-em. of spring cereals.
	(3)	Kills some grasses till 2" high, some dicots till 5" high.
	(4)	Seedling weeds pre-em. and post-em. of ornamental bulbs, corms
Chlorbromuron		Annual weeds in potatoes, soybean, carrots, parsnips
Chlortoluron		Wild oat, blackgrass, dicot seedlings in some winter cereals
Fenuron[5]		As pellets, for woody plant control in non-crop situations
Metobromuron		Annual grasses, dicots, in potatoes
Metoxuron		Blackgrass, annual grasses, mayweed in some winter cereals
Monolinuron		Many seedling weeds in potatoes and post-planting of leeks.
		Annual grasses and seedling dicots in carrots, onions
Chloroxuron		Annual weeds post-harvest in strawberries; seedling weeds in container plants

(1) Not all of these are registered in some countries.
(2) Major urea herbicides in the USA.
(3) Linuron acts on emerged seedlings via both soil and foliage.
(4) In mixtures with other herbicides.
(5) Fenuron is relatively soluble (0.3 per cent) and not soil-sorbed, so it washes down.

Abbreviations: em., emergence; dicot., dicotyledonous weeds; esp., especially.
Sources of information: MAFF (1980); Anderson (1977); Klingman and Ashton (1975); Fryer and Makepeace (1978).
For latin names of weeds, see Index.

on soil constituents determines not only persistence, leaching characteristics and availability for root uptake but also influences the rate of degradation by micro-organisms. Microbial action is largely limited to that fraction of a dose of herbicide that is present in soil water. It is therefore mainly determined by the aqueous concentration which results from an equilibration of the forces of adsorption on, and desorption from, soil particles (section 11.5). Adsorption characteristics within the series are partly determined by the following molecular variants:

(1) In the series fenuron, monuron, diuron adsorption increases with increasing chlorine content.

(2) Methoxy compounds such as linuron are more strongly adsorbed than are their dimethyl analogues.

(3) Brominated compounds are more strongly adsorbed than are their chlorinated analogues.

Adsorption of ureides on soil constituents is normally greatest in organic soils and for ureides of low solubility. This is well illustrated by the work of Rahman *et al.* (1978) who studied the adsorption of five soil-acting herbicides from different chemical groups, using Horotiu sandy loam soils with organic matter varying from 8 to 19.3 per cent. **Linuron** was one of the substances investigated. For all five herbicides some two or three times as much was required for a similar level of weed control in the soil of highest organic content compared to that with the lowest organic content. There was also a good inverse correlation between adsorption and aqueous solubility. In this study, the activity of trifluralin (section 13.8) was the most affected by changing levels of organic matter, terbacil (section 13.5) the least so, with linuron activity occupying an intermediate position.

Urea herbicides in solution or suspension readily enter plant roots but the extent of translocation of different substances varies greatly. Autoradiographic techniques have shown, for example, that **chloroxuron** and **neburon** tend to remain in the root system whereas **monolinuron**, **fluometuron** and **metobromuron** are largely translocated to the leaves. Once within leaves, some tend to stay near the leaf veins whereas others enter cells of the mesophyll. Since movement is apoplastic, it is accelerated by factors that increase the rate of transpiration, including higher temperatures, lower humidities and opening of stomata. Gross *et al.* (1979) have recently investigated the uptake and distribution of **chlortoluron**, a ureide used for pre- and post-emergence weed control in cereals. This substance resembles diuron, but has a methyl group in the C-4 position (figure 13.1). Wheat plants placed in a nutrient solution containing 3.1 ppm of ^{14}C-labelled chlortoluron absorbed 32 per cent of the total radioactivity within four days. Rapid translocation ensued, 80 per cent of the absorbed ^{14}C soon being found in shoots.

About 20 per cent of a dose of a urea herbicide within a plant normally becomes bound within two weeks. Binding is apparently to proteins, for it can be released *in vitro* by the action of proteolytic enzymes. It is still not clear whether this bound material represents herbicides held back from translocation, herbicide bound at its site of action, or material inactivated by some defence mechanism.

13.2 UREA HERBICIDES: METABOLISM AND MODE OF ACTION

Metabolism of urea herbicides follows the pattern established in chapter 3, although the relative importance of routes of degradation depends on both the organism and the ureide. Hydrolysis of the (apparently vulnerable) amide linkage normally plays only a minor role, although Ross and Tweedy (1974) obtained some evidence that chlorbromuron was hydrolysed to a substituted aniline by micro-organisms.

Figure 13.2 illustrates the probable major degradative routes for **diuron** and **linuron**. The metabolism of monuron follows a similar pattern (Menzie, 1978) and the scheme probably applies to several other ureides also. Two major mechanisms operate, namely, N-demethylation (or, where applicable, N-demethoxylation), and ring hydroxylation. For other ureides, alternative mechanisms occasionally exist.

The mechanism of N-demethylation is described in section 3.4. Its oxidative nature is confirmed by the fact that intermediate hydroxymethyl derivatives or their glucose conjugates have been detected in plants. The reaction is probably NADPH-dependent and catalysed by microsomal mono-oxygenases. Ureides such as linuron appear to be metabolised by removal of the methoxy group and its replacement by a hydrogen atom (i.e. oxidation followed by elimination of the elements of formic acid). This probably happens in preference to the removal of the methyl group (figure 13.2). A further example is **monolinuron**, which has

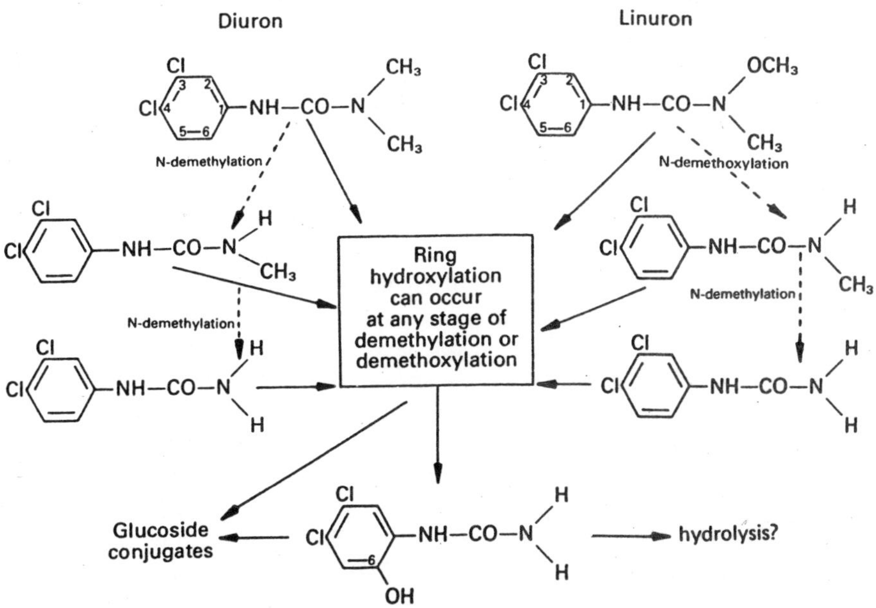

figure 13.2 Probable metabolic routes of diuron and linuron

been studied by Schuphan and Ebing (1978), who found that both N-demethylation and N-demethoxylation occurred in three crop plants. They also detected the formation of the glucoside of a hydroxymethyl intermediate.

Another type of mono-oxygenase activity results in ring hydroxylation. Diuron and linuron and their demethylated (or demethoxylated) derivatives usually hydroxylate in the C-6 position. In contrast, data summarised by Menzie (1978) indicate that monuron tends to hydroxylate in the C-2 position. A third (and for the ureide group, unusual) type of oxidation occurs with **chlortoluron** (Gross *et al.*, 1979). For this substance the main route of metabolism in wheat plants involves the oxidation of the C-4 methyl group first to $-CH_2OH$ (i.e. to a benzyl alcohol derivative) and then to carboxyl. In mature plants, the conjugates are mainly derivatives of the carboxyl oxidation product. Young wheat plants cannot oxidise the methyl group beyond the benzyl alcohol stage and the conjugates contain this product. These observations imply that a conjugate is not necessarily to be regarded as a terminal residue, for, as the plant matures, the benzyl alcohol conjugates must be dismantled to allow further oxidation to occur, with subsequent re-conjugation. N-demethylation also occurs but it is a minor pathway of metabolism of chlortoluron in wheat plants.

Differences in metabolism, with or without differences in uptake and translocation, are major factors contributing towards the selective responses of various crops and weeds to particular ureides. For example, Smith and Sheets (1967) found that cotton, which is resistant to diuron, degraded this substance much more rapidly than did soybean, which is sensitive to it. Geissbühler *et al.*, (1975, p. 271) list species which are resistant to, or tolerant of, seven urea herbicides, and in almost all cases metabolic factors seem to be in part responsible for differences in susceptibility. In the case of the triazines (section 13.4) it is likely that the mono-N-dealkylation metabolic intermediates possess diminished but protracted activity and may play a part in determining differential phytotoxicity. The mono-N-dealkylated ureides also possess diminished toxicity but their significance to differential herbicidal action in this group of compounds is uncertain. Another possible cause of selective toxicity is differential responsiveness to poisons of the sites of action in the thylakoids and this is considered shortly.

The ureide family is one of several groups of herbicides known to affect the light reaction of photosynthesis. It will be recalled that green plants contain two light-induced electron excitation systems called photosystems I and II (figure 12.7). The second of these is responsible for the production of oxygen and supplies electrons which move, via a chloroplastic electron transport system, to a positive moiety formed on excitation of photosystem I. As electrons move along the transport system, one molecule of ATP is probably formed from ADP for each electron entering the transport system. Photosystem I produces the high energy electrons that ultimately reduce $NADP^+$ to NADPH, and the latter is responsible for the reduction of carbon dioxide to sugar. Photosystem II is activated by near-red rather than far-red light and is absent from photosynthetic bacteria.

The pioneering work of Wessels and Van der Veen (1956) showed that diuron inhibited by 50 per cent the Hill reaction in isolated chloroplasts (section 12.4) at concentration below 1 μM. Later work demonstrated that photosynthetic bacteria are not readily affected by ureides (Hoffman *et al.*, 1960). Moreover, ureides do not affect (artificially induced) cyclic photophosphorylation but they do stop non-cyclic photophosphorylation, the electrons for which originate from photosystem II. Similarly, they inhibit photosynthesis in those organisms where oxygen is evolved as a by-product. All these data suggest that ureides (and probably triazines and uracils also) inhibit reactions at one or both ends of photosystem II. The position marked A in figure 12.7 (i.e. the reducing end of the system) is currently the favoured locus of their attack. For further details reference should be made to the useful reviews by Moreland (1977) and by Dodge (1977).

As was seen earlier, there is considerable evidence that the different susceptibilities of plants to equal levels of absorbed urea, triazine or uracil herbicides can often be explained in terms of rates or routes of detoxication. This may not, however, be the complete answer for, contrary to some earlier work on triazines (Moreland *et al.*, 1959; Moreland and Hill, 1962) and other herbicides (van Oorschot, 1976) there is now a certain amount of evidence that the site of action may be in some way altered in resistant strains of several species. Santakumari and Das (1978) have shown that the Hill reaction in isolated chloroplasts from pre-treated plants (of several types) was less susceptible to inhibition by diuron than was the reaction in chloroplasts from untreated plants. Moreover, Radosevich *et al.* (1979), working with resistant (R) and susceptible (S) biotypes of *Senecio vulgaris* (groundsel), have demonstrated that 60 times the concentration of diuron is needed to inhibit the Hill reaction by 50 per cent in thylakoid membranes from R plants than in those from S biotypes. Naturally, it does not follow that this cause of the appearance of resistant strains which can occur on contact with herbicide can also explain the variation in tolerances of different species of (hitherto unexposed) weeds in nature.

It is still not clear whether differences in the susceptibilities of sites of action in R and S biotypes represent differences in the ease of penetration of thylakoid membranes from the two biotypes or whether they are a function of configurational changes at the active sites within these membranes. In any case, the exact nature of the action of ureides which leads to inhibition of the light reaction is uncertain; if, however, as seems probable, ureides and triazines have a similar mode of action (section 13.4) it is probable that lipid peroxidation leads to the disintegration of chloroplastic membranes (Anderson and Krinsky, 1973).

13.3 TRIAZINES: CLASSIFICATION, STRUCTURE AND USES

Most herbicides based on the symmetrical triazine nucleus have alkyl-substituted amino groups in positions 4 and 6 (figure 13.3) and either a chlorine group or a methylthio group in position 2. Some exceptions are listed in the detailed review by Esser *et al.*, (1975). The compounds containing a chlorine atom have names terminating in *-azine* and those containing a methylthio group end in *-etryn(e)*.

Triazine herbicides are solids with low vapour pressures at room temperature and with aqueous solubilities which mostly vary within the range of 5–500 ppm.

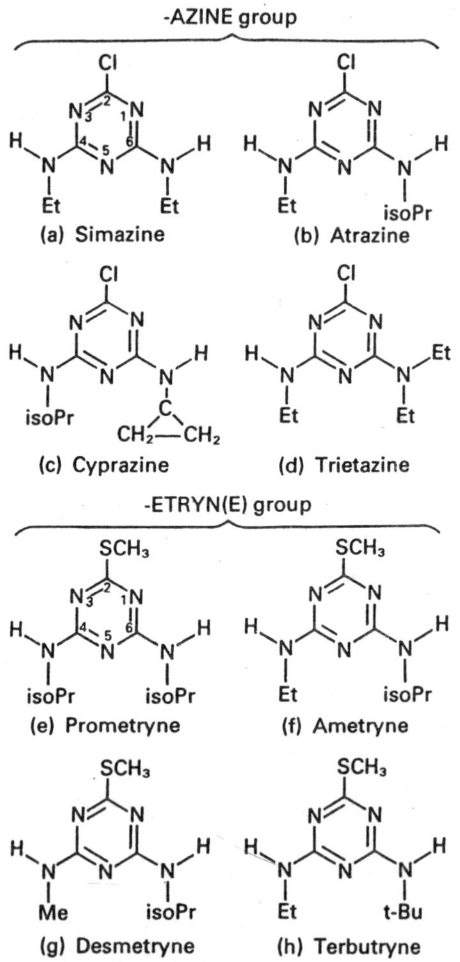

figure 13.3 Some triazine herbicides. ((a) 2-chloro-4,6-bis(ethylamino)-1,2,4-triazine; (b) 2-chloro-4-ethylamino-6-isopropylamino-1,2,4-triazine; (c) 2-chloro-4-*iso*propylamino-6-cyclopropylamino-1,2,4-triazine; (d) 2-chloro-4-ethylamino-6-diethylamino-1,2,4-triazine; (e) 2-methylthio-4,6-bis(isopropylamino)-1,2,4-triazine; (f) 2-methylthio-4-ethylamino-6-*iso*propylamino-1,2,4-triazine; (g) 2-methylthio-4-methylamino-6-*iso*propylamino-1,2,4-triazine; (h) 2-methylthio-4-ethylamino-6-*tert*-butylamino-1,2,4-triazine)

Simazine and **cyprazine** are amongst the less soluble, **ametryne** and **desmetryne** amongst the more soluble. Solubility in suitable organic solvents is insufficient to allow ready formulation as emulsifiable liquids. They are therefore usually marketed as wettable powders comprising up to 80 per cent of active ingredient supplemented with detergents and, sometimes, diluents such as clay. Some are also formulated as granules. Most are of low toxicity to mammals (table 1.2).

Most triazine herbicides are applied to soil and absorbed by the roots of seedling weeds although some possessing higher aqueous solubility may also enter through foliage (section 12.12). Soil-acting members are normally applied before, or immediately after, the emergence of the crop and are used to control seedling weeds. Some crops are tolerant of specific triazines. However, for most crops, apparent selectivity in the field largely depends on the fact that the crop is large-seeded, or consists of bulbs, corms, etc., and in consequence is more deeply placed in the soil than are the seedling weeds. In addition, several triazines are used at higher application rates for non-selective and long-term weed control in non-crop situations. Some of the varied uses of soil-acting triazines are shown in table 13.2.

The effectiveness of triazine herbicides is dependent on soil type and rainfall (section 11.5). For a similar herbicidal effect, much less is normally needed in sandy soils and when rainfall is high than in organic soils and when rainfall is low during the first few weeks after application. The toxicity of ureides to the crop also tends to be much higher in sandy soils than in organic soils, especially when rainfall is high. The effect on phytoxicity of soil organic matter is illustrated by the work of Rahman *et al.* (1978) who showed that the equitoxic dose of **atrazine** increased nearly threefold as the content of organic matter increased from eight to 19.3 per cent. The adsorption curves (section 11.5) which relate amount of atrazine adsorbed by soil constituents to the applied dose have been shown by Rao and Davidson (1979) to be non-linear as the external concentration increases from zero to the aqueous solubility limit. For the four soils examined, the adsorption constant was much less variable when calculated in terms of soil organic matter than when it was based on total soil mass.

The use of ^{14}C-labelled triazines has shown that uptake from nutrient solutions occurs readily in both resistant and susceptible plants. Moreover, radioactivity can usually be detected in shoots within an hour of application to roots. Upward movement is in the transpiration stream and factors which favour high transpiration usually favour rapid movement of triazines. Such factors include elevated temperature and reduced environmental humidity (section 11.4). However, work of this kind does not prove that unchanged triazines accumulate in leaves. Additional experiments involving partition into solvents followed by chromatographic separation of the extracted substances are needed to distinguish between unchanged herbicide and its metabolites. This and similar work has demonstrated that most triazines are quite rapidly metabolised to both water-soluble and chloroform-soluble metabolites. Often, but not always, plants that are able to metabolise triazines rapidly prove to be relatively tolerant of them.

Table 13.2 Uses of some triazine herbicides

Herbicide	Uses
Simazine	Total weed control on paths. Control of germinating grass and broadleaved weeds before planting, pre-em. or early post-em. Seedling weeds, in beans, sweet corn, in established rhubarb and hops. Around a variety of fruit bushes, canes and trees. Seedling weeds in alfalfa, pineapples, sugar cane
Atrazine	Weed seedlings in maize, pineapples, sorghum, sugar cane, raspberries, roses, young forest trees
Cyanazine	Annual dicot seedlings and annual meadow grass in peas. Annual weed seedlings pre-em. of potatoes (in mixtures with linuron). Post-em. dicot weed control in cereals (in mixtures with MCPA or with mecoprop)
Prometryne	Germinating annual weeds pre-em. of peas, potatoes. Post-em. many vegetables including carrots, celery. In cotton, pre-em. and directed post-em.
Ametryne	Germinating annual weeds pre- and post-em. potatoes. Ditto in bananas, sugar cane; in pineapple soon after planting. Some foliage activity on emerged weeds
Cyprazine	Newly-emerged weeds in maize. Germinating weeds pre-em. of peas, potatoes (often in mixtures with linuron or simazine)
Desmetryne	Fat hen and other seedlings in many brassicas. When used post-em., the crop must have three true leaves
Terbutryne	Blackgrass and dicot seedlings pre-em. of early drilled winter wheat, barley. Seedling weeds in sorghum, peas, potatoes. Filamentous algae in water courses

Abbreviations and sources of information: see footnote, table 13.1.

13.4 TRIAZINES: METABOLISM, SELECTIVITY AND MODE OF ACTION

The three major routes by which triazines are metabolised will first be described and then the probable relevance of each to plant tolerance will be considered. It should be remembered that any discussion about the importance to selectivity of the existence of different routes of metabolism is only meaningful in relation to the protection afforded against the effects of equal concentrations of a herbicide actually entering the plant, not to the amount applied per square metre of soil surface. Soil factors that influence availability are discussed in sections 11.2, 11.5 and 13.1.

First, **chlorinated triazines** undergo non-enzymic but catalysed hydrolysis, the chlorine atom on C-2 being replaced by a hydroxyl group (figure 13.4). Some plants, including maize, contain non-enzymic catalysts that accelerate this reaction. Several related catalysts have been recognised but the best known is benzoxazinone. (Benzoxazinone, or DIMBOA, is 2,4-dihydroxy-7-methoxy-1,4-benzoxazine-3-one glucoside.) The hydroxy triazine so formed has no herbicidal action, so that fraction of an absorbed dose of a triazine metabolised in this way is totally inactivated. **Methylthio** triazines are not metabolised at a significant rate by this route.

Secondly, some plants possess one or more glutathione-S-transferases (section 3.5) which allow glutathione to conjugate directly with herbicidal triazines of the **chlorotriazine** family. The enzymes so far investigated do not conjugate glutathione with hydroxy or dealkylated derivatives of chlorotriazines, nor with triazines of the methylthio family. All glutathione conjugates are herbicidally inactive, so this pathway also leads to total detoxication of that fraction of an internal dose of triazine that is metabolised by this route. It is of interest that Guddewar and Dauterman (1979) have achieved a 61-fold purification of a glutathione-S-transferase (from mouse liver) which is capable of conjugating 2-chlorotriazines. It is a dimer each of the two units of which have a molecular weight of about 23 000.

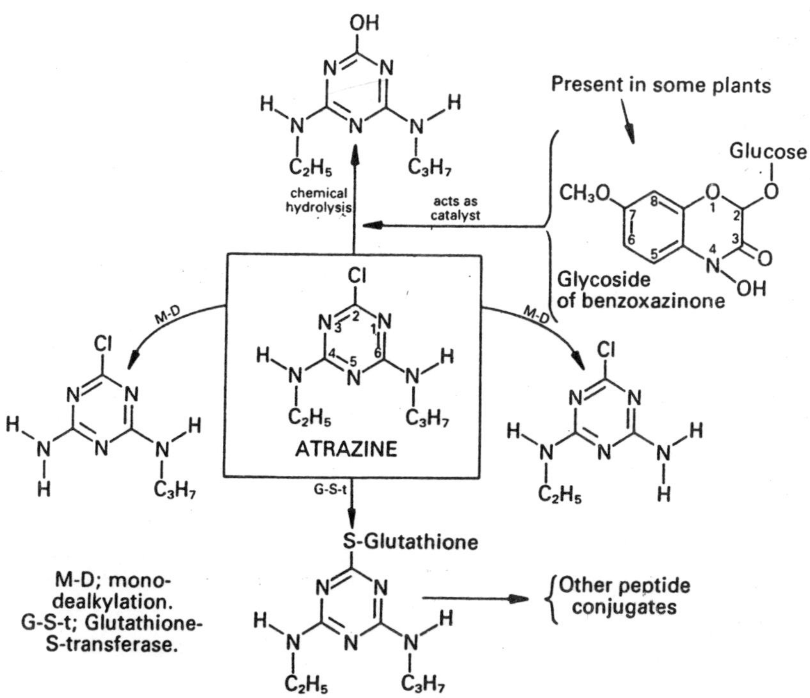

figure 13.4 Atrazine metabolism: three possible routes

Thirdly, the secondary amine groups on C-4 or C-6 of any triazine may undergo N-dealkylation (section 3.4). Hydroxy-derivatives of chlorotriazines can undergo similar changes. In 'mixed amine' herbicides such as **atrazine** (figure 13.4) the ethyl group is removed from $-NHC_2H_5$ more rapidly than is the isopropyl group from $-NH.CH(CH_3)_2$. In suitable cases, further N-dealkylation can lead eventually to molecules containing only primary amine groups. Monodealkylated triazines are less effective herbicides than are the parent compounds but they do retain some activity; consequently, that fraction of a total dose metabolised by this route is not entirely inactivated. These partly dealkylated intermediates do not seem to conjugate rapidly with glutathione, so their activity, although limited, often tends to be rather long-lasting.

In relation to the importance to selectivity of the three routes of metabolism described above, little correlation has been observed between tolerance of chlorotriazines and the presence of benzoxazinone. Maize, for example, is resistant and wheat is not, yet both contain catalysts of the benzoxazinone type. Conversely, Hamilton (1964) found that a maize mutant with a low benzoxazinone content was no less resistant to atrazine than were ordinary lines of maize. On the other hand, there may be a partial correlation between the ability of a plant to dealkylate a triazine and its resistance to it. Species such as peas and cotton that are of intermediate tolerance may owe that tolerance, in part, to the presence of this system (Shimabukuro and Swanson, 1970). It will be recalled that the partially inactivated herbicides may persist for some while within the plant.

In contrast, there is evidence that an efficient glutathione-S-transferase system is well developed in many plants tolerant of herbicide that has entered them. The system is well-developed in maize, sugar cane and sorghum, all of which are tolerant, but is poorly-developed in wheat, barley and pea, which are of low or moderate tolerance (Frear and Swanson, 1970). Naturally, if plants well-endowed with the transferase system also possess high hydrolytic activity, the latter degradative route may well play a significant additional role in ensuring rapid detoxication of chlorinated triazines.

The mode of action of triazines resembles that of ureides (section 13.2). Briefly, electrons from photosystem II are prevented from entering the electron transport system of the chloroplast, probably as a result of the herbicide blocking the pathway at position A in figure 12.7 (p. 276). It has been suggested that this leads to the formation of an active form of oxygen which gives rise to peroxides and that these probably attack unsaturated membrane lipids and destabilise chloroplastic membranes (Dodge, 1977).

As in the case of the ureides, there is conflicting evidence as to whether the photochemical activity in chloroplasts from sensitive and resistant species (or biotypes of the same species) is inhibited by similar or very different concentrations of triazines. The work described in section 13.2 is therefore also relevant to triazines, and indeed some of it was done on atrazine and ametryne. Radosevich and Devilliers (1976) reported that the uptake and internal distribution of atrazine by atrazine-sensitive (S) and atrazine-resistant (R) biotypes of *Senecio vulgaris* were similar but that the photochemical activity of isolated chloroplasts from the S biotype was inhibited by atrazine concentrations that had no effect on that of chloroplasts from the R strain. They later showed (Radosevich *et al*., 1979) a 3200-fold difference in ametryne concentrations having quantitatively similar effects on the Hill reaction in S and R chloroplasts from *Senecio vulgaris*. Moreover, in the case of atrazine, the lack of susceptibility of the R strain was so profound that no reliable ratio could be recorded. Radosevich (1977) obtained similar results using chloroplasts from S and from R strains of *Amaranthus retroflexus* and of *Chenopodium album*.

Many investigators have reported that triazines have effects on plants other than on activities in the chloroplasts, although some symptoms could be consequential upon shortage of energy or of photosynthates. Amongst the more recent reports Churchill and Klepper (1979) have shown that ametryne greatly reduces nitrate reductase activity and concluded that accumulation of nitrite contributes to herbicidal damage in wheat plants (plants absorbing ammonium salts from the nutrient solution, or absorbing no nitrogen at all, were less sensitive to ametryne in the nutrient solution than were plants absorbing nitrate ions). Renosto *et al*. (1979) found that 10^{-4}M atrazine inhibited sulphate ion uptake. This effect was not only more evident in barley and wheat than in atrazine-resistant maize, but the inhibition in maize was reversed by washing the atrazine out. Ashton and Glenn (1979), in a wide-ranging investigation, examined the effects of triazines on RNA synthesis, protein synthesis, lipid synthesis and photosynthesis in isolated leaf cells of red kidney bean plants (*Phaseolus vulgaris*). The effects of propazine, prometone and prometryne were very similar; in general, in 100 μM solutions photosynthesis was more readily affected than RNA synthesis or lipid synthesis, with protein synthesis least affected. At 1 μM, photosynthesis and RNA synthesis were inhibited, but lipid synthesis was stimulated by all three of these triazines.

13.5 URACIL AND PYRIDAZINE HERBICIDES

Both of these small groups of soil-acting herbicides contain substances whose molecules are based upon six-membered heterocyclic rings with two nitrogen atoms. Structures of selected members of the two groups are shown in figure 13.5. The basic ring structure of the uracils is that of pyrimidine and this provides an alternative name for the group.

The **uracil** herbicides were introduced in 1961, the three commonest members in current use being **bromacil, terbacil** and **lenacil**. They are of relatively low selectivity and are often used for total weed control in non-crop situations. However, each has also found uses in the presence of selected crops which either have deep root systems or show some tolerance (table 13.3). Bromacil is used for weed control in citrus and pineapple as well as in such cane fruits as raspberry

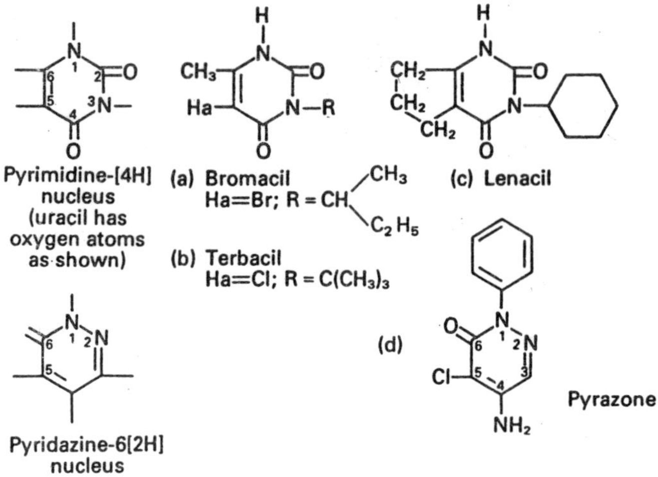

figure 13.5 Uracil (pyrimidine) and pyridazine herbicides. ((a) 5-bromo-3-*sec*-butyl-6-methyl uracil; (b) 5-chloro-3-*tert*-butyl-6-methyl uracil; (c) 3-cyclohexyl-5,6-trimethylene uracil; (d) 4-amino-5-chloro-1-phenyl-6 [2H]-pyridazinone)

and loganberry. Lenacil is used in Britain pre-drilling, pre-emergence and post-planting of specified crops (MAFF, 1980). Although various limitations exist, the latter crops can include members of the beet family, certain bulbs and certain bush and cane fruits (table 13.3).

Terbacil is used against annual and some perennial weeds in established sugar cane, citrus, apple, asparagus and strawberry; again, various precautionary restrictions exist. Experimentally, Doughty (1978) found that up to 90 per cent control of *Agropyron* growing amongst highbush blueberry plants is possible. When the application was done while the blueberries were still dormant little phytotoxicity was evident. Millhollon (1977) reported that sugar yields from sugar cane grown without cultivation but kept relatively weed free with terbacil and hand hoeing, were normally little lower than the yield obtained using the two cultivations of standard cultural practice. Arnold and Aldrich (1979) found mixtures of terbacil with other compounds to be very effective in the control of bermudagrass (*Cynodon dactylon*) and purple nutsedge (*Cyperus rotundus*) growing in well-established pecan (*Carya*) and peach. Various terbacil mixtures were reported by the same authors to provide some control of yellow nutsedge, dogfennel, camphorweed and large crabgrass.

Uracil herbicides are normally formulated as wettable powders but granules are also available. At standard application rates of 2–4 kg/ha most uracils persist in many soils at phytotoxic levels for upwards of six months. Bromacil is of moderate aqueous solubility (750 ppm) and of medium-term persistence and for these reasons has a limited soil-sterilising action. The adsorption constants determined by Rhodes *et al.* (1970) indicate that the two more soluble uracils,

Table 13.3 Uses of uracil (pyrimidine) and pyridazine herbicides

Substance	Uses
Bromacil	At low rates of application seedling weed control in citrus, cane fruit and pineapple. At high rate of application can bring about soil sterilisation and total weed control in non-crop situations. Some perennial weeds controlled by high application rates(1). Some brushwood species controlled by very high application rates
Terbacil	Annual weed control in strawberry, peaches, citrus, apples. Ditto in sisal and sugar cane. Some perennial weeds controlled by high application rates(1)
Lenacil	Pre-em. control of seedling weeds in sugar beet, mangolds. Post-planting for strawberries, bulbs, some ornamentals
Pyrazon(e)	Seedling weeds pre-em. sugar beet, fodder beet, red beet, mangolds

(1) Higher-than-normal concentrations are not necessarily compatible with use in crop situations.

Abbreviations and sources of information: see footnote, table 13.1.

bromacil and terbacil, are not so strongly adsorbed on to (Keyport silt loam) soils as are monuron, diuron or chloroneb, and in consequence rain is able to wash them down in soil to the roots of weeds or even to the deeper roots of the crop plants. Gardiner *et al.* (1969) showed that bromacil washed down into soil zones lower than three inches below the surface more rapidly than did terbacil. This is in conformity with the positions of these two uracils in the mobility classification of Helling (1971), for terbacil lies between groups 3 and 4 and bromacil is in group 4 (group 1 comprises compounds of low mobility and group 5 compounds of great mobility; see section 11.5). The rate at which bromacil leaches under favourable conditions into ground water has been studied by Hebb and Wheeler (1978); it took three months from application before the first bromacil was detected there and one month later the concentration rose to its maximal value of 1.25 ppm.

Bromacil appears to be metabolised in a rather similar way by enzyme systems present in plants and animals (Menzie, 1974). The structures of several metabolites (isolated from the urine of rats that had been fed bromacil for one month) are such that it is likely that microsomal mono-oxygenases are involved in their formation (figure 13.6a). Thus hydroxylation occurred on the C-6 methyl group and in two different positions in the sec-butyl group attached to the N-3

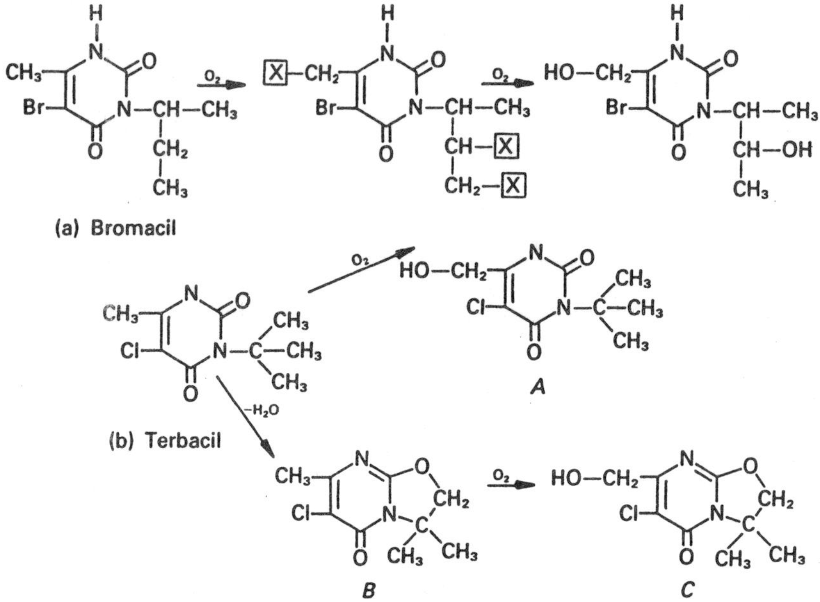

figure 13.6 Metabolism of bromacil and terbacil. (a) Any one of the three positions, X, can be hydroxylated in different metabolites (the other two being hydrogen). (b) *A* is the main primary metabolite in the dog; in one study *C* was the main metabolite in alfalfa

nitrogen atom. The metabolism of terbacil appears to be essentially similar except that a complication occurs in the form of oxidative cyclisation involving the tertiary butyl group on the N-3 nitrogen atom (figure 13.6b). The main metabolite in alfalfa was found by Rhodes (1977) to be metabolite C, but in work with animals quoted by Menzie (1974) metabolite A predominated.

It is very probable that the primary mode of action of the herbicidal uracils is the same as that of the ureides and triazines (sections 13.2 and 13.4), namely, that they interfere with electron transport at the reducing end of photosystem II. Concentrations of the order of 1–5 μM inhibit the Hill reaction of isolated chloroplasts by 50 per cent, whereas Jordan *et al.* (1966) have demonstrated that concentrations of this order do not affect the dark growth of tobacco callus tissue supplied with an organic source of energy. Moreover, Hoffman (1971) found that the toxic effects of monuron and bromacil to photosynthesising *Chlorella* were additive, indicating the probability of a common mode of action. Moreover, monuron-resistant strains of *Euglena* were also resistant to bromacil, while in the dark neither resistant nor sensitive strains of *Euglena* were damaged by bromacil. It should perhaps be added that, despite the similarity of herbicidal uracils to the pyrimidines present in DNA and RNA, bromacil does not become incorporated into DNA and no mutagenic effects have been recorded.

Pyrazone (figure 13.5) is the best-known **pyridazine** herbicide and is of particular value for the pre-emergence control of annual weeds in various kinds of beet (Fryer and Makepeace, 1978). In certain soils it is also sometimes mixed with di-allate for use as a treatment before sowing beet in ground heavily infected with wild oat (*Avena fatua* and *A. ludoviciana*). A second member of the group is **norflurazon(e)**. This is used to control annual grasses in cotton (Hawtree, 1980).

In contrast to most uracil herbicides, pyrazone binds somewhat strongly to constituents of organic and heavy soils. Its availability at different depths in the soil depends on soil type and its use on sandy soils is often inadvisable (section 11.5). Upon reaching the roots of plants it is translocated to the leaves. For a given amount absorbed it accumulates in some plants and is inactivated in others, a difference that probably accounts, at least in part, for its selectivity. Beet can be damaged when the transpiration rate is very high since in this circumstance uptake may proceed faster than inactivation. The principal method of inactivation probably involves direct conjugation with glucose. In soil it is broken down rather more quickly than are the uracil herbicides; its reported half-life of about a month in a loam soil means that its concentration often (but not necessarily) falls below the phytotoxic level within three or four months.

Pyridazine herbicides resemble the uracils in that they powerfully inhibit the Hill reaction in isolated chloroplasts. They are therefore believed to affect electron transport in the chloroplasts. Perhaps for this reason a symptom of their action under field conditions is the development of severe chlorosis of green tissues. Pyrazone, as well as entering plant roots, has some foliar action on seedling weeds.

13.6 SOIL-ACTING PHENYLCARBAMATES

Esters of phenylcarbamic acid were originally introduced as a possible means of controlling monocotyledonous weeds in broadleaved crops, i.e. to complement the action of the very successful phenoxyalkanoic acids (section 12.1). This early promise was, in fact, not fulfilled, for although seedling grasses succumb, established grasses usually do not. It was, however, soon evident that some crops are moderately resistant to carbamates at dosages, or under conditions, which destroy seedling grasses and some other weeds.

Phenylcarbamic acid esters are compounds in which a phenyl (or substituted phenyl) group has replaced an amino hydrogen atom in aminoformic acid (equation (13.1a)). They can equally well be termed carbanilates, for they can be regarded as aniline in which a carboxyl group has replaced an amino hydrogen atom. The structures of several important members of the group are shown in figure 13.7. They should be distinguished from the aryl esters of alkylcarbamic

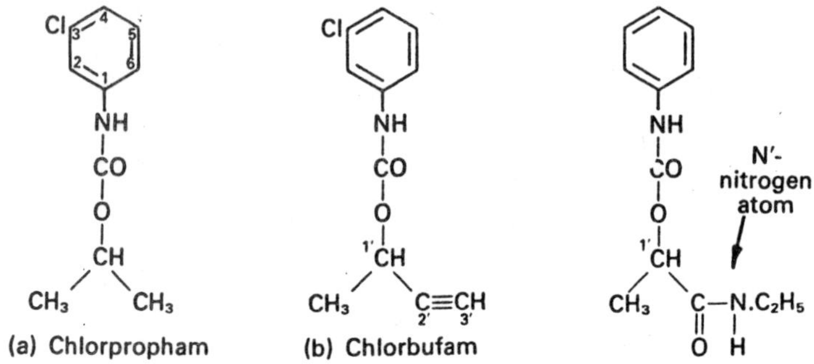

(a) Chlorpropham (b) Chlorbufam (c) Carbetamide

figure 13.7 Some herbicidal carbamates. Propham resembles (a) but lacks the C-3 chlorine; Swep resembles (a) but has a second chlorine on C-4. ((a) Isopropyl-N-(3-chlorophenyl) carbamate; (b) 1′-(methylprop-2′-ynyl)-N-(3-chlorophenyl) carbamate; (c) 1′-(N′-ethylcarbamoyl)ethyl-N-phenyl carbamate)

acid (equation (13.1b)) and phenylamides (equation (13.1c)) which have other uses in crop protection.

(a) Phenyl—NH—COO—R (e.g. propham)	R—esters of phenylcarbamic acid	(13.1)
(b) Alkyl—NH—COO—Aryl (e.g. carbaryl)	Aryl esters of alkylcarbamic acid	
(c) Phenyl—NH—CO—R (e.g. propanil)	Not esters; phenyl-substituted amides	

The phenylcarbamates used as soil-acting herbicides are solids with moderately low vapour pressure, although appreciable amounts may volatilise when application is made to a soil surface. They are soluble in organic solvents such as acetone and benzene. Their solubility in water is low but variable. Examples are chlorpropham, 90 ppm; chlorbufam, 540 ppm; carbetamide, 3500 ppm.

Soil-acting carbamates are frequently formulated as emulsifiable concentrates but wettable powders and granular formulations are also marketed. Turner *et al.* (1978) showed that when a conventional emulsifiable concentrate and micro-encapsulated chlorpropham were compared by applying them to the surface of bare silt loam under otherwise identical conditions, the rate of volatilisation from the encapsulated formulation was initially only one third of that from the emulsifiable concentrate. Moreover, the encapsulated formulation allowed the chlorpropham to persist longer, 20 per cent of the active ingredient being still present after 50 days compared to two per cent for the emulsifiable concentrate. Dawson (1979) reported that when clay granules containing chlorpropham were put on the surface of moist soil, a considerable proportion gradually moved into the atmosphere.

Phenylcarbamates are adsorbed on to soil colloids and especially on to soil organic matter. In this respect they resemble the ureides and the same considerations apply as were outlined in sections 11.5 and 13.1. In general the less soluble members including **chlorpropham** tend to be more strongly adsorbed and to wash down more slowly than those that are more soluble. This, it will be recalled, is the rationale underlying **positional selectivity**, for large-seeded crops such as peas and beans are planted deeply compared to the location of most of the competitive weed seedlings (section 11.2). In addition, the extent of adsorption can affect both their herbicidal efficiency and the time that they persist at herbicidal concentrations. Chlorpropham is in group 2 (low mobility) of the classification of Helling (1971). This scheme is discussed in section 11.5. Toxicity to crop plants is similarly influenced by the extent of adsorption on to soil constituents. McGrath and McCormack (1979) have shown that the phytotoxicity to sugar beet of chlorpropham (and also of atrazine and linuron) is strongly and negatively correlated with the amount and type of soil organic matter, but that soil pH and soil texture are of minor importance. They found that for their samples a simple test for potential crop damage could be based on the fact that light absorbance of aqueous sodium hydroxide extracts is highly correlated with phytotoxicity.

Herbicidal effectiveness is usually greatly increased by working these substances into the soil to an appropriate depth, either by mechanical means or by irrigation. In loam soils the half-life of many carbamates is of the order of two to five weeks when used at normal herbicidal levels, but this range is greatly influenced by soil temperature, by the type and numbers of soil microflora and by the level of rainfall in the period shortly after application. Substances termed **extenders** exist which increase the longevity of propham and chlorpropham in soil. One such substance is 4-chlorophenyl-N-methylcarbamate. It probably acts by inhibiting bacterial hydrolases (esterases and amidases) that would otherwise detoxify these herbicides (Anderson, 1977).

Some selective herbicidal uses of soil-acting carbamates are summarised in table 13.4. Chlorpropham, the most widely used member of the group, is extensively employed in Europe for the control of established chickweed (*Stellaria media*). In addition, it controls germinating weeds in many crops, some of which are listed in the Table. It is frequently used in conjunction with other soil-acting substances, including diuron, fenuron, linuron and propham (MAFF, 1980). In the United States it is registered for the control of annual grasses and of some annual broadleaved weeds in about twenty crops (Klingman and Ashton, 1975). An important non-herbicidal use of chlorpropham is to prevent sprouting in ware potatoes.

Carbamates readily enter through roots and also through the subterranean shoot. Once within plants, apoplastic translocation is usually rapid, although some tolerant species may owe this tolerance, at least in part, to restricted move-

Table 13.4 Some uses of carbamate herbicides

Substance	Uses
Chlorpropham	Established chickweed in bulbs, fruit bushes, dormant strawberries[1]. Germinating weeds and established chickweed pre-em. broad beans, field beans, peas. Weed seedlings in alfalfa, blueberries, cranberries, soybeans. Post-harvest inhibition of sprouting of potato tubers
Propham	Germinating weeds, esp. wild oats, pre-em. peas, sugar beet. Germinating weeds pre-em. sugar beet, fodder beet, mangolds Weed seedlings in alfalfa, lentils, safflower
Chlorbufam	Germinating weeds pre-em. sugar beet, some vegetables and some bulbs[1]
Carbetamide	Winter treatment of land later to be sown with brassicas, to control annual grasses, chickweed, speedwells
Swep	In Asia, against annual grasses in transplanted rice

(1) Often in mixtures with other herbicides.

Abbreviations and sources of information: see footnote, table 13.1.

ment caused by limited binding to root components. Swep, for example, appears to be incorporated as the intact carbamate into the macromolecular structure of the plant. In other cases rapid metabolism in the roots of tolerant species may prevent extensive upward movement, either by direct detoxication or by the formation of metabolites which are bound to root components. More generally, there is considerable unanimity in the belief that differential metabolism is responsible for selective phytotoxicity, the enzymes in tolerant plants being more active, or more strategically located, or more capable of catalysing alternative routes of degradation, than the enzyme systems in sensitive species.

Some of the observations in the preceding paragraph are illustrated by the work on soybean plants by Still and Mansager (1971, 1973). These workers demonstrated that chlorpropham was changed first to polar products and then to insoluble residues, the proportion of the latter increasing with time. Polar metabolites were rapidly formed in roots and did not move upwards to the shoot and were later bound to unknown root constituents. The apparently vulnerable ester and amide linkages do not appear to be rapidly attacked in plants, for the main polar products identified were 2- and 4-hydroxypropham and their O-glucosides.

Data relating to the metabolism of chlorpropham and propham by animals and by micro-organisms have been collated by Menzie (1974, 1978). In rats up to 30 per cent of an oral dose of chlorpropham was split by hydrolysis to give chloroaniline and its N-acetylated derivative (figure 13.8). However, 4-hydroxy-chlorpropham was a major product in most animal species investigated by various authors; both this substance and its N-acetylated hydrolysis product (3-chloro-4-hydroxy-acetanilide) are found in rat urine in the form of their glucuronide and sulphate conjugates (Grunow *et al.*, 1970). In addition, some authors have suggested that the isopropyl moiety can be hydroxylated and that the new $-CH_2OH$ group can then be converted to carboxyl.

The carbamates can interfere with several plant systems, amongst them being photosynthesis, mitosis and oxidative phosphorylation. Inhibition of the Hill reaction in isolated chloroplasts requires rather high concentrations of propham and chlorpropham, the concentration for 50 per cent inhibition (the I_{50}), being variously reported by different workers to be as low as 93 μM or as high as 1200 μM. The available data have been assembled and critically assessed by Corbett

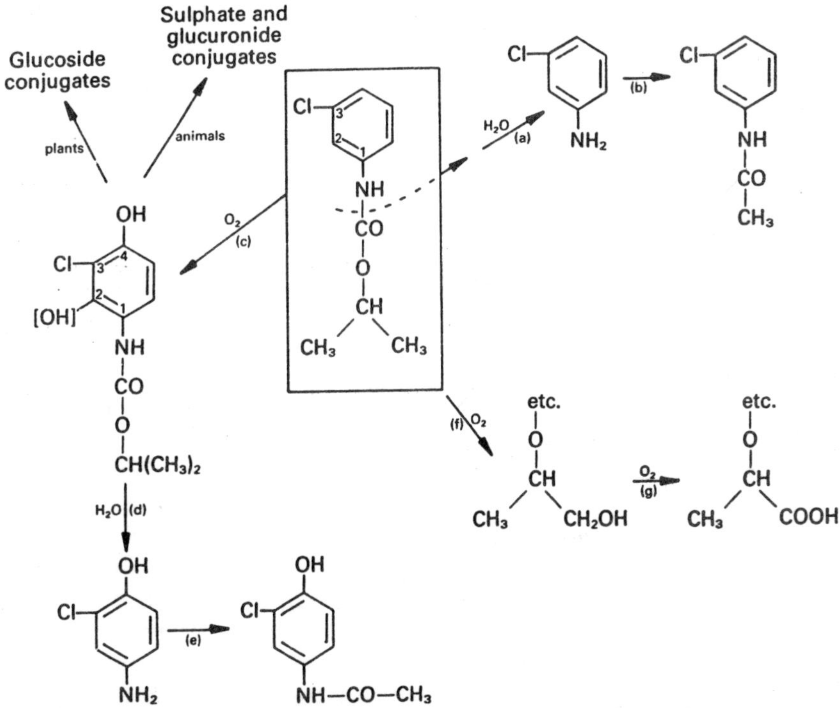

figure 13.8 Metabolism of chlorpropham. Chlorpropham (boxed) can be hydrolysed (a) and the product acetylated (b); in one study 30% was so metabolised by rats. Hydroxylation in the 4- or 2-position (c) is common in plants and in animals; the product(s) can be hydrolysed (d) and acetylated (e). There have been reports that the isopropyl group can also be oxidised (f) and (g)

(1974). It seems likely that under field conditions their primary action is to disrupt growth by acting as mitotic poisons.

A very characteristic effect of carbamates on germinating rye grass or wild oats is to stunt growth and to cause the development of a bulbous shoot apex. Cells of affected apices contain swollen cells with giant nuclei. The electron micrographs of Hepler and Jackson (1969) show that 55 μM propham affects within two hours the organisation of the microtubules of the spindle fibres. These comprise globular protein (tubulin) and their function is to separate the daughter chromatids. For comparison, the lowest recorded I_{50} for inhibition of the Hill reaction by propham is 180 μM. Hartmann and Schroeder (1979) described the abnormalities in cell division and of differentiation which followed the action of chlorpropham on winter rape (*Brassica napus*). The meristematic state becomes protracted and cell vacuolation and differentiation are delayed. Mitotic division is disturbed and the presence of nuclear components can be detected in the general cytoplasm.

It has also been suggested that carbamates inhibit gene de-repression, thus preventing the onset of new metabolic changes rather than affecting existing ones. Cartwright (1976) has reviewed the evidence for the various possible effects of carbamate and other herbicides on cell division.

13.7 THIOLCARBAMATES

Although chemically related to the carbamate herbicides, the thiolcarbamates differ from them in respect to their physical properties and their method of application to the soil. In addition, it is possible that their phytotoxic mode of action is different. These herbicides are by preference termed *thiol*carbamates rather than *thio*carbamates since the sulphur atom is invariably in a position corresponding to the oxygen of the –OH group and not to that of the C=O group of the parent carbamic acid, $H_2N{-}C({=}O){-}OH$. Formulae and names of some members of this rather large group appear in figure 13.9.

Most thiolcarbamates are yellowish liquids with vapour pressures ranging from 10^{-1} to 10^{-3} mm Hg at 20°C. Most of them are therefore more volatile than herbicidal carbamates and a few of them are, compared with most pesticides, very volatile indeed. This property greatly influences how they must be used in agriculture if good field results are to be achieved. In view of the importance of volatility it is essential that the vapour pressures of thiolcarbamates should be known accurately and some are being redetermined. For example, Grover *et al.* (1978) reported that the vapour pressure of triallate is 1.08×10^{-4} mm Hg at 20°C and can be calculated at other temperatures using equation (13.2) where T is the temperature on the absolute scale and P is the vapour pressure (mm Hg). Their aqueous solubilities are quite variable, ranging from four to 900 ppm.

$$\mathrm{Log}_{10}P = 11.05 - \left(\frac{4401}{T}\right) \qquad (13.2)$$

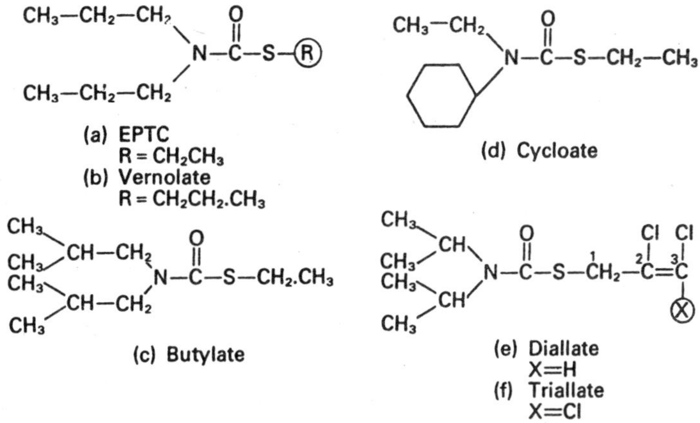

figure 13.9 Some herbicidal thio(l) carbamates. ((a) S-ethyl-N,N-dipropyl-thiolcarbamate; (b) S-propyl-N,N-dipropyl thiolcarbamate; (c) S-ethyl-N,N-diisobutyl-thiolcarbamate; (d) S-ethyl-*N*-ethyl-N-cyclohexyl-thiolcarbamate; (e) S-(2,3-dichloroallyl)-N,N-diisopropyl-thiolcarbamate; (f) S-(2,3,3-trichloroallyl)-N,N-diisopropyl-thiolcarbamate)

Thiolcarbamates normally only kill weed seeds at the time of germination but they have the advantage that they attack troublesome grasses such as wild oat (*Avena fatua* and *A. ludoviciana*), black-grass (*Alopercurus myosuroides*) and barnyardgrass (*Echinochloa crus-galli*). In addition, some of them can be successfully used against certain established perennial weeds, a paradox that is explained by the fact that they can inhibit the development of buds on underground parts of the plant, the consequence of which is to restrict future growth.

As well as being volatile, thiolcarbamates are somewhat unstable chemically and not strongly adsorbed on to soil constituents. In consequence they are ideal substances to perform the often tricky task of delivering a 'short, sharp, shock' – and then partially disappearing before a more or less tolerant crop is drilled or planted a week or so later (table 13.5). Nevertheless, this same volatility also dictates the manner of their application, for they must normally be **worked into the soil** to a suitable depth, and not merely applied to the soil surface. Granular formulations may sometimes be exceptions, while application in irrigation water achieves the same end in a different way.

Ethyl dipropyl thiolcarbamate, **EPTC** (figure 13.9a) is the most volatile member of the group (10^{-1} mm Hg at 20°C) and its rapid incorporation into the soil is especially necessary. This should be done within 15 min of application. When it is being done prior to sowing wheat, two passes should be made at a wide angle using a straight-toothed harrow (Fryer and Makepeace, 1978). In Britain, it is worked into the soil to a depth of 15 cm before planting potatoes. The use of rotary cultivators helps to cut grass rhizomes into short lengths and improves the performance of the chemical. Usually, potatoes can be planted within two weeks of treatment.

Table 13.5 Uses of herbicidal thiolcarbamates

Substance	Uses
EPTC	Soil-mixed before planting potatoes and sowing wheat (couch grass, bent grasses and wild oats). Soil-mixed prior to maize, some legumes, cotton, sunflower (nutsedges and some other weeds)
Triallate	Soil-mixed before drilling barley, beans, carrots, peas (blackgrass, wild oats). Post-planting but pre-em. deeply drilled wheat, barley, beans(1)
Diallate	Soil-mixed before drilling brassicas, beet (blackgrass, wild oats). In USA, annual seedlings in lentils, peas, flax
Cycloate	Soil-mixed 3 weeks before drilling beet, mangolds (wild oats and other annual weeds)(2)
Pebulate	Soil-mixed before planting beet, tomatoes and before transplanting tobacco (annual grasses, nutsedges, hairy nightshade)
Vernolate	Soil-mixed before planting or transplanting peanuts, soybeans, sweet potatoes, tobacco (annual grasses, nutsedges)
Butylate	Soil-mixed before sowing maize (annual weeds, including grasses, nutsedges)
Molinate	Soil-mixed before planting rice (barnyardgrass, annual grasses)
Benthiocarb	Pre- or post- planting rice (barnyardgrass)

(1) Often mixed with lenacil.

(2) The seed must be below the depth of incorporation of herbicides.

Abbreviations and sources of information: see footnote, table 13.1.

In conjunction with a herbicidal antidote (see below) EPTC has proved useful for the control of troublesome weeds in maize. Moyer and Dryden (1979) reported that good control of wild oat (*Avena* spp.) was obtained with little subsequent crop damage when soil-incorporated before seeding. In a triple mixture containing butylate it controlled green foxtail (*Setaria viridis*). Hemp dogbane (*Apocynum cannabinum*) can be controlled for up to two months after soil incorporation prior to sowing maize (Schultz and Burnside, 1979). Similarly Ogg and Drake (1979) found that a mixture of EPTC and an antidote gave 95 per cent control of barnyardgrass (*Echinochloa crus-galli*) as well as control of fat hen (*Chenopodium album*), redroot pigweed (*Amaranthus*) and yellow foxtail (*Setaria lutescens*) when incorporated into soil before planting maize; no loss of yield, quality or succulence of maize was evident. Used alone, EPTC has been reported to give useful control of the yellow nutsedge, *Cyperus esculentus*, in ground in which alfalfa was to be grown (Keeley *et al.*, 1979).

Diallate and **triallate** (figure 13.9f) are also somewhat volatile and need to be incorporated into soil as soon after application as possible. Diallate is used pre-planting of beet and brassicas, particularly for the control of black-grass (*Alopecurus myosuroides*) and of wild oat (*Avena fatua* and *A. ludoviciana*). Triallate is frequently used to control the same two weeds in cereals, although in this case the cereals must be drilled deeply and not undersown with grasses (Fryer and Makepeace, 1978). In appropriate circumstances it can be used both before and after drilling wheat and barley; in the case of spring wheat, for example, triallate can be incorporated between 3 days and 3 weeks before drilling. Zemanek and Kovar (1978) reported that triallate and TCA showed synergistic action, with the result that the application rate of triallate could be halved yet the toxicity to wild oats maintained.

Butylate (figure 13.9c) is often mixed with atrazine and is primarily useful for weed control in maize. When mixed with an antidote it gives good but temporary control of hemp dogbane (*Apocynum cannabinum*); its use by Moyer and Dryden (1979) in a triple mixture to control green foxtail (*Setaria viridis*) was referred to earlier. Viable tubers of yellow nutsedge (*Cyperus esculentus*) were reduced by 96 per cent in two years when used on ground to be double cropped with barley and maize (Keeley *et al.*, 1979). **Vernolate** is frequently employed as a component of a mixture containing trifluralin. Mixed with an antidote, Ogg and Drake (1979) found that it did not reduce yield or quality of maize but greatly reduced the incidence of barnyardgrass (*Echinochloa crus-galli*) and other weeds.

Molinate and **benthiocarb** are employed to control germinating weeds in rice. Molinate is rapidly leached from soil and it also breaks down quickly. In consequence, little normally remains in soil a month after application. Both substances are used against barnyardgrass. Molinate applied into floodwater after emergence of rice seeded into water may damage the crop; a safer procedure is to apply after flooding dry-seeded rice (Smith, 1970).

A typical effect of thiolcarbamates upon susceptible plants is to cause their leaves to have very thin cuticles (Schultz and Burnside (1979) considered that this reduction in thickness may assist the absorption of foliar herbicides later applied to surviving hemp dogbane plants). Still *et al.* (1970) found that the thickness of weed epicuticle can be reduced by as much as 80 per cent. In consequence, leaves of plants affected by thiolcarbamates tend to be dark green and shiny. It has been suggested that the shortage of cuticular wax may be but one consequence of a more general interference with lipid biosynthesis, and especially with the synthesis of unsaturated fatty acids (Wilkinson and Smith, 1975). If so, interference with growth is also to be anticipated, since unsaturated fatty acids are essential components of phospholipids in the lipoprotein membranes that form the fabric of living cells.

It is of interest that compounds are known which can diminish the toxic effect of thiolcarbamates in maize and wheat (crops other than these two do not seem to benefit to any significant degree). Such compounds are variously known as **'herbicidal antidotes'** 'protectants' (in Britain) or 'safeners' (in USA). An example is N,N-diallyl-2,2-dichloroacetamide (often called R25788 in USA literature). This protects maize from soil-applied butylate, EPTC and vernolate and, as a seed treatment, protects wheat from triallate (Anderson, 1977). Stephenson *et al.* (1979) synthesised 18 substances and tested them for their antidote potential against the effects on maize of six thiolcarbamate herbicides. Twelve of these substances showed significant antidote activity when combined with one or more of the six herbicides. When studied in quartz sand nutrient culture, the synthetic dichloroacetamides were more active as antidotes than were the mono- or trichloro analogues. In soil, however, the standard antidote, N,N-diallyl-2,2-dichloroacetamide, was the most effective for five of the herbicides, probably because of the greater ease with which this substance can be taken up from the soil. The mode of action of herbicide antidotes is uncertain.

Rather little detailed information exists about the precise routes of degradation of thiolcarbamates in plants, in soil and in animals. In plants, early work demonstrated that resistant species often metabolised EPTC faster than susceptible ones and that the sulphur from the EPTC was in part incorporated into cystine, methionine and cysteic acid. Later work with EPTC labelled with ^{14}C on the ethyl group (figure 13.9a) showed that the radiolabel entered various carbohydrates and amino acids. Both types of work imply that a radical breakdown of EPTC probably takes place in plants.

Metabolism by micro-organisms is probably a major factor determining the rate of disappearance of thiolcarbamates from soils, for in sterile soils they persist much longer. Nevertheless, factors considered elsewhere (section 11.5) also contribute. Volatilisation is probably greater from wet soils than from dry ones and persistence is longer in soils of high organic content than in those of a lower carbon content, while loss from damp soils is much more influenced by a raised temperature than is loss from dry soils. Above all, loss from soil is greatly diminished if a volatile thiolcarbamate such as EPTC is incorporated to a depth of at least two inches. In all probability the main degradative route involves hydrolysis of the thioester linkage to give a mercaptan, an amine and carbon dioxide (equation (13.3)).

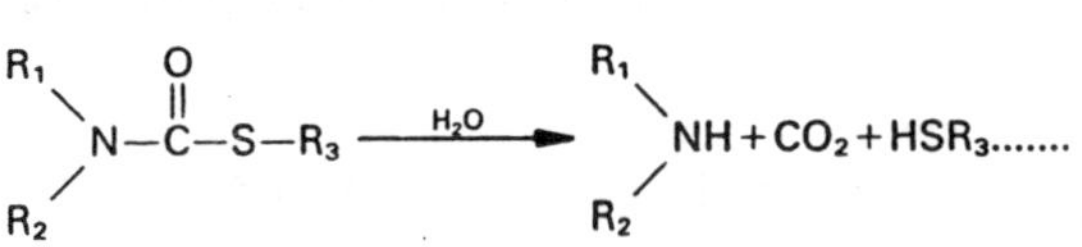

(13.3)

In animals, hydrolytic breakdown of EPTC to a mercaptan is probably a major metabolic route, the thiol then being converted to propanol which may be metabolised by the normal pathways of intermediate metabolism (e.g. converted to propionic acid). Less unstable carbamates, such as cycloate, however, probably form sulphoxides and sulphones, after which glutathione-S-transferase cleaves the molecule. The rather incomplete evidence has been collated by Menzie (1978).

Most thiol carbamates are of low acute toxicity to animals, their LD_{50}s mostly lying within the range 1400–2000 mg/kg body weight. This, when considered in conjunction with their low chemical stability, suggests that residue hazards are unlikely to exist. It should, however, be added that various pesticides containing chloroallyl, chlorovinyl or similar groups have been shown to be mutagenic in tests in which *Streptomyces coelicolor* and *Salmonella typhimurium* were test organisms (Carere *et al.*, 1978). Of the various types of pesticides tested by these authors, dichlorvos and triallate caused mutations in *Salmonella* while diallate and triallate were mutagenic in *Streptomyces*. They concluded that the presence of chlorinated vinyl or chlorinated allyl groups in pesticide molecules could cause the substances to induce point mutations in the two genera.

13.8 NITROANILINE AND NITROPHENOL DERIVATIVES

The formulae of selected members of these two groups of nitrophenol-based weedkillers appear in figure 13.10 and their principal field uses are summarised in table 13.6.

Trifluralin is a typical member of the nitroaniline group. It is a solid of low aqueous solubility (1 ppm). Although its acute toxicity to mammals is extremely low (oral LD_{50}, 10 g/kg for rats) fish can be harmed by it. It is normally formulated as an emulsifiable liquid which should be **incorporated into the soil** to an appropriate depth within 30 min of applying it. In Britain, its main use is to remove weeds from land into which cabbage, cauliflower, kale and sprouts are to be planted. It is also used for pre-plant or pre-sowing weed control in ground that is to carry beans (broad, field and runner), carrots, parsnips, cotton, soybeans, peanuts and sunflowers. It has found considerable use for the pre-planting control of annual grass weeds in cotton, for the activity of dinitroanilines is much less influenced by soil type than is that of the ureides (Hawtree, 1980). In addition, it has been used post-emergence in sugar beet (MAFF, 1980).

Trifluralin persists in soils from a few weeks to a few months at normal application rates. In common with other volatile compounds it is lost more

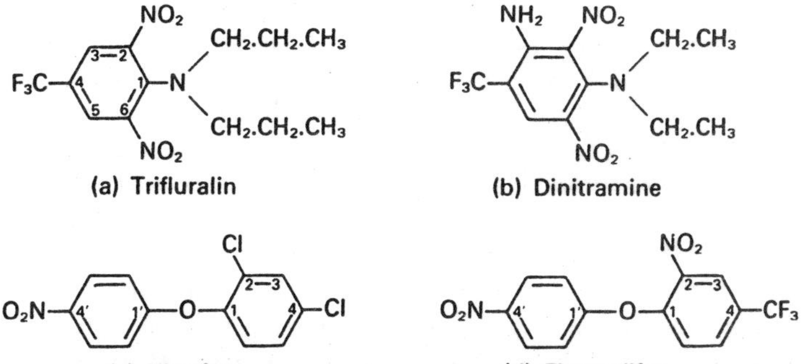

(a) Trifluralin (b) Dinitramine (c) Nitrofen (d) Fluorodifen

figure 13.10 Nitroaniline and nitrophenol derivatives. ((a) 2,6-dinitro-4-trifluoromethyl-N,N-dipropylaniline; (b) 2,6-dinitro-3-amino-4-trifluoromethyl-N,N-diethylaniline; (c) 2,4-dichlorophenyl-4′-nitrophenyl ether; (d) 2-nitro-4-trifluoromethylphenyl-4′-nitrophenyl ether) rapidly from moist soils than from dry ones because it is more readily desorbed from soil constituents (section 11.5). Ratnayake and Audus (1978), who studied the effects of several herbicides on the oxidation of ammonia and nitrite in soil perfusates containing nitrifying organisms, reported that trifluralin is relatively more toxic to *Nitrobacter* than to *Nitrosomonas*.

Most of the routes by which trifluralin is metabolised are predictable (chapter 3). In carrots, for example, N-dealkylation is a major degradative pathway, although in addition, the nitro-groups can be converted to amino groups and the $-CF_3$ group can be converted to carboxyl (Probst and Tepe, 1969). Soil bacteria probably degrade it similarly, except that double N-dealkylation and reduction of both nitro groups can sometimes occur. In animals the $-CF_3$ group may not be readily modified; otherwise the primary metabolic changes are the same as in plants. Metabolic data have been collated by Menzie (1974) and by Naylor (1976). More recently the metabolism of trifluralin by rat liver microsomes has been studied by Nelson *et al.* (1977). These investigators confirmed that aliphatic hydroxylation, N-dealkylation and reduction of one nitro group had taken place to give three different metabolites. In addition, cyclisation had occurred to give a benzimidazole derivative. The distribution and degradation of trifluralin in an aqueous ecosystem have been studied by Kearney *et al.* (1977).

Seedlings poisoned with trifluralin are severely stunted and lateral roots do not develop properly. In the case of grasses the leaves may not fully emerge from the coleoptile. All these symptoms suggest possible interference with **nuclear division**. This conclusion was confirmed by Lignowski and Scott (1972) who showed that 1 ppm trifluralin inhibits mitosis in the root tips of wheat seedlings, and by Swanson (1972) who observed fragmented chromosomes in the meristems of plants treated with various nitroanilines. Young and Camper (1979) also concluded that the mode of action involved disruption of mitosis. Using tobacco callus tissue, the inhibitory effects of trifluralin were studied using Feulgen staining and light microscopy. Compared to untreated tissue a marked decrease occurred in the number of cells present at any one time in the anaphase and telophase states. They also observed arrested metaphases and multinucleate cells. The effects were similar to those brought about by 10^{-4}M colchicine but occurred in the presence of trifluralin concentrations much lower than this.

Other nitroanilines with similar general properties and uses to trifluralin include **dinitramine**, **benfluralin**, **nitralin**, **isopropalin** and **profluralin**. Of these, **benfluralin** resembles trifluralin except that the amino nitrogen atom carries an ethyl and a butyl group instead of two propyl groups, while dinitramine has two ethyl groups on the amino nitrogen atom and also an amino group in position 3 of the benzene ring (figure 13.10a). Of the compounds listed, only dinitramine appears in the (British) List of Approved Products for 1980, in which it is recommended for the control of blackgrass, annual meadow grass and broad-leaved weeds prior to drilling kale, swedes, turnips and runner beans (MAFF, 1980).

Nitrofen and **fluorodifen** (figure 13.10 and table 13.6) are nitrophenyl

Table 13.6 Uses of some nitroaniline and nitrophenol derivatives

Group and substance	Uses
(a) *Nitroanilines*	
Trifluralin	*Soil-mixed* before planting brassicas, beans and some other vegetables (annual weeds). In USA for weeds in many crops, esp. cotton, soybeans
Dinitramine	As above, before planting cotton, soybeans, sunflowers, groundnuts (annual grasses, some broadleaved weeds)
(b) *Nitrophenyl ethers*	
Nitrofen	Applied to soil surface (*not soil-mixed*) pre-em. some brassicas, carrots, onions, celery and sugar beet (germinating grasses and some other seedlings)
Fluorodifen	As above, but pre-em. rice, cotton and some larger-seeded legumes in warm climates

Abbreviations and sources of information: see footnote, table 13.1.

ethers. An important difference between members of this group and both the nitroanilines and the thiolcarbamates is that mixing with soil reduces their effectiveness. The reason for this is that uptake is primarily through the *emerging shoot* rather than through the roots.

Nitrophenyl ethers are solids of low aqueous solubility (1–2 ppm). They are strongly adsorbed on to soil colloids, they are not appreciably volatile and they do not readily leach from soil. At normal rates of application they remain herbicidally effective for one to four months. They are not very toxic to mammals.

Uses of nitrofen and fluorodifen are listed in table 13.6; both are principally pre-emergence weedkillers, although post-emergence uses do exist. In either case, however, their action is primarily one of contact, because little translocation occurs. They are employed for the control of germinating weeds (both mono-and dicotyledons) in a wide range of crops, including several vegetable crops, rice, large-seeded legumes and cotton.

Relatively few studies have been carried out on the routes by which trifluralin is metabolised and less work still has been done on other nitrophenyl ether herbicides. A major route in some plants probably involves reduction of the 2-nitro group of fluorodifen (figure 13.10d) to amino followed by rupture of the ether linkage to give nitrophenol and 2-amino-4-trifluoromethylphenol. These two substituted phenols are then transformed to unknown products. Small amounts of the 4′-amino- and of the 2,4′-diamino derivatives of fluorodifen have also been detected (Eastin, 1971). Whether the selectivity of fluorodifen and other nitrophenyl ethers depends upon differential uptake, distribution or metabolism is unknown.

The mode of action of nitrofen has been more studied than that of other members of the group. Light is a requirement for its herbicidal activity, or at least accentuates its toxic effects following foliar contact. In a recent study Lambert *et al.* (1979) found that non-cyclic electron flow in the chloroplasts of microalgae was not affected by concentrations up to about 10 μM, i.e. electrons flowed from photosystem II, down the electron transport system to the oxidised moiety of photosystem I (figure 12.7). However at concentrations below 10 μM both non-cyclic and cyclic photophosphorylation were severely inhibited. Nitrofen appears to compete with ADP in the ATP-synthetase reaction and, at higher concentrations, uncouples photophosphorylation.

It must, however, be pointed out that while the mode of action just described may be highly relevant to foliar-applied members of the group, simple deprivation of light-generated ATP can hardly explain the failure of weed seeds to germinate properly. Such seedlings are killed long before food supplies in the cotyledons would be exhausted in untreated plants. This is a criticism which, in the author's view, can be levelled at other groups of pre-emergence herbicides that are undoubtedly powerful inhibitors of photosynthesis *in vitro*. In the present instance, the concentration of nitrofen which inhibits the Hill reaction and that which uncouples oxidative phosphorylation in mitochondria are of comparable magnitude (Corbett, 1974, p. 87) and it is relatively easy to speculate that both mitochondrial and chloroplastic sources of ATP are denied to the weeds.

13.9 MISCELLANEOUS SOIL-ACTING HERBICIDES

Of the many soil-acting herbicides not considered in earlier sections, three further types merit brief consideration. Formulae of selected substances are given in figure 13.11 and some of their agricultural uses are listed in table 13.7.

Diphenamid and **propachlor** (figure 13.11a,b) are two examples of amides that are principally used for pre-emergence weed control. Other members of the group include **alachlor, butachlor, napropamide, propyzamide** and **CDAA** (2-chloro-N,N-diallyl acetamide). Two other amides, pentanochlor and propanil, are primarily foliage- rather than soil-acting (section 12.12). All these compounds are amides (R.CONH$_2$), but the majority have a substituted aromatic ring attached to the nitrogen atom, (R.CONH.phenyl), and are then usually classified as anilides.

Diphenamid (which is not an anilide) is used both before planting and before the emergence of a variety of crops (table 13.7) its particular value being that it preferentially controls annual grass seedlings. It is moderately strongly adsorbed on to constituents of clay soils and herbicidal concentrations sometimes persist for six months. Experiments with plants in culture solutions have shown that it quite readily enters plant roots, from where it moves upwards and tends to accumulate in leaves. The major metabolic product in several plants is N-methyl-2,2-diphenylacetamide, which is formed from diphenamid by N-demethylation (compare table 3.1).

Table 13.7 Uses of some miscellaneous soil-acting herbicides

Group and substance	Uses
(a) *Amides* or *Anilides*	
Diphenamid (amide)	Pre-em. tomatoes, potatoes, groundnuts, ornamentals (seedling grasses)
Propyzamide (amide)	Pre-em. and/or post-em. several crops, including lettuce, field beans. Also in some established fruit bushes. To control chickweed and seedling weeds but members of the Compositae and Leguminosae are resistant
Propachlor (anilide)	Pre-em. some bulbs, brassica crops, ornamentals (annual seedlings). In USA, pre-em. maize, cotton, sorghum (many seedling weeds)
Alachlor (anilide)	Pre-em. cotton, maize, peanuts, soybeans, sugar cane (seedling grasses and some broadleaved weeds)
(b) *Nitriles*	
Dichlobenil	Soil-mixed (unless granular) in non-crop situations, against grass and other weeds (seedling and established). In established fruit, rose, forestry trees and in some canes and bushes. Kills aquatic weeds through soil contact
Chlorthiamid	As dichlobenil (in which form it acts). Granular forms used against established couch grass, ground elder, nettles and thistles

Abbreviations and sources of information: see footnote, table 13.1.

A characteristic effect of diphenamid is that it allows seeds to germinate but they die before they emerge through the soil. At lower concentrations it powerfully inhibits root development. It reduces the uptake of several cations but possibly of greater interest is the fact that it appears to alter the internal distribution of calcium and magnesium. Nashed and Illnicki (1968) found that the shoots of cabbage plants contained less, and the roots more, calcium than did untreated plants.

Alachlor and **butachlor** are somewhat similar in structure to **propachlor** (figure 13.11b) and all three of these anilides are used for pre-emergence weed control in various crops. Butachlor is often employed for annual weed control in rice. All three are adsorbed on to soil colloids and consequently do not readily leach from soils. They normally remain herbicidally active for 2–3 months. Propachlor is rapidly metabolised in most plants to give water-soluble derivatives. Two of these have been identified by Lamoureux *et al.* (1971) and are conjugates

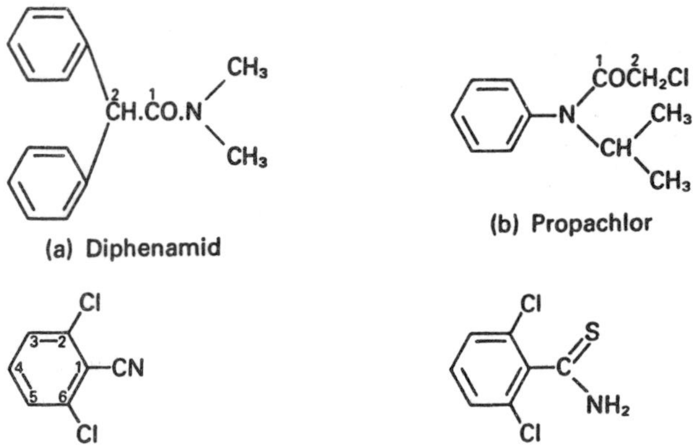

(a) Diphenamid (b) Propachlor (c) Dichlobenil (d) Chlorthiamid

figure 13.11 Other nitrogen-containing herbicides. ((a) N,N-dimethyl-2,2-diphenylacetamide; (b) 2-chloro-N-isopropyl-acetanilide; (c) 2,6-dichlorobenzonitrile; (d) 2,6-dichlorothiobenzamide)

in which the propachlor is attached to glutathione (or to a peptide formed by the partial breakdown of glutathione).

Symptomatic of the toxic action of propachlor and alachlor is a stunting of growth and an inhibition of root development. The significance of these symptoms is uncertain, but it has been suggested that these amides interfere with protein synthesis. Alternatively, they may interfere with lipid biosynthesis by preventing the incorporation of malonic acid into fatty acids (Mann and Pu, 1968). They do not affect the Hill reaction at normal concentrations (Chandler *et al.*, 1972).

Dichlobenil (figure 13.11c) is a soil-acting benzonitrile derivative. Structurally, it is similar to bromoxynil and ioxynil (section 12.10) but its field use is quite different from that of these foliar-applied herbicides. **Chlorthiamid** (figure 13.11d) is a thioamide but it and dichlobenil are appropriately considered together since chlorthiamid actually breaks down to give dichlobenil after application. The difference between these two substances relates to their method of application because dichlobenil is somewhat volatile (vapour pressure, about 10^{-3} mm Hg, at 25°C) and it is rapidly lost from some soils unless it is either irrigated into the soil or applied in the form of granules. Chlorthiamid, on the other hand, is much less volatile and it is not necessary to work it into the soil.

In Britain, both substances are marketed as granules and both are used for non-selective weed control. In addition, chlorthiamid is recommended for the control of troublesome perennial weeds, established couch grass, ground elder (*Aegopodium podagraria*), nettles (*Urtica* spp) and thistles (*Cirsium* spp; and *Carduus nutans*) on ground not immediately to be used for crops (MAFF, 1980). This use may at first seem surprising since chlorthiamid and dichlobenil, like many other soil-acting compounds, only kill seedlings. However, they also severely inhibit the development of buds on subterranean parts of perennial plants and so prevent them from spreading. They have no effect on dormant tubers. In some crops (usually well-established fruit trees or bushes) it is possible at appropriate application rates to obtain selective control of some perennial species (Fryer and Makepeace. 1978).

Dichlobenil is strongly adsorbed on lignin and humus, with the result that it does not readily leach in soils with a high organic content. It is this property which enables it to be employed for selective weed control in some deep-rooted trees and shrubs. At the lower concentrations used for pre-emergence weed control it may disappear in six to nine weeks but at higher concentrations it may persist for six months or longer.

Apoplastic movement of dichlobenil in plants following root uptake is restricted because of its affinity for lignin. The symptoms of its toxic action have been compared to those caused by boron deficiency, for in both cases there is an inhibition of growth followed by blackening of tissues. Sometimes a melanin-like pigment has been observed deposited in the region of the apical meristem and in the phloem. Leaves tend to be dark because of an elevated chlorophyll content while cell division in active tissues may be greatly diminished, the nuclei appearing granular and the chromosomes blotchy. Its precise mode of action remains uncertain; as well as its probable effects on cell division, it has been suggested that it may restrict the movement of the products of photosynthesis.

Much of the loss of dichlobenil from both plants and soil can sometimes be by volatilisation, but metabolism also occurs. In plants, it was originally reported that the nitrile group was hydrolysed so that 2,6-dichlorobenzoic acid was the final product. More recently, some doubt has been cast upon this observation. It now seems more likely that the major metabolic route, as is so often the case, is a microsomal oxidation. This results in the formation of two isomeric hydroxydichlorobenzonitriles from which two corresponding hydroxydichlorobenzamides appear to be formed by hydrolysis. Each of these four products may then, in plants, be conjugated with glucose.

REFERENCES

Anderson, W. P. (1977). *Weed Science: Principles*, West Publishing Co., New York

Anderson, S. M. and Krinsky, N. I. (1973). *Photochem. Photobiol.*, **18**, 403

Arnold, C. E. and Aldrich, J. H. (1979). *Weed Sci.*, **27**, 638

Ashton, F. M. and Glenn, R. K. (1979). *Pestic. Biochem. Physiol.*, **11**, 201

Carere, A., Ortali, V. A., Cardamone, G. and Morpurgo, G. (1978). *Chem. Biol. Interact.*, **22**, 297

Cartwright, P. M. (1976). *Herbicides: Physiology, Biochemistry, Ecology*, 2nd edn, Vol. 1 (ed. L. J. Audus), Academic Press, London, chapter 2

Chandler, J. M., Croy, L. I. and Santelmann, P. W. (1972). *J. agric. Fd Chem.*, **20**, 661

Churchill, K. and Klepper, L. (1979). *Pestic. Biochem. Physiol.*, **12**, 156
Corbett, J. R. (1974). *The Biochemical Mode of Action of Pesticides*, Academic Press, London
Dawson, J. H. (1979). *Weed Sci.*, **27**, 274
Dodge, A. D. (1977). *Herbicides and Fungicides: Factors Affecting their Activity* (ed. N. R. McFarlane), Chemical Society, London, Special Publication No. 29, Alden Press, Oxford
Doughty, C. C. (1978). *Weed Sci.*, **26**, 488
Eastin, E. F. (1971). *Weed Sci.*, **19**, 261
Esser, H. O., Dupuis, G., Ebert, E., Vogel, C. and Marco, G. J. (1975). *Herbicides: Chemistry, Degradation and Mode of Action*, 2nd edn, Vol. 1 (ed. P. C. Kearney and D. D. Kaufman), Marcel Dekker, New York, p. 129
Frear, D. S. and Swanson, H. R. (1970). *Phytochemistry*, **9**, 2123
Fryer, J. D. and Makepeace, R. J. (1977). *Weed Control Handbook*, 6th edn, Vol. 1: *Principles*, Blackwell, Oxford
Fryer, J. D. and Makepeace, R. J. (1978). *Weed Control Handbook*, 8th edn, Vol. 2: *Recommendations*, Blackwell, Oxford
Gardiner, J. A., Rhodes, R. C., Adams, J. B., Jr. and Soboczenski, E. J. (1969). *J. agric. Fd Chem.*, **17**, 980
Geissbühler, H., Martin, H. and Voss, G. (1975). *Herbicides: Chemistry, Degradation and Mode of Action*, 2nd edn, Vol. 1 (ed. P. C. Kearney and D. D. Kaufman), Marcel Dekker, New York, chapter 3
Green, M. B., Hartley, G. S. and West, T. F. (1977). *Chemicals for Crop Protection and Pest Control*, Pergamon Press, Oxford
Gross, D., Laanio, T., Dupuis, F. and Esser, H. O. (1979). *Pestic. Biochem. Physiol.*, **10**, 49
Grover, R., Spencer, W. F., Farmer, W. J. and Shoup, T. D. (1978). *Weed Sci.*, **26**, 505
Grunow, W., Böhme, C. and Budczies, B. (1970). *Fd Cosmet. Toxic.*, **8**, 277
Guddewar, M. B. and Dauterman, W. C. (1979). *Pestic. Biochem. Physiol.*, **12**, 1
Hamilton, R. H. (1964). *Weeds*, **12**, 27
Hartmann, W. and Schroeder, H. (1979). *Flora (Jena)*, **168**, 529 (quoted from abstract)
Hawtree, J. N. (1980). *Outl. Agric.*, **10**, 184
Hebb, E. A. and Wheeler, W. B. (1978). *J. environ. Qual.*, **7**, 598
Helling, C. S. (1971). *Soil Sci. Soc. Amer. Proc.*, **35**, 737
Helper, P. K. and Jackson, W. T. (1969). *J. Cell Sci.*, **5**, 727
Hoffman, C. E. (1971). *Proc. 2nd Internat. IUPAC Congress Pestic. Chem.*, **5**, 65
Hoffman, C. E., Hersch, R. T., Sweetser, P. B. and Todd, C. W. (1960). *Proc. 14th N. E. Weed Conf., USA*, p. 16
Jordan, L. S., Murashige, T., Mann, J. D. and Day, B. E. (1966). *Weeds*, **14**, 134
Kearney, P. C., Isensee, A. R. and Kontson, A. (1977). *Pestic. Biochem. Physiol.*, **7**, 242
Keeley, P. E., Thullen, R. J., Miller, J. H. and Carter, C. H. (1979). *Weed Sci.*, **27**, 463
Klingman, G. C. and Ashton, F. M. (1975). *Weed Science: Principles and Practices*, John Wiley, New York
Lambert, R., Kunert, K.-J. and Böger, P. (1979). *Pestic. Biochem. Physiol.*, **11**, 267
Lamoureux, G. L., Stafford, L. E. and Tanaka, F. S. (1971). *J. agric. Fd Chem.*, **19**, 346
Lignowski, E. M. and Scott, E. G. (1972). *Weed Sci.*, **20**, 267

MAFF (1980). *List of Approved Products for Farmers and Growers*. For full reference, see references, p. 334
Mann, J. D. and Pu, M. (1968). *Weed Sci.*, **16**, 197
McGrath, D. and McCormack, R. J. (1979). *Ir. J. agric. Res.*, **18**, 89
Menzie, C. M. (1974, 1978). *Metabolism of Pesticides: Updates I and II*. For full reference see p. 119
Millhollon, R. W. (1977). *Weed Sci.*, **25**, 431
Moreland, D. E. (1977). *Am. Chem. Soc. (ACS) Symposium Series No. 37: Pesticide Chemistry in the 20th Century* (ed. J. R. Plimmer), American Chemical Society
Moreland, D. E., Genter, W. A., Hilton, V. L. and Hill, K. L. (1959). *Plant Physiol.*, **34**, 432
Moreland, D. E. and Hill, K. L. (1962). *Weeds*, **10**, 129
Moyer, J. R. and Dryden, R. D. (1979). *Can. J. Plant Sci.*, **59**, 383
Nashed, R. B. and Illnicki, R. D. (1968). *Proc. 22nd N.E. Weed Control Conf., USA*, p. 500
Naylor, A. W. (1976). *Herbicides: Physiology, Biochemistry, Ecology*, 2nd edn, Vol. 1 (ed. L. J. Audus), Academic Press, London
Nelson, J. O., Kearney, P. C., Plimmer, J. R. and Menzer, R. E. (1977). *Pestic. Biochem. Physiol.*, **7**, 73
Ogg, A. G., Jr. and Drake, S. (1979). *Weed Sci.*, **27**, 608
van Oorschot, J. L. P. (1976). *Herbicides: Physiology, Biochemistry, Ecology*, 2nd edn, Vol. 1 (ed. L. J. Audus). Academic Press, London, p. 322
Probst, G. W. and Tepe, J. B. (1969). *Degradation of Herbicides* (ed. P. C. Kearney and D. D. Kaufman), Marcel Dekker, New York, chapter 9
Radosevich, S. R. (1977). *Weed Sci.*, **25**, 316
Radosevich, S. R. and Devilliers, O. T. (1976). *Weed Sci.*, **24**, 229
Radosevich, S. R., Steinback, K. E. and Arntzen, C. J. (1979). *Weed Sci.*, **27**, 216
Rahman, A., Dyson, C. B. and Burney, B. (1978). *N.Z. Jl exp. Agric.*, **6**, 69
Rao, P. S. C. and Davidson, J. M. (1979). *Water Res.*, **13**, 375
Ratnayake, M. and Audus, L. J. (1978). *Pestic. Biochem. Physiol.*, **8**, 170
Renosto, F., Nardi, S. and Ferrari, G. (1979). *Pestic. Biochem. Physiol.*, **11**, 243
Rhodes, R. C. (1977). *J. agric. Fd Chem.*, **25**, 1066
Rhodes, R. C., Belasco, I. J. and Pease, H. L. (1970). *J. agric. Fd Chem.*, **18**, 524
Ross, J. A. and Tweedy, B. G. (1974). *Bull. Environ. Contam. Toxic.*, **10**, 234
Santakumari, M. and Das, V. S. R. (1978). *Pestic. Biochem. Physiol.*, **9**, 119
Schultz, M. E. and Burnside, O. C. (1979). *Agron. J.*, **71**, 723
Schuphan, I. and Ebing, W. (1978). *Pestic. Biochem. Physiol.*, **9**, 107
Seiler, J. P. (1979). *Pestic. Biochem. Physiol.*, **12**, 183
Shimabukuro, R. H. and Swanson, H. R. (1970). *Weed Sci.*, **18**, 231
Smith, J. W. and Sheets, T. J. (1967). *J. agric. Fd Chem.*, **15**, 577
Smith, R. J., Jr. (1970). *Weed Sci.*, **18**, 467
Stephenson, G. R., Bunce, N. J., Makowski, R. I., Bergsma, M. D. and Curry, J. C. (1979). *J. agric. Fd Chem.*, **27**, 543
Still, G. G., Davis, D. G. and Zander, G. L. (1970). *Plant Physiol.*, **46**, 307
Still, G. G. and Mansager, E. R. (1971). *J. agric. Fd Chem.*, **19**, 879
Still, G. G. and Mansager, E. R. (1973). *Pestic. Biochem. Physiol.*, **3**, 87
Swanson, C. R. (1972). *Herbicides, Fungicides, Formulation Chemistry* (ed. A. S. Tahori), Proc. 2nd Internat. IUPAC Congress, Vol. 5, p. 87
Turner, B. C., Glotfelty, D. E., Taylor, A. W. and Watson, D. R. (1978). *Agron. J.*, **70**, 933
Wessels, J. S. C. and van der Veen, R. (1956). *Biochim. Biophys. Acta*, **19**, 548

Wilkinson, R. E. and Smith, A. E. (1975). *Weed Sci.*, 23, 90
Young, L. W. and Camper, N. D. (1979). *Pestic. Biochem. Physiol.*, 12, 117
Zemanek, J. and Kovar, J. (1978). *Ochr. Rostl.*, 14, 143 (in Czech; quoted from abstract)

Bibliography: Background, General or Wider Reading, and Sources of General Information(see Section 1.2)

BACKGROUND READING

1. Roberts, M. B. V. (1976). *Biology: A Functional Approach*, 2nd edn, Nelson & Sons, Sunbury, Middlesex. (For those approaching pesticides from the non-biological disciplines)
2. Edwards, N. A. and Hassall, K. A. (1980). *Biochemistry and Physiology of the Cell*, 2nd edn, McGraw-Hill, Maidenhead. (A knowledge of biochemical systems and control mechanisms is essential to an understanding of the mode of action of pesticides; any introductory text will suffice but this book was written with the needs of non-specialists specifically in mind)
3. Parke, D. V. (1968). *Biochemistry of Foreign Compounds*, Pergamon Press, Oxford. (For those expecting to specialise in the metabolism of pesticides this book presents the subject in the context of general drug metabolism).
4. Graham, F., Jr. (1970). *Since Silent Spring*, Hamish Hamilton, London. (A timely reminder that problems still occur and that pesticides have their critics)
5. Gunn, D. L. and Stevens, J. G. R. (eds.) (1976). *Pesticides and Human Welfare*, Oxford University Press. (Individual experts contribute specialist chapters as such diverse topics as the economics of pesticide usage, pesticides in developing countries and legal considerations)
6. Perring, F. H. and Mellanby, K. (eds.) (1977). *Ecological Effects of Pesticides*, Academic Press, London. (Contributions to a symposium held at the Royal Geographical Society, London, in September, 1976)
7. Wilson, C. L. and Loomis, W. E. (1967). *Botany*, 4th edn, Holt, Rinehart Inc., New York

General and wider reading

Many of these works are more specialised than the present book, with individual chapters written by specialists. Some of our readers will wish to concentrate on particular topics and it is to such works as these that they should turn after having considered the principles and general discipline of pesticide chemistry.

8. Matthews, G. A. (1979). *Pesticide Application Methods*, Longmans, London. (This deals with engineering subjects that are not the concern of the present book)

9. Cremlyn, R. J. (1978). *Pesticides: Preparation and Mode of Action*, John Wiley, Chichester. (A useful account of the manufacture and formulation of pesticides)
10. Brown, V. K. (1980). *Acute Toxicity in Theory and Practice*, John Wiley, Chichester. (For those concerned with toxicity tests relating to the safety of pesticides; this book stresses some of the problems)
11. Corbett, J. R. (1974). *Biochemical Mode of Action of Pesticides*, Academic Press, London. (A useful survey, the compounds being classified according to their mode of action)
12. Matsumura, F. (1975). *Toxicology of Insecticides*, Plenum Press, New York. (A chemical and pharmacological approach, including metabolism, mode of action and effects upon wildlife)
13. Wilkinson, C. F. (ed.) (1976). *Insecticide Biochemistry and Physiology*, Heyden, London. (A valuable contribution but not a textbook in the sense that it does not attempt a progressive development of the subject. Each chapter is written on a selected topic by a specialist writer)
14. White-Stevens, R. (ed.) (1971, 1976, 1977). *Pesticides in the Environment*, Vols 1 (parts 1 and 2) 2, and 3, Marcel Dekker, New York. (Not a textbook but a highly detailed survey, sections of which will be of value to some present readers when they know in which direction they wish to specialise)
15. Siegel, M. R. and Sisler, H. D. (eds.) (1977). *Antifungal Compounds*, 2 Vols, Marcel Dekker, New York. (Probably the best of the few works which survey fungicides in depth)
16. Audus, L. J. (ed.) (1976). *Herbicides: Physiology, Biochemistry, Ecology*, 2nd edn, 2 Vols, Academic Press, London. (Not a textbook in the sense mentioned under 7 above. Individual contributors deal with selected topics in depth)
17. Kearney, P. C. and Kaufman, D. D. (eds.) (1975, 1976). *Herbicides: Chemistry, Degradation and Mode of Action*, 2nd edn, 2 Vols, Marcel Dekker, New York. (Probably the most authoritative text on the subject up to 1975)

Reference books

Some of these link proprietary names with the common names used in the text (see also Appendix B). Some describe the agricultural uses of various types of pesticides and others relate to analysis, to safety precautions and to law.

18. Gunther, F. A. and Gunther J. D. (eds.) (1962–1980). *Residue Reviews*. A series of 74 volumes so far. Springer Verlag, New York. (Specialist writers cover a wide range of topics, including metabolism, environmental effects, methodology of analysis and reviews of individual pesticides)
19. Zweig, G. (ed.) (1963–1980). A series of ten volumes so far. *Analytical Methods for Pesticides, Plant Growth Regulators (and Food Additives)* Academic Press, New York. (Specifically for chemists)
20. MAFF (1980 and yearly). *Ministry of Agriculture, Fisheries, and Food (U.K.), Agricultural Chemicals Approval Scheme: List of Approved Products for Farmers and Growers*, H.M.S.O., London. (Published February each year, naming currently Approved products, their uses and hazards, with cross references to commercial products)
21. Martin, H. and Worthing, C. R. (eds.) (1976). *Insecticide and Fungicide Handbook*, 5th edn, Blackwell, Oxford. (A useful summary of available chemicals, application methods and uses)

22. Fryer, J. D. and Makepeace, R. J. (1977, 1978). *British Weed Control Handbook*, Blackwell, Oxford. (A valuable compendium of Principles (1977, 6th edn) and Recommendations (1978, 8th edn). A source of information on available chemicals, their uses and methods of application)
23. BAA (1978). *British Agrochemicals Association: Directory*, Alembic House, 93 Albert Embankment, London. (A list of commercial formulations, their active ingredients and the names of suppliers)
24. *The Nanogen Index* (1975, with Updates). Compiled by K. Packer and published by Nanogens International, P.O. Box 487, Freedom, Ca., USA
25. Worthing, C. R. (ed.) (1979). *The Pesticide Manual: A World Compendium*, 6th edn, British Crop Protection Council, 74, London Rd., Croydon, Surrey, CR0 2TB. (Formulae, nomenclature, properties, uses, toxicology, analysis. One page or less devoted to each substance)
26. ENTOMA (1977). *Entomological Society of America: Pesticide Handbook*, 27th edn, 4603, Calvert Road, Box AJ, College Park, Maryland 20740, USA
27. NACA (1970). *Nat. agric. Ass. News. 28*, (3). (A list of tolerances compiled from the official tolerance orders published in the Federal Register, Food and Drugs Administration, USA. Final list in the volume published February, 1970)
28. Edson, E. F., Sanderson, D. M. and Noakes, D. N. (1966). *Wld Rev. Pest Control*, 5, 143. (A list of oral and cutaneous LD_{50}s for pesticides in use in 1966)

Appendix A: Conversion Tables (Metric/Non-Metric, and Miscellaneous)

METRIC/NON-METRIC CONVERSIONS

1 hectare (ha) = 2.471 acres
1 acre = 0.405 ha

1 kilogram (kg) = 2.205 pounds (lb)
1 lb = 0.454 kg

1 kg/ha = 0.892 lb/acre
1 lb/acre = 2.471 lb/ha
1 lb/acre = 1.121 kg/ha

1 litre (1)/ha = 0.0892 imperial gallons (gal)/acre
1 gal = 10 lb water
1 gal/acre = 2.471 gal/ha
1 gal/acre = 11.21 1/ha

1 ha = 100 metres (m) x 100 m = 10 000 m^2

1 m^2 ≙ 1.2 square yards

1 m = 1.094 yards
1 yard = 0.914 m

MISCELLANEOUS

1000 nanometres (nm) = 1 micrometre (μm)

1000 μm = 1 millimetre (mm)

1000 mm = 1 metre (m)

Similarly, nanogram (ng), microgram (μg), milligram (mg), gram (g),
μg/g = mg/kg = parts per million (ppm)

In this book,

one billion = 10^9 (a thousand million)

1 imperial gallon (gal) = 1.2 US gallons

1 US gallon = 0.834 gal

1 US gallon/acre = 2.47 US gallons/ha = 9.35 1/ha

1 cupful ≙ 8 fluid ounces ≙ 0.5 pints

1 cupful ≙ 16 tablespoonfuls ≙ 48 teaspoonfuls

Appendix B: Some Proprietary Products Containing Active Ingredients Mentioned in the Text

Thousands of proprietary formulations exist and only a few can be included. Excluded are trade names that reflect common names (e.g., Smith's Malathion) or are serial numbers (e.g., Brown's B237). Where after such exclusions several proprietary products still remain a maximum of three have been chosen at random. Brackets imply that more than one active ingredient is present.

SECTION A. INSECTICIDES AND RELATED COMPOUNDS

Common name	Trade names	Common name	Trade names
Aldicarb	Temik	Diazinon	Basudin; Sarolex
Aldrin	Aldrex; Soildrin	Dichloropropene	Telone; (D-D)
Allethrin	Pynamin	Dichlorvos	Vapona; Nogos; Nerkol
Azinphos-methyl	Gusathion; Guthion	Dicofol	Kelthane; (Childion)
BHC	Gammexane; Gamma-Col	Dicrotophos	Bidrin; Ektafos
Bromophos	Nexion; Brophene	Dieldrin	Dilstan; Octalox
Carbaryl	Sevin; Murvin	Diflubenzuron	Dimilin
Carbofuran	Yaltox; Faradan	Dimethoate	Rogor; Cygon; De-Fend
Chlordane	Sydane; Octa-Klor	Dinitrocyclohexylphenol	Dinex
Chlorobenzilate	Acarben; Folbex	Disulfoton	Disyston; Parsolin
Chlorbenside	Orthocide; Chlorocide	Endosulfan	Thiodan; Malix
Chlordimeform	Fundal; Galecron	Endrin	Mendrin
Chlorfenvinphos	Birlane; Sapecron	Fenchlorphos or Ronnel	Korlan; Trolene; Etrolene
Chlorfenson	Ovotran	Fenitrothion	Dicofen; Folithion; Agrothion; Sumithion
Coumaphos	Co-Ral; Muscatox	Formothion	Anthio; Aflix
Cyhexatin	Plictran	Haloxon	Loxon
DDD	Rhothane	Malathion	Malastan; Cython; Karbophos; Chemathion
DDT	Deestan; Vitanol; Gesarol		
Demeton	Systox		
Demeton-methyl	Metasystox; Duratox; Demetox; Azotox		

Common name	Trade names	Common name	Trade names
Menazon	Sayphos; Saphizon	Phosalone	Zolone; Rubitox
Metaldehyde	Slugit; Slugoids	Phosphamidon	Dicron; Dimecron
Metalkamate	Bux	Pirimicarb	Pirimor; Aphox
Methiocarb	Draza; Mesurol	Pirimiphos-methyl	Actellic
Methomyl	Lannate	Propoxur	Baygon; Blattenex
Methoxychlor	Marlate; Moxie	Sulfotep	Dithio; Bladafume
Methyl parathion	Dalf; Metron; Nitrox	Tar Oil	Carbo-Craven, Mortegg
Mevinphos	Phosdrin	Tetradifon	Tedion; Duphar
Mirex	Dechlorane	Tetramethrin	Neopynamin
Niclosamide	Fenasal; Lintex	Toxaphene	Alltox; Phenatox
Nicotine	Black Leaf 40; XL All	Triazophos	Hostathion
Oxamyl	Vydate	Trichlorphon	Dipterex; Anthon
Parathion	Phoskil; Thiophos	Vamidothion	Kilvar; Trucidor
Permethrin	Ambush		
Phorate	Thimet; Granutox		

SECTION B. FUNGICIDES

Common name	Trade names	Common name	Trade names
Benomyl	Benlate	Maneb	Manzate; Tubothane; Dithane M22
Binapacryl	Morocide; Ambox	Mercuric oxide	Santar
Bupirimate	Nimrod	Mercurous chloride	Cyclosan; M-C; 4% Calomel
Captafol	Difolatan; Sanspor	Metalaxyl	Ridomil; (Fubol)
Captan	Orthocide	Nabam	Dithane; (X-Spor)
Carbendazim	Defosal; Bavistin	Organomercury compounds	Agrosan D; Trosan; Ceresan; Panogen M
Carboxin	Vitavax; (Murganic)	Phosetyl-Al	Aliette
Chlorthalonil	Bravo; Termil; Daconil	Pyrazophos	Afugan
Copper compounds	Burcop; Wetcol; Cuprosana	Quintozene	Botrilex; Bras-sicol; Folosan
Cycloheximide	Acti-dione	Streptomycin	Agrimycin; Agri Strep
Dichlofluanid	Elvaron; Euparen	Silvex	Kuron; Kurosal
Dichlone	Phygon	Sulphur	Elosal; Thiovit
Dicloran	Allisan; Botran; Ditranil	Tecnazene	Fusarex; Hytec
Dimethirimol	Milcurb	Thiabendazole	Storite; Tecto
Dinocap	Karathane; Crotothane Mildex	Thiophanate-methyl	Mildothane; Cercobin
Dodemorph	Meltatox	Thiram	Tripomol; Fernasan; Arasan; Pomarsol
Dodine	Cyprex; Melprex	Tridemorph	Beacon; Calixin
Ethirimol	Milstem; Milgo	Triforine	Saprol
Fenoprop	Kuron; Kurosal	Zineb	Dithane Wettable; Tiezene
Fentin	Farmatin; Du-ter; Tubotin; (Brestan)		
Folpet	Folpan; Phaltan		
Hexachlorobenzene	Anti-Carie		
Imazalil	Fungaflor		
Mancozeb	Dithane 945; Karamate N		

SECTION C. HERBICIDES

Common name	Trade names
Alachlor	Lasso
Ametryne	Evik; Gesapax
Aminotriazole	Weedazol; Azolan
Asulam	Asulox
Atrazine	Gesaprim; Weedex-4G; Aatrex
Barban	Carbyne; Oatax
Benfluralin	Balan; Quilan; Benefin
Benthiocarb	Saturn
Bromacil	Hyvar X; Phytar
Bromoxynil	Buctril; (Certrol-E)
Butachlor	Machete
Butylate	Sutan
Carbetamide	Legurame
Chlorthiamid	Prefix
Chloramben	Amiben; Vegiben
Chlorbromuron	Bromex; Maloran
Chloroneb	Demosan; Tersan
Chlortoluron	Dicurane
Chloroxuron	Norex; Tenoran
Chlorpropham	Furloe; Herbon CIPC; Mirvale 50 AC
Cycloate	Ro-Neet; Eurex
Cyprazine	Outfox
2,4-D	Chloroxone; Shell D; Palormone D; Salvo; Planotox; Weedone
Dalapon	Dowpon
Dazomet	Basamid; Mylone
Desmetryne	Semeron
2,4-DB	Butoxone; Butyrac; Embutox
Diallate	Avadex
(Dicamba)	(Phenoxylene Super); (Hyban); (Herbitox)
Dichlobenil	Casoron
Dichlorprop	Fernoxone; Cornox-RK64; Polygon Plus
Dinitramine	Cobex
Dinoseb	Supersevtox; Morto; Haulmex; Premerge
Diphenamid	Enide; Dymid
Diquat	Reglone
Diuron	Karmex; (Dexuron)
EPTC	Eptam; Eradicane 6E
Fluometuron	Cotoran; Lanex
Fluorodifen	Soyex; Preforan
Glyphosate	Roundup
Ioxynil	Totril; Actrilawn; (Actril C)
Lenacil	Venzar
Linuron	Afalon; Lorox; (Lanslide)
MCPA	Agricorn; Agroxone M; Phenoxylene Plus; Shell M; Weedar
MCPB	Can-Trol; Tropotox
Mecoprop	Clovotox; Methoxone M
Metham-Na	Vapam; Vitafume
Metobromuron	Patoran
Metoxuron	Deftor; Dosanex; Dosaflo
Molinate	Ordram; Hydram
Monolinuron	Arresin; (Gramonol)
Monuron	Telvar; Karmex W
Neburon	Kloben
Nitralin	Planavin
Nitrofen	Tok-E; Nip; Tokkorn
Paraquat	Gramoxone; Weedol
Pentanochlor	Herbon Solan
Phenmedipham	Betanal
Picloram	Tordon; (Atladox HI)
Prometryne	Caparol; Gesagard
Propachlor	Ramrod; Herbon Orange
Propanil	Rogue; Stam F
Propham	Triherbide IPC; Chem Hoe; (Herbon Pink)
Pyrazone or Chloridazon	Pyramin; Alicep; Trojan SC
Simazine	Weedex S2G; Simadex; Princep
2,4,5-T	Silvapron T; Phortox; Vegatrol; Trioxone
Terbacil	Sinbar
Terbutryne	Prebane; Clarosan
Triallate	Avadex BW; Far-Go
Trietazine	Gesafloc; (Aventox); (Bronox)
Trifluralin	Treflan; Trident; Tristar

Common name	Trade names	Common name	Trade names
2,3,6-TBA	Trysben; (Camtox); (Cambilene)	Vernolate	Vernam
TCA	Varitox; Tecane		

Disclaimer: The object of this table is for information, not trade, and proprietorship of registered products is not provided. The list has been compiled with care from published information, but the author, editor and publisher disclaim further responsibility for the accuracy of its contents. Suggestions for improvement in possible later editions would be welcomed.

Index